EVAN M. MELHADO

Jacob Berzelius

The Emergence of His Chemical System

David
with regards
Evan

Berzelius at the age of 37. Water-colour by L. H. Roos af Hjelmsäter 1816.

EVAN M. MELHADO

Jacob Berzelius

The Emergence of His Chemical System

ALMQVIST & WIKSELL INTERNATIONAL
STOCKHOLM — SWEDEN

THE UNIVERSITY OF WISCONSIN PRESS

First published by
Almqvist & Wiksell International, Stockholm, Sweden
ISBN 91-85286-21-4 (Bound)
ISBN 91-85286-22-2 (Paper)

Also published as "Studies and Sources published
by The Swedish History of Science Society No. 34"

Published in North and South America by
The University of Wisconsin Press
114 North Murray Street, Madison, Wisconsin 53715
ISBN 0-299-08970-3
LC 81-7593

Library of Congress Cataloging in Publication Data

Melhado, Evan M., 1945
Jac. Berzelius, the emergence of his chemical system.
Bibliography
Includes index.
1. Berzelius, Jöns Jacobs, Friherre, 1779—1848.
2. Chemists—Sweden—Biography. I. Title.
QD22.B5M43 540'.92'4 [B] 81-7593
ISBN 0-299-08970-3 AACR2

Printed in Sweden by
Almqvist & Wiksell, Uppsala 1981

TO THE MEMORY OF
STEN LINDROTH

Contents

On citation of sources

Each source, primary or secondary, cited in a chapter, preface, or introductory section of the text is cited fully the first time the source is invcked in that chapter, preface, or introductory section. Full citation is given even if the work was cited fully in a prior chapter, though the elaborate bibliographic discussion given some sources is not repeated; the reader seeking bibliographic details is referred to the notes containing them. Sources invoked repeatedly in a chapter are assigned a short title when first cited in the chapter; thenceforth, they are cited by author and short title. However, repeated references to works by Berzelius, whether or not the he was the sole author, are given by short title only. Some books are cited only in abbreviated form. The abbreviations and the full citations of the works they represent appear in the list of abbreviations preceding the bibliography.

Quotations from sources are verbatim, though occasional corrections of orthography, typography, or punctuation are made (and sometimes noted by square brackets). Emphasis is that of the originals unless stated otherwise. In many quoted sources, italics, capitals, and similar means of emphasis were used not so much to stress a point as to serve a stylistic or graphic function. Emphases of this kind are routinely and silently suppressed.

Berzelius's publications appeared primarily in English, French, German, and in his native Swedish. The choice of which version to exploit in this study rests on two criteria. First, except where unavoidable, Swedish texts of Berzelius's writings are not used (though they are cited wherever a Berzelian text that appeared in a Swedish version is first introduced). Second, every effort is made to employ the published version that was either the original or the closest non-Swedish version to the original. (Where they exist, Swedish versions were not necessarily the originals. Berzelius often composed his papers in French or German—though not in English—publishing Swedish versions later.) Except in obvious cases, the reader is spared discussion of the evidence by which the proximity of a text to its original is judged.

Cross references refer the reader to the page on which the note *number* of the referenced note or quotation appears (though the relevant portion of the referenced note or quotation may appear on a preceding or following page).

Preface

This, the bicentennial year of the birth of Jöns Jacob Berzelius (1779—1848), is a timely occasion for the completion of a monograph about his chemistry. The need for new investigations of his science has been acute. Anders Lundgren's study, *Berzelius och den kemiska atomteorin,* which also marks the bicentenary, is the first recent investigation of Berzelius's chemistry that has not relied on narratives dating from the late-nineteenth century. The sole comprehensive analysis of Berzelius's chemistry, H. G. Söderbaum's *Berzelius' Werden und Wachsen, 1779—1821,* appeared at the turn of the century, motivated by the fiftieth anniversary of the chemist's death. Söderbaum's book capped a tradition of historical writing that accorded Berzelius a prominent place in the history of chemistry and that colored subsequent accounts, including Partington's discussion and works by Prandtl and Jorpes.[1]

The nineteenth-century authors were mindful of the towering position that Berzelius enjoyed in his day. The *Namenregister* of Hermann Kopp's classic *Die Entwickelung der Chemie in der neueren Zeit* (Munich: 1873) and that of Ernst von Meyer's *Geschichte der Chemie von den ältesten Zeiten bis zur Gegenwart* (Leipzig: 1889) list more entries under Berzelius's name than under any other.[2] Moreover, both authors, especially Meyer, organized the core of their books, devoted to the atomic theory and organic chemistry, around what they saw as the pivotal position of Berzelius. Söderbaum's book improves on these accounts, providing focussed development of their themes and invoking a historiographical novelty that led to new insights. In its main lines, however, the picture he presents of Berzelius is not dramatically different from those of Kopp and

[1] The present monograph is a descendant of the author's doctoral dissertation, Evan M. Melhado, "Jac. Berzelius: foundations and development of his chemistry" (Doctoral dissertation, Princeton University: 1977). Lundgren's book, *Berzelius och den kemiska atomteorin,* Acta Universitates Upsaliensis. Studier i idé- och lärdomshistoria, Vol. II, utgivna av institutionen för idé- och lärdomshistoria vid Uppsala Universitet (Uppsala: 1979), addresses the manner in which Berzelius gradually came to terms with a corpuscular ontology for the phenomena of chemical proportions. Söderbaum's book (Leipzig: 1899), which is Vol. III of *Kahlbaum's Monographien aus der Geschichte der Chemie,* is henceforth cited as Söderbaum, *Werden und Wachsen;* for its motivation, see Kahlbaum's "Verbemerkung," p. v. For Partington's discussion, see his *History,* IV, Ch. 5. Wilhelm Prandtl's study is *Humphrey Davy, Jöns Jacob Berzelius: zwei führende Chemiker aus der ersten Hälfte des 19. Jahrhunderts,* Grosse Naturforscher, Vol. III, H. W. Frickhinger, ed. (Stuttgart: 1948). J. Erik Jorpes study is *Jac. Berzelius: his life and work* (Berkeley: 1970).

[2] Kopp's book is henceforth cited as Kopp, *Entwickelung;* Meyer's as Meyer, *Geschichte.* For another nineteenth-century view of Berzelius's theories, see C. W. Blomstrand, *Die Chemie der Jetztzeit vom Standpunkte der electrochemischen Auffassung aus Berzelius Lehre entwickelt* (Heidelberg: 1869), esp. pp. 5—12.

Meyer. Söderbaum's book codified the nineteenth-century image of Berzelius, which is still our own.

However, Berzelius also appears in historical discussions devoted to events following his death, particularly the evolution of organic chemistry in the late nineteenth century. In these studies, Berzelius's position is not central, but it seems significant. Works by participants in the emergence of structural organic chemistry, such as Wurtz and Ladenburg, by their opponents, such as Hermann Kolbe, and by non-participating observers suggest that Berzelius remained influential in the development of organic chemistry despite the fall of the Berzelian system in the years just prior to Berzelius's death.[3]

After reading about Berzelius in both kinds of discussions, I became persuaded that Berzelius's achievements were immensely important, but I was unclear about just what made them important. In view of the picture of Berzelius, described shortly, that emerges from the work of Meyer, Kopp, and Söderbaum, it seems unlikely that he should have merited all the attention they paid him. Moreover, the persevering influence of Berzelius suggested by treatments of organic chemistry in the latter part of the nineteenth century seemed to me inconsistent with the sharp temporal constraints that, as I note below, students of Berzelius place on his historical role. These shortcomings of available accounts of Berzelius led me to undertake this study, which attempts a new assessment of his achievements by means of a genetic analysis of his thought.

Discussions of Berzelius's role in the early nineteenth century often share a pattern of organization. First, Berzelius's work is given an overall characterization, usually by singling out one aspect of his science as its fundamental core and unifying theme. Next, his varied achievements are covered in a succession of topical discussions, pursued in rough chronological order. In each case, the discussion emphasizes the steady acquisition of positive results, whether of theoretical or practical import, and views them in the light of the initial characterization. Repeated reference to the characterization serves several purposes. It keeps in view the well-known synthetic character of Berzelius's thought, otherwise obscured by the topical treatment, and it points out how Berzelius's numerous achievements rested on a few central ideas. As pillars of Berzelius's science, the characteristic ideas were also the targets of later attacks by a new generation of chemists. Accordingly, these narratives prepare the reader for an account of the

[3] See Albert Ladenburg, *Vorträge über die Entwicklungsgeschichte der Chemie in den letzten Hundert Jahren* (1st ed., 1869; the present author has consulted the 2nd ed., Brunswick: 1887); (Ladenburg was a student of Kekulé, Wurtz, and Friedel;) Charles Adolph Wurtz, *Histoire des doctrines chimiques depuis Lavoisier jusqu'à nos jours,* Dictionnaire de chimie pure et appliquée, Discours préliminaire (Paris: 1868). For Kolbe's views, see the sources listed by Partington, *History,* IV, pp. 503—504, text and notes. An informative study by a non-participating observer (a partisan of Kolbe), is Albrecht Rau's *Die Theorien der modernen Chemie* (3 pts., Brunswick: 1877—1884).

overthrow of Berzelian science in the 1840's and for an assessment of the durability and influence of Berzelius's achievements.

Meyer's, Kopp's, and Söderbaum's studies provide examples of this pattern. Beginning with a few pages of overall characterization, Meyer goes through a succession of topical discussions: atomic theory and chemical proportions, electrochemical theory and dualism, early currents opposing dualism, dualism in organic chemistry. This approach sets the stage for his next subject, the work of those younger chemists who destroyed Berzelius's system. Meyer singled out the atomic theory and chemical proportions as the fundamental, unifying characteristic of Berzelius's science. Kopp proceeded similarly, but found the underlying characteristic in his theoretical views. Though Söderbaum differed significantly from Kopp and Meyer in his choice of characteristic theme and in his approach to making that choice, he nevertheless insisted on finding something characteristically Berzelian to which Berzelius's other achievements could be referred, namely Berzelius's appropriation of Lavoisier's insistance on the central position of oxygen in chemical theory.[4]

In addressing the durability and influence of Berzelius's achievements, authors of nineteenth-century studies of Berzelius were mindful of the stark contrast they saw between Berzelius's domination of the chemical profession prior to the 1840's and the complete destruction of his elaborate system just before mid-century. Berzelius seemed to them to bridge two periods of upheaval, the revolution centering on Lavoisier and the shift from Berzelian science to new paths of research, in organic chemistry. Accordingly, Berzelius was readily viewed as a consolidator of revolutionary gains, not as an innovator. Especially for Kopp and Söderbaum, Berzelius's significance lay chiefly in his reflecting the commitments of a post-revolutionary generation and in his realizing the implications of revolutionary principles. Berzelius dominated his time only because he was its best spokesman.[5] In this light, the fall of the Berzelian system could be

[4] Meyer's discussion of Berzelius, in the context of early-nineteenth-century chemistry, is in his *Geschichte,* pp. 165—230. Kopp's account of Berzelius is less localized. Berzelius figures most prominently in Kopp, *Entwicklung,* pp. 311—333, 418—432, 498—517, 544—581, 600—631. For Meyer's characterization, see his *Geschichte,* pp. 165—171; for Kopp's, see his *Entwicklung,* pp. 311—317; for Söderbaum's, see his *Werden und Wachsen,* esp. pp. 91—95.

[5] Kopp expresses his opinion thus (*Entwickelung,* p. 312): "An seinen [Berzelius's] Namen knüpft sich nicht, wie an den seines Vorgängers Lavoisier, die Erinnerung an den Umsturz eines vorher herrschend gewesenen Systems und den Aufbau eines neuen, und nicht einmal wie an den seines Zeitgenossen H. Davy . . . die Erinnerung an die Widerlegung gewisser Lehren, welche als fundamentale betrachtet waren, in dem Sinne des Nachweises, dass sie keineswegs allgemein gültige seien. Wohl aber knüpft sich an den Namen Berzelius's die Erinnerung, wie dieser grosse Forscher in ruhigem Anschluss an das vorher schon in der Chemie für wahr gehaltene dieses Gebiet des Wissens nach jeder Seite hin erweiterte und überall vervollständigte und befestigte; wie er selbständig, und mit grösserer Schärfe und im weiterem Umfang als irgend einer seiner Zeitgenossen die Beweissführung für die wichtigste allgemeine Lehren unserer Wissenschaft gab; wie er das empirisch Erkannte vollständig und geordnet zusammenzufassen wusste, welche als den Anforderungen der damaligen Zeit entsprechend von den Meis-

portrayed as another revolution, in which new forces overturned the beliefs articulated in the characteristic ideas of Berzelius's chemistry. The image of an achievement having significance for only one generation is reinforced in these narratives by noting the insistent conservatism of Berzelius in his last years. Its author inflexible, its leading tenets felled by telling blows, the system of Berzelius inevitably crumbled to make way for new developments.[6]

Accounts of the kind given by Meyer, Kopp, and Söderbaum left me dissatisfied. One reason for my dissatisfaction was that the choice of the theme considered as the characteristic core of Berzelius's chemistry differs from author to author. Except in Söderbaum's study, the choice is made in terms of hindsight, and the relations between the characteristic ideas with the remaining topics are left incompletely elaborated. Moreover, differing judgments of what was characteristically Berzelian implied different identifications of the beliefs and commitments that he represented for his generation and differing judgements of what novelities occasioned the fall of Berzelius's system and led chemists along new paths. Söderbaum made a significant contribution to the understanding of Berzelius by insisting that the characteristic core be determined by observing the construction of the system, not by selecting a single theme from the completed structure. However, because he found the characteristic core almost wholly in precedents provided by Lavoisier, Söderbaum reinforced the perception of Ber-

ten anerkannt wurden." Cf. also this remark (pp. 313—314): "'Berzelius' theoretische Ansichten schlossen sich, namentlich auch was die näheren Bestandtheile in complicirter zusammengesetzten Verbundungen betraf, zunächst an die vorher schon in Geltung gewesenen an, und als eine Vervollständigung dieser Lehren über die Zusammensetzung der Körper erschien, was er bezüglich der Letzteren feststellte und als der atomistischen Begründung dieser Lehren, was er in den elektrochemischen Theorie zu finden und zu geben glaubte."

Söderbaum offered a similar opinion of Berzelius's originality (*Werden und Wachsen,* pp. 132—133): "Berzelius war, streng genommen, in höherem Grade Systematiker als Theoretiker. Ein origineller Theoretiker in dem Sinne, dass er neuen und ungewöhnlichen Ideen nachjagte, in der bewussten oder unbewussten Absicht, dadurch die Aufmerksamkeit der Mitwelt auf seine Person zu lenken, wurde er niemals. Für diejenigen, welche derartige 'Originalität' als den höchsten Exponent der wissenschaftlichen Produktion betrachten, muss Berzelius immer auf sehr niedriger Stufe stehen. Seine Stärke in Bezug auf die theoretische Behandlung der Wissenschaft lag vielmehr, wie übrigens schon von einigen älteren Biographen angedeutet worden ist, in seiner ungemeinen Fähigkeit, von einander unabhängige Theorien zu einem harmonischen Ganzen, zu einem einheitlichen Lehrgebäude zusammenzufügen. Nicht ohne Ursache nannte ihn Liebig 'unsern ersten Baumeister.' Und obwohl der Grund, auf welchem er sein Gebäude aufführte, und der im wesentlichen dessen Plan bestimmte, schon von Lavoisier gelegt war, entwickelte Berzelius bei der Vollendung des Systems eine so selbständige Thätigkeit in der Bearbeitung, Anordnung und Verschmelzung des, wenn auch manchmal von andern Forschern zuerst hervorgebrachten Materials, dass er es wohl verdient, neben Lavoisier als einer der grössten Förderer der antiphlogistischen Sauerstofftheorie genannt zu werden. Lavoisier hatte seine Ideen zum Siege geführt, Berzelius hat den Sieg organisiert."

Meyer, more conscious of the novelty and importance of Berzelius's work in the atomic theory and theory of proportions, did not insist on the Swede's lack of originality.

[6] For the fall of the system, see Söderbaum, *Werden und Wachsen,* pp. 92—97, 133—135; Meyer, *Geschichte,* esp. pp. 212—230; and Kopp's chapter entitled "Bestreitung älterer chemischer Lehren um 1840," in his *Entwickelung,* pp. 582—631.

zelius as an unoriginal thinker.[7] In this study, I follow Söderbaum's example of taking a genetic approach to Berzelius's science, in the belief that the identification and assessment of the significant themes and novelties of his chemistry could be made more reliably by watching them develop rather than by selecting them after dissection of the developed system. However, as I note below, rather than attempting to characterize Berzelius's science by citing a few leading ideas, I try to observe the functions performed by the system. Moreover, my work does not support the view that Berzelius's science lacked novelty and originality.

That view is another reason for my dissatisfaction with older accounts. The immense prestige and influence of Berzelius seemed inadequately explained by denial of his originality and dismissal of his achievements as transient phenomena. This description was hard to reconcile with my awareness that Berzelius, more than any other figure, taught the chemical profession that chemistry must be conducted on a rigorous stoichiometric foundation and showed how this was to be done. This study documents that one of Berzelius's principal concerns, especially insofar as complex compounds were concerned, was to deploy a combination of stoichiometric principles and analytical techniques to discriminate specific compounds from one another and from mixtures. The study also credits Berzelius with recognizing before other chemists that progress in organic chemistry depended on first gaining access to specific organic compounds composed in definite proportions, instead of the animal and vegetable products that were obtained from analysis by extraction and distillation. These materials, Berzelius recognized, were commonly mixtures of similar compounds.

It was also hard for me to see how Berzelius's impact could be dismissed as limited to his own generation when it was in terms of his ideas that the concept of organic radicals was transformed, partly by Berzelius and partly by his successors, from a mere speculation into a serviceable entity (as noted at the conclusion of this study). My disquiet with claims for Berzelius's limited impact was intensified by what I had read of Berzelius in the other kind of nineteenth-century historical discussion in which he appears, treatments of the later evolution of organic chemistry up to the foundation of the structural theory. Discussions of Berzelius in relation to later developments reveal that the original battle between Berzelius and the succeeding generation was reenacted. Though by Berzelius's death his system was a shambles, something Berzelian was still present and remained influential. Whatever it was, it had acquired a new standard bearer in Hermann Kolbe (1818—1884), who forced the victors over Berzelius, and especially their successors, to fight new battles. Though Kolbe employed some particular suggestions developed by Berzelius as part of his last-ditch

[7] For Söderbaum's choice of characteristic core and its justification, see his *Werden und Wachsen*, pp. 67—69, 77—89, 90—101; cf. also pp. 127—133, and see the passage from Söderbaum's book quoted above, in n. 5.

defense, notably the idea of conjugation, Kolbe's approach cannot be reduced to the invocation of a few specific notions. He clearly believed himself to be representing something much broader, and to be opposing an equally broad anathema, the chemistry of Gerhardt and his successors.

I cannot yet describe the Berzelian mantle that Kolbe assumed. However, I found some indication of the issues in which Berzelius's influence was felt in the themes debated by Kolbe and his opponents. The focal point of their discussion was whether the multiplicity of reactions exhibited by organic compounds could be reconciled with a univocal description of the manner in which the elements of compounds were united. Accordingly, they asked whether the nature or the position of atoms (or groups of atoms) determined the properties and reactions of organic compounds; whether individual atoms or groups of atoms (radicals) should have been considered primary; and, whichever was primary, what characteristics the primary units possessed. Gaining insight into the Berzelian attitudes that informed the debate became one of my longer term goals. An understanding of Berzelius's own ideas seemed to me a necessary prelude to an assessment of his later influence. This need reinforced my interest in replacing older studies of Berzelius's chemistry.

This study, then, is intended to serve two functions. For the present, it is intended to replace older discussions of the nature of Berzelius's chemistry. For the future, it is intended to undergird an investigation of Berzelius's influence in the later development of organic chemistry. With the former function primarily in mind, I pursue the development of Berzelius's ideas only through the full articulation of the Berzelian system, that is, up to about 1820.[8] Leaving the subsequent evolution of his ideas for later investigation of his influence on the development of organic chemistry, I close this study by suggesting how organic chemistry made a beginning in the nineteenth century in terms of Berzelius's views.

In offering a corrective to older accounts of Berzelius, the study examines the role of both precedent and contemporary belief in shaping Berzelius's ideas and in directing the course of his research. Part I of the study explores the background to Berzelius's work, and Part II examines the development of Berzelius's ideas in the light of the background. Though the study confirms Söderbaum's belief in the importance of Lavoisier's ideas in the work of Berzelius, it shows that some of Berzelius's most important contributions depended upon his departing significantly from just those aspects of Lavoisier's chemistry that were *non*-re-

[8] There is considerable justification for concluding a study of the development of Berzelius's thought around the year 1820, as Söderbaum remarks in his preface (*Werden und Wachsen*, pp. viii—ix). Most significant for the present study is that, by 1820, Berzelius had completed the construction of his system, changing it later only in the details, not in the fundamentals.

volutionary. In particular, Part I shows that Lavoisier's chemistry was organized according to one of the principal motifs of late-eighteenth-century chemical thought, that of two-component theories. Part II shows that several of Berzelius's leading theoretical tenets, which were essential to his ability to specify compounds by stoichiometric means, rested on his departure from this precedent. Part II also points out that this innovation was devised in terms of the new phenomena presented by the Voltaic pile, which Berzelius was among the first to explore. Another departure from Lavoisier that is exhibited in Part II has already been suggested. Though Berzelius owed much to Lavoisier's ideas about organic matter, Berzelius took a significant step beyond Lavoisier in emphasizing the existence of a fundamental specific level in organic chemistry, the level of specific compounds composed in definite proportions.

This study differs from prior accounts in more than its enhanced assessment of Berzelius's originality. It suggests that Berzelius's science is best understood not by finding out what were its characteristic ideas, but by inquiring how Berzelius's system functioned in helping him solve certain problems. The study argues that the leading question that Berzelius had in view was how to invoke stoichiometry to distinguish specific compounds from one another and from mixtures, and it suggests that the Berzelian system was an apparatus designed to perform this function in the greatest possible number of cases.[9] Part II describes the evolution of Berzelius's techniques of specification from his efforts to describe complex salts as specific compounds and then explores his deployment of the system to solve the much more difficult problems of discriminating species of naturally occurring minerals and of organic compounds. In preparation for this exploration, Part I examines the background to the deployments in mineralogy and organic chemistry and notes the differing roles of stoichiometry and natural-historical characterization in discrimination of kinds in each field. In Part II, Berzelius's studies in mineralogy and organic chemistry are examined, and his efforts to apply stoichiometric criteria in both are investigated.

The deployment in mineralogy clearly exhibits the functioning of Berzelius's system as an apparatus for discrimination of species of complex compounds. The organic deployment was less fruitful, in that up to 1820 Berzelius had succeeded chiefly in identifying the obstacles to discrimination of organic species and offering some explanations of the origins of the obstacles. After examining the organic deployment, the study notes that Berzelius urged organic chemists to pursue the same goal that had inspired him: obtaining specific compounds and endeavoring to discriminate them by a combination of stoichiometric principles and analytical techniques. The study concludes by pointing out how

[9] The main stoichiometric motif explored in this study is Berzelius's use of stoichiometric laws to control analytical results. In this connection, Berzelius's role in the history of the atomic theory is not centrally important and is given minimal treatment.

organic chemists, mindful of this goal, made a beginning in their science in the 1830's by applying Berzelius's system to discriminate newly isolated specific organic compounds in a fashion fully commensurate with Berzelius's own successes in inorganic chemistry and mineralogy.

Many persons and institutions aided me during my studies of Berzelius, which I began more than a decade ago. My work benefited from the insights of Thomas S. Kuhn, under whose direction this investigation began as a doctoral dissertation. Other persons who read drafts of this study are Richard W. Burkhardt, Jr., Jeanne E. Combs, Gerald L. Geison, Charles C. Gillispie, Sten Lindroth, Anders Lundgren, L. Lee Melhado, and R. Steven Turner. I am grateful for their commentary. Elaine R. Ryks and Jeanne E. Combs, my research assistants, facilitated my work in many other ways. For welcoming me into her Swedish classes, I am indebted to Rochelle Wright. David L. Ransel also helped me with Swedish sources, and Gerald Day provided assistance with Latin texts. The staffs of the libraries of Washington University and the University of Illinois aided me for extended periods. For typing the final draft of this monograph, I am grateful to Nancy Edwards.

My work has been supported by a University of Illinois Faculty Summer Fellowship; by the University of Illinois Research Board, which provided funds for a research assistant; and by an appointment as a Resident Associate at the University of Illinois Center for Advanced Study, where Daniel Alpert, Peggy Harris, and the Center staff facilitated my activities. I also enjoyed the cooperation of the faculty and staff in the Department of Chemistry at Washington University and the Department of History and the School of Chemical Sciences at the University of Illinois.

For sustaining the costs of publication of this study in the *Lychnos-Bibliotek*, I am grateful to the Research Board and the School of Chemical Sciences at the University of Illinois and to the Swedish Natural Science Research Council and the Swedish Research Council for the Humanities and Social Sciences.

My greatest debt is to my wife, L. Lee Melhado, who long shared Berzelius's presence in our household.

Evan M. Melhado

Urbana, Illinois

13 November 1979

PART I

The Background to Berzelius's Achievements

Introduction

Historical study of eighteenth-century chemistry has been motivated chiefly by the need to explain the events of the revolution centering on Lavoisier. However, by selecting themes for study in the light of their relevance to the revolution, historians have ignored many of the remaining trends.[1] Those trends are of interest not only in their own right, but also in their role as precedents for novel nineteenth-century developments that remain inexplicable (or even unobservable) without eighteenth-century referents. Especially in the early nineteenth century, chemical innovations are likely to have ties with eighteenth-century antecedents previously little noted or to represent breaks with neglected motifs that survived the revolution intact.[2]

Berzelius's achievements are a prominent example of early-nineteenth-century innovations requiring analysis in relation to themes of the previous century that were peripheral to the revolution. Without the appropriate background, the novelties of his chemistry remain as invisible as they were to the nineteenth-century authors discussed in the preface, who regarded Berzelius's work as only the expression of revolutionary achievements. The three chapters forming the first part of this two-part study analyze the eighteenth-century background to Berzelius's achievements. The first two chapters of Part I examine the chemical

[1] A major exception to this generalization lies in some of the pioneering studies by Hélène Metzger. See particularly her *Les doctrines chimiques en France du début du XVIIe à la fin du XVIIIe siècle* (1923; rpt., Paris: 1969) and her *Newton, Stahl, Boerhaave et la doctrine chimique* (1930; rpt., Paris: 1974). The most significant recent corrective to the neglect of many eighteenth-century chemical traditions is Arnold Thackray, *Atoms and powers: an essay on Newtonian matter-theory and the development of chemistry* (Cambridge, Mass.: 1970); henceforth cited as Thackray, *Atoms and powers*.

[2] The emergence of stoichiometry provides an example of a nineteenth-century development arising out of eighteenth-century traditions that survived the chemical revolution. Motivated in part by what Thackray (*Atoms and powers*, Ch. 7) calls the "force-quantification program" of the late-eighteenth-century affinity theorists, stoichiometry began to apear without benefit of the Daltonian atomic theory. For a general discussion of pre-Daltonian stoichiometry, see Partington, *History*, III, Ch. 14. For the quantification of affinities as one motive for Berzelius's stoichiometry see Anders Lundgren, *Berzelius och den kemiska atomteorin*, Acta Universitatis Upsaliensis. Studier i idé- och lärdomshistoria, Vol. II, utgivna av institutionen för idé- och lärdomshistoria vid Uppsala Universitet (Uppsala: 1979), esp. pp. 75, 91, 97.

Dalton's work itself exhibits both continuity with some eighteenth-century traditions and a break with others. For example, as Thackray argues (*Atoms and powers*, pp. 241—242), part of Dalton's achievement was to make explicit and to exploit certain prevailing eighteenth-century assumptions about microscopic material particles and their modes of combination. However, Dalton's ability to do this rested on his break with eighteenth-century affinity theory, as Thackray also points out (*Atoms and powers*, pp. 257—259).

background of Berzelius's work. The main function of the chapters is to elucidate one of the principal explanatory techniques of chemistry in the eighteenth century, that of two-component theories. These theories explained both the existence of generic groups of chemicals, like metals or acids, and the differences among members of the groups. Membership in the genus was taken to result from the presence, in all its members, of a generic ingredient. A property-bearing principle, the generic ingredient stamped its properties upon all members of the genus. The genus of combustibles, for example, consisted of all substances that contained the principle of combustibility, phlogiston. The specificity of the different members of the genus was explained by asserting that each member contained a different specific ingredient. The two-component theories were hierarchical, the genus subsuming all its members or species and the generic ingredient being the important one insofar as membership in the genus was concerned. The emphasis of two-component theories was biased toward the generic level, and contemporaries focussed their attention on generic groups and generic properties at the expense of the specific. A major novelty of Berzelius's chemistry was to replace hierarchical two-component theories, especially of salts, with an approach that gave each component equal status. For Berzelius, the acidic and basic constituents of salts were species having opposing properties and equivalent status, and they owed their acidity or basicity to the character of their radicals, not to the presence of generic principles of acidity or basicity.

This novelty must be analyzed against the background of the work of Lavoisier and his contemporaries, especially Bergman and Guyton de Morveau. Their principal achievement, insofar as two-component theories are concerned, was a thorough articulation of that approach in its applications to acids, salts, and organic compounds. The first of these applications invoked the concept of "radicals," which emerged in the work of Bergman and Guyton and was acquired from the latter by Lavoisier. Inorganic radicals, for Lavoisier, were the entries in his list of elements and served as the specific components of acids. By combining with oxygen, the principle of acidity, radicals were transformed into acids. Like its predecessors, this theory emphasized the generic at the expense of the specific. Despite his offering a list of diverse, specific elements, Lavoisier was little concerned with the properties of the radicals themselves. It was their acidification that attracted his attention. Acids, in turn, were the generic components of salts, and bases were the specific ingredients. Here, too, the properties of the specific components were all but ignored. There was no term "basicity" referring to a set of positive properties opposed to those of acids. A base was merely a passive substrate for the coagulation of an acid into a typically crystalline neutral salt. Berzelius replaced the two-component theory of acidity with one that emphasized the specific and made the hitherto generic property of acidity a function of the specific components, the radicals.

Lavoisier also deployed the concept of radicals in a two-component theory of organic chemistry, modelled on that in inorganic chemistry, as shown in Chapter III. With appropriate changes, corresponding to his innovations in inorganic chemistry, Berzelius followed this precedent as well. However, he identified, for the first time, a fundamental requirement that would have to be met in organic chemistry before any further advance in that field could be achieved. He pointed out that most organic substances then known were generic mixtures of similar compounds, and that they had to be separated into species before anything else, experimental or conceptual, could profitably be done with them. Once again, he emphasized the importance of the specific, a consideration scarcely visible in Laviosier's account of organic bodies.

The culmination of the eighteenth-century two-component tradition in the work of Bergman, Guyton, and Lavoisier is the subject of Chapter II. To establish the antecedents for their achievements, Chapter I offers a selective overview of eighteenth-century chemistry. It treats the early part of the century in terms of the reaction among chemists against the attacks launched by the mechanico-corpuscular philosophy on the qualitative chemistry prevailing up to the seventeenth century, and it examines efforts to reconstruct chemical science. Two central desiderata of a reconstructed chemistry were an explanation of specificity of qualities and an account of specificity of reactivities. Central figures in these developments were G. E. Stahl and his partisans and Herman Boerhaave. Their work is shown to have provided the two leading sources for a French Stahlian school that emerged after mid-century under the banner of Newtonianism. The leader of the new school was Buffon, and his principal protégés were P. J. Macquer, A. Baumé, and Guyton. The chapter concludes with an examination of the intellectual milieu established by the Newtonian Stahlians, setting the stage for Chapter II, which analyzes the work of Guyton and Bergman—the latter was a fellow traveller with the French group—and which explores Lavoisier's debts to them.

In a chemistry that emphasized the specific level, it was essential to define chemical species and to establish criteria for the discrimination of species from one another and from mixtures. In the eighteenth century, these tasks had been addressed as much by mineralogy as by chemistry, and even chemical approaches to them frequently were provided in a mineralogical context. The various mineralogical traditions of the eighteenth century were defined largely by their differing approaches to the taxonomy and discrimination of kinds of inorganic matter, and they constitute another little noticed source of antecedents for the novelties of early-nineteenth-century chemistry. The third chapter of Part I provides an examination of two mineralogical traditions insofar as they bore on Berzelius's innovations.

The mineralogical traditions were connected with Berzelius's work in three main ways. First, Berzelius deployed his chemical achievements in mineralogy, consciously attaching himself to one of the major antecedent mineralogical traditions, Haüy's crystallography. Because the problem of the speciation of inorganic matter had been so long prominent in mineralogy, the character of Berzelius's innovations becomes particularly apparent when his success in mineralogy is measured. Second, the traditions of mineralogy provided Berzelius with sources for his own ideas, especially regarding the application of stoichiometry to define chemically complex species. Last, eighteenth-century mineralogy provided Berzelius with a foil against which he developed his ideas further. This was the Wernerian school of mineralogy, which approached the subject along lines closely allied to the procedures of the other two branches of natural history, botany and zoology.

Besides mineralogy, organic chemistry was the other area in which Berzelius attempted a deployment of his chemical achievements, and with the same goal: achieving a rigorous discrimination of species by invoking stoichiometric considerations. The third chapter of Part I also establishes the background to this deployment. The background was similar to that in mineralogy, in that organic matter was characterized by natural-historical methods (though in less formal fashion than in mineralogy). However, organic chemistry in the eighteenth century differed from mineralogy in at least one major respect. In organic chemistry, there was no conception of a specific level as fundamental and no discrimination of generic mixtures from single compounds. If Berzelius enjoyed less success in subordinating organic chemistry to stoichiometric principles than he did in mineralogy, he still achieved a major conceptual breakthrough, which he urged chemists to acknowledge. In his view, recognition of a fundamental specific level—that of the single, discrete compound, composed in definite proportions—and the experimental isolation of species were essential preconditions for further research in organic chemistry. The third chapter notes the prevailing natural-historical approach to organic matter in the late eighteenth century, develops certain precedents provided by Lavoisier for Berzelius's views of organic matter, and suggests the absence of a fundamental species concept in late-eighteenth-century organic chemistry.

CHAPTER I

Some Chemical Traditions of the Eighteenth Century

1. Particles and principles

The corpuscular matter-theory that flourished during the second half of the seventeenth century undermined the science of chemistry in two main ways. Corpuscularism denied a secure ontological status to the level of material organization accessible to the laboratory chemist, and it offered conceptual techniques too spare to account for the phenomenological details of chemistry.[1] In either the kinematic corpuscularism of Boyle (1627—1691), Descartes (1596—1650), and their partisans, or the dynamic version of Newton (1642—1727) and his followers, all matter was considered to be always and everywhere fundamentally the same. For Boyle and Newton, matter was built up in a hierarchy of composition from primitive solid particles, *prima* or *minima naturalia,* of a universally self-identical matter. Newtonians affirmed the uniformity of matter in terms of physics by claiming that all the primitive solid particles had the same inertial constant. Cartesians assured the homogeneity of all matter by identifying it with space. Though Cartesians denied in principle the existence of *minima,* they did not insist on the ultimate divisibility of matter that their theory entailed. In all corpuscular theories, the chemical level was very remote from these ontologically secure microscopic particles; ordinary chemicals represented only the most complex level of particulate aggregation. Accordingly, no chemicals were elementary. Indeed, in the extreme versions of corpuscularism, such as Boyle's, nothing was elementary but the *minima,* for in principle, by rearrangement of the aggregates, any thing might have been transmuted into

[1] For the denial of secure status to laboratory chemicals, see Thomas S. Kuhn, "Robert Boyle and structural chemistry in the seventeenth century," *Isis,* 43 (1952), 12—36, and Arnold Thackray, *Atoms and powers: an essay on Newtonian matter theory and the development of chemistry* (Cambridge, Mass.: 1970), Ch. 6. In the same chapter, Thackray also discusses the inadequacy of corpuscularians' explanations of the qualitative particulars of chemistry. Kuhn's article henceforth is cited as Kuhn, "Structural chemistry"; Thackray's book, as Thackray, *Atoms and powers.* Much of the substance of this section of this chapter is based principally on these two sources, which are cited further only infrequently in the section. For the general development of the corpuscularism of the seventeenth and eighteenth centuries, see also Marie Boas [Hall], "The establishment of the mechanical philosophy," *Osiris,* 10 (1952), pp. 413—541, and Hélène Metzger, *Les doctrines chimiques en France du début du XVIIe à la fin du XVIIIe siècle* (1923: rpt., Paris: 1969). The latter is cited henceforth as Metzger, *Doctrines chimiques.*

any other thing. Though other versions of corpuscular matter-theory were not quite so uncompromising in denying elements as was Boyle's, they were nevertheless inhospitable to the concept of elements. The chemical practitioner, concerned with laboratory chemicals and macroscale material changes, could find in a thoroughgoing corpuscularism no firm conceptual foundations for dealing with the phenomena that his science studied, in particular for undergirding the techniques of analysis and synthesis.

Furthermore, he could find no sufficient basis to account for the qualitative specificity of chemistry. Making a frontal assault upon the qualitative explanations and essentialism of the past, corpuscularism had no tolerance for the qualitative explanations of the "Chymists" and "Peripateticks" whose ideas Boyle attacked. Rather, it insisted that all sensible, or secondary qualities of matter were derived from the minimal attributes or primary qualities of matter: the size, shape, motion, position, and (with Newton) the forces between the corpuscles. To explain the sensible qualities of bodies, the "Chymists"—alchemists or iatrochemists—invoked the alchemical concept of principles. As a material bearer of qualities, such as color, flammability, and fusibility, a principle conferred its properties or qualities on a body by entering into it as an ingredient. In the Aristotelian matter-theory of the "Peripateticks" or Schoolmen, the concept of an element played much the same role as that of principle did in the thought of the "Chymists." Among the latter, moreover, the distinction between principles and elements was rarely drawn in the seventeenth century. The five principles of the iatrochemist, which included two Aristotelian elements, water and earth, and the Paracelsan *tria prima,* mercury, sulfur, and salt, were all typically identified as elementary in the seventeenth century. Accordingly, the terms "element" and "principle" were commonly considered synonymous. They were considered synonymous by the corpuscularians, too, particularly Boyle, who denied both their status as elements and their property-bearing role. Those seeking to build a chemical science of matter had thus been deprived of their chief conceptual pillars—the idea of elementary or stable chemicals and the endowment of stable chemicals with secondary qualities—by those seeking to build a physical science of matter.

They way had been prepared for this corpuscularian attack by the impact of analytic and synthetic techniques on concepts of material combinations. The union of principles to form a homogenous body had been termed "mixtion," and the body so formed, a "mixt." The concept of mixtion is ancient, having been devised by Aristotle to account for certain material changes. Though in Aristotle's eyes major qualitative transformations occasioned by certain material unions were inadequately explained by mere particulate juxtaposition, as proposed by the ancient atomists, the recoverability of the ingredients seemed to deny the status of true generation to such changes. "Mixtion" was a new category, vaguely midway between mixture and generation, which Aristotle suggested to accommodate the compounding of ingredients into a new entity from which they could be recovered.[2] Until the seventeenth century, Hooykaas points out, the mixt was regarded hylomorphically, as a unitary entity.[3] In Scholastic terms, it was said that the mixt had a single form even though its ingredients were recoverable, because the presence of the ingredients in the mixt was potential, not actual. By emphasizing the recoverability of the ingredients and the reproducibility of efforts to make mixts, successful analyses and syntheses helped bring the early-seventeenth-century iatrochemists, such as Sala, Scaliger, Van Helmont, and Sennert, back to a view of the mixt as formed by particulate juxtaposition. The particles were those of the principles, that is, principles were considered collections of quality-bearing atoms. The iatrochemists, though adherents of a qualitative chemistry, had begun to espouse a corpuscularism that, when stripped of its qualitative aspects, entailed the destruction of qualitative chemistry.

Despite this contribution of the analytic and synthetic arts to the rise of corpuscularism, the conceptual foundations of analysis and synthesis were also destroyed by the later corpuscularian attack. A qualitative atomism was compatible with analysis and synthesis, but a mechanical atomism was not. Because, in its extreme versions, corpuscularism granted no secure status to qualitatively distinguishable chemicals, no chemicals could rightly be deemed the ingredients of others. In brief, because there were no elements, there were no analyses. Moreover, standard analytical techniques, most of which were means for using heat, created the products of analysis, rather than separated the supposed preexisting ingredients. Fire, the instrument of material change, did its work microscopically by destruction, construction, and rearrangement of corpuscular aggregates.[4] Matter was not analyzed by analytical techniques, but transmuted. Though these destructive implications of corpuscularian postulates were rarely fully elicited in the late seventeenth or early eighteenth century, the weight of corpuscularian arguments was sufficient to make analysis a long standing and pressing problem of chemical theory and practice. This question is explored further below in section 5 and in Chapter III.

The corpuscularians most destructive of chemical theory and practice were those of thinkers with largely physical interests. John Keill (1671—1721), John Freind (1675—1758), and other rather thoroughgoing Newtonian physicalists of

[2] For the origins of the term "mixt," see Robert P. Multhauf, *The origins of chemistry* (New York: 1967), pp. 70—71.

[3] For the points in the remainder of this paragraph, see these articles by R. Hooykaas: "The discrimination between 'natural' and 'artificial' substances and the development of the corpuscular theory," *Arch. int. d'hist. sci.*, 1 (1947—1948), 640—651, esp. 645—649; "De natuurlijke klassificatie der chemische substanties," *Chemisch Weekblad*, 33 (1936), 599—610, esp. 602—603; and "Die Elementenlehre der Iatrochemiker," *Janus*, 41 (1937), 1—28, esp. 19—20. This last is cited henceforth as Hooykaas, "Elementenlehre."

[4] For the instrumental role of fire, see Kuhn, "Structural chemistry," 29—31.

the early eighteenth century attended little to the needs of the chemist; accordingly, their matter theories were the more arid and the less useful chemically. On the other hand, Boyle and Nicolas Lémery (1645—1715), and others working with chemical issues in mind offered some means, however modest, for dealing with the specificity of laboratory phenomena. Despite his emphatic affirmation that any material transmutation was possible in principle, Boyle, and after him the still alchemically inclined Newton, suggested the utility of a hierarchy of relatively stable corpuscular aggregates. Each level in the hierarchy corresponded to a particular level of chemical organization, that is, to certain chemicals that were stable, contained lower level ingredients, and entered as ingredients into higher level aggregates. Moreover, unlike those seeking to calculate microscopic attractions according to Newtonian force laws, Boyle and especially Lémery tried to refer specific properties and reactivities to the characteristics, typically geometrical and mechanical, of these relatively stable corpuscles of particular chemicals. Boyle's belief that the springiness of the air was a result of the springy particulate texture of its component particles is a famous example. Similarly, on page after page of his immensely popular text, Lémery detailed the attributes of microscopic particles—typically citing the primary ones of size, shape, and porosity, but lapsing occasionally into requiring secondary ones, like brittleness—necessary to account for the properties of macroscopic aggregates.[5]

The victory of the physicalist approach to matter theory, then, was not total. Nevertheless, the reinvigoration of the chemical approach during the first half of the eighteenth century was slow, and it was achieved by compromising with corpuscularism, not by ignoring it. Because the Newtonian version of corpuscularism gradually eclipsed the Cartesian, all chemical systems of the eighteenth century were constructed to accommodate physicalist foundations of matter theory with the requirements of the practicing chemist.[6] Principles and qualities were restored to chemistry, but only in relation to an underlying corpuscular ontology. The two chief early architects of reconstructed chemistries were Stahl and Boerhaave.

[5] For Newton and his early followers, see Thackray, *Atoms and powers,* Chs. 2 and 3. For Boyle's use of a hierarchy of aggregates, see Kuhn, "Structural chemistry," 24—26, and Marie Boas [Hall], *Robert Boyle and seventeenth-century chemistry* (Cambridge: 1958), pp. 99—102. On Lémery, see Metzger, *Doctrines chimiques,* pp. 281—338.

[6] This argument was first developed by Metzger in her classic—and, so far as Stahl and Boerhaave are concerned, still unsurpassed—study, *Newton, Stahl, Boerhaave et la doctrine chimique* (1930; rpt., Paris: 1974). Much of the present chapter is based on this work, to which repeated reference is made. It is cited henceforth as Metzger, *NSB.* Thackray pursued Metzger's point in his *Atoms and powers,* esp. Ch. 6.

2. Elements

Perhaps the most successful attempt to restore a qualitative chemistry in the light of corpuscularism was that of Georg Ernst Stahl (1660—1734) and his school.[7] Stahl's chemistry rested on a qualitative atomism. He inoked a set of principles, which served as secure and elementary chemicals, and he assumed that they consisted of quality-bearing atoms.[8] Stahl recognized only three elements, earth, water, and air. Air was not a principle, however, in that it did not enter into mixtion. The Stahlians said little about water, but much about earth. In fact, there were three kinds of earth: the vitrifiable, or hypostasis of matter; the inflammable, or phlogiston; and the mercurial or metallic earth. In the theory of Stahl, then, all bodies were composed of water and earth.[9]

Though given secure status, these principles were still removed from the chemical level of organization. Stahl's principles could not be obtained by analysis but always existed in combination. Their existence was therefore not directly demonstrable, but was inferred by reflection upon the chemist's experience.[1] Chemicals were interpreted in terms of a hierarchy of composition, beginning with the principles and reaching the level of even heterogeneous aggregates.[2] The simplest level of composition was that of the "mixt" (also called a "secondary" or "principiated principle") formed by the union of principles. Only a few chemicals were secondary principles, notably the noble metals, gold and silver, and the universal acid (sulfuric). The noble metals were each composed of the three earths, but in different proportions, and the acid was composed of the vitrifiable earth and water. A "compound" or "secondary mixt" was formed by the union of secondary principles, either with one another or with principles. Examples of secondary mixts formed from two secondary principles were the regulus of antimony, which was composed of a vitrifiable, arsenical substance and the inflammable earth; and spirit of wine, which was composed of a rarified oil and a subtle vegetable acid. An example of a secondary mixt formed from a secondary principle and a principle was sulfur, composed of the universal acid and the phlogistic earth. The union of two or more secondary mixts gave rise to a "supercompound," such as antimony (the trisulfide), formed

[7] For Stahl and his school, see Metzger, *NSB*, pp. 93—188. Two groups of Stahlians are discussed in the present chapter: first, Stahl himself and his immediate German followers; second, a group of late-eighteenth-century French Stahlians. The second group itself had two parts, an earlier group unsympathetic to Newton and a later group working under the banner of Newtonianism. The earlier French Stahlians are mentioned only in the introduction to section 6. The later French Stahlians are mentioned frequently, however. To avoid confusion in references to the original school of Stahl and the later French Stahlians of Newtonian bent, the former is called "the Stahlians" or "the school of Stahl"; the latter, "the French Stahlians," "the Newtonian Stahlians," or "the French Newtonians."

[8] *Ibid.*, pp. 118—120.

[9] *Ibid.*, p. 130.

[1] *Ibid.*, pp. 119—120.

[2] For the Stahlian hierarchy of mixtion, see *ibid.*, pp. 121—129. The remainder of this paragraph is based on pp. 121—122.

from sulfur and a metallic regulus, and ores, containing arsenic and stones. At any given level in the hierarchy of mixtion, a substance was formed of lower-level ingredients that endured in it, that is, they were present potentially, as evidenced by their recoverability. Analysis consisted in the recovery of ingredients lower in the hierarchy. Supercompounds represented the extreme of complexity in mixtion, and secondary principles represented the extreme of simplicity. The secondary principles were therefore the lowest level of qualitative organization accessible to the chemist. They, and not the principles, represented for the Stahlians the first term of synthesis and the last term of analysis.

Aggregation was the other kind of material combination.[3] Only by accumulation of qualitatively distinguished particles of mixts or compounds into massy aggregates were chemicals made accessible to manipulation in the laboratory. Pure substances, consisting of a single chemical combination, were homogeneous aggregates; the impure, heterogeneous. Though this distinction between mixtion and aggregation was gratifying theoretically, the Stahlians were able to make little practical use of it, and Lavoisier fused the two concepts, as Metzger points out.[4] In the context of late-eighteenth-century French chemistry, however, Lavoisier's reluctance to discuss the particulate level of organization was exceptional. The vocabulary distinguishing the two was a standard feature of French chemistry and crystallography after mid-century. It may be illustrated here with the words of one of Stahl's disciples, Johann Juncker:

> Le corps mixte diffère du corps agrégé en ce que les parties du mixte sont des parties constituantes, tandis que celles de l'agrégé sont des parties intégrantes; en ce que le mixte est plus petit, qu'il est plus difficile à dissoudre et que chacun de ces atomes ne se ressemblent pas, tandis que ceux du corps agrégé sont homogènes et en plus grand nombre.[5]

When aggregated, the particles of a mixt body formed a sensible mass, which could be manipulated in the laboratory. The particles of the mixt that formed the aggregate were the integrant parts. The constituent parts were the particles of the ingredients of the mixt. The distinction between integrant and constituent parts is roughly similar to the modern distinction between the molecules (integrant parts) of a substance and the atoms of the molecules (constituent parts). As shown below in section 6, the notions of integrant and constituent parts were complemented late in the century by the corresponding notions of affinity of aggregation and affinity of composition. The corresponding crystallographic concepts of integrant and constituent molecules are pointed out in Chapter III.

Herman Boerhaave (1668—1738) developed the other major early eighteenth-century response to the corpuscularian attack.[6] Highly sensitized to the problems of material ontology raised by the corpuscularians, Boerhaave attempted no boldly systematic reconstruction of chemistry comparable in its degree of articula-

[3] For aggregates, see *ibid.*, pp. 124—129.

[4] *Ibid.*, p. 129.

[5] Quoted in *ibid.*, p. 124.

[6] On Boerhaave, see *ibid.*, pp. 191—305.

tion to that of the Stahlians.[7] He articulated brilliant theoretical fragments, however, thereby violating his own cautionary remarks about the need to constrain chemical theory.[8] As Metzger remarks, "La doctrine, d'ailleurs, dans son ensemble, se présentera comme une juxtaposition plutôt que comme une synthèse d'éléments aprioriques et d'éléments empiriques."[9] Consequently, despite Boerhaave's denial that analysis led to anything that truly merited the name of element or principle—a point of view he believed supported by the failure of synthesis[1]—he nevertheless included in the theoretical portions of his famous *Elementa Chemiae* separate and widely read treatises on each of the four traditional elements. Reflecting the opposing tendencies of his thought, he concluded the last of the four treatises by remarking,

> On a vu que les corps qui passent communément pour être des éléments sont mêlés avec une infinité d'autres choses et que, par là même, ils sont très-composés; mais que, si on peut parvenir à les séparer artificiellement de toute autre matière et à les avoir seuls, alors, ils surpassent tous les autres en simplicité. Qu'y a-t-il de plus simple que le feu, l'air, l'eau et la terre? Mais, en même temps, qu'y a-t-il, dans un autre sens, de plus composé? J'ai tâché de ne rien affirmer sur ces éléments qui ne fût solidement démontré; et lorsque j'ai en quelque raison de douter, j'en ai averti et je ne me suis point hâté de tirer des conclusions.[2]

Boerhaave's attitude toward other, non-elementary chemicals was similar, a circumstance that obscured the distinction between his elements and any other reagents, most of which were categorized as "menstrua" because they operated by processes taking place in solution. In all cases, his chemistry rested (at least in principle) on atomism, in particular, on the claim that to any stable chemical corresponded a stable, microscopic corpuscle, itself complex, but remaining undisturbed during reactions.[3]

Boerhaave also considered the distinctions between the particulate and the aggregate level of organization. Both he and Stahl appear to have offered this distinction in reaction against the corpuscularians' technique of attributing numerous geometrical and mechanical properties to the particles of bodies to account for macroscopic properties. In the eyes of Boerhaave and Stahl, such descriptions had no demonstrable foundation.[4] Boerhaave denied, for example,

[7] *Ibid.*, pp. 206—207.

[8] *Ibid.*, pp. 201, 207—208, 209.

[9] *Ibid.*, p. 293.

[1] *Ibid.*, pp. 206—207.

[2] Quoted in *ibid.*, pp. 279—280. Metzger also points out (*ibid.*, p. 226) that Boerhaave nevertheless considered elementary fire to be compound, if undecomposable.

[3] *Ibid.*, pp. 224, 263—264, 286, 287—288.

[4] For Stahl's rejection of descriptions of microscopic particles, see *ibid.*, pp. 102—106; for Boerhaave's, pp. 48—49, 290—291. Boerhaave did not refrain totally from offering such descriptions. He affirmed, for example (pp. 53, 227), that the fluidity, as opposed to the elasticity of the air resulted in part from the ease with which its particles could slide past one another. He also attributed (p. 227) the properties of elementary fire to its round, polished, and elastic particles. In a comparison (pp. 249—250) of the properties of air and fire, Boerhaave compared the attributes of the particles of these two substances. He also claimed (pp. 268—269, 291) that particle shapes and sizes had an important role in producimg solubility effects.

that the elasticity of air resulted from the structure of its individual particles,[5] and that the fluidity of water was a consequence of the character of its particles.[6] For Boerhaave, these and other properties of bulk matter were products of aggregation and the forces responsible for aggregation, as noted in the next paragraph. By taking this position, Stahl and Boerhaave deprived themselves of one of the corpuscularians' principal explanations of specificity in chemistry. The nature of their replacement is considered below in section 3.

Unlike the Stahlians, Boerhaave was able to make effective use of the distinction between particulate and aggregate properties.[7] He developed an account, to which Stahl offered nothing comparable, of the aggregate properties of substances in terms of two opposing agents that operated upon them: the mutual attractions of the particles of the mass, which tended to consolidate it, and the activity of fire, which tended to separate the particles and dilate the mass.[8] The fluidity of water, for example, derived from the action of fire, for when that was diminished, water froze. In this view, "ice is the natural state of water," that is, the state to which the mutual attractions of its particles would always have brought it had it not been for the intervention of fire.[9]

Throughout much of the eighteenth century the dual attitude toward the simplicity of the elements, or indeed the stability of any chemicals, remained a constant theme. Wherever chemicals were made ontologically secure, even if only by assumption, they were nevertheless taken to be complex. The chemical level of organization was considered remote from its ultimate foundation in minimal material corpuscles. This view prevailed even as analysis increasingly proved able to extract, reproducibly, recognized ingredients of substances, and as synthesis was successful in producing expected products from the ingredients so obtained. The much discussed late-eighteenth-century "negative-empirical" definition of the elements, culminating in Lavoisier's *Traité,* rested on these circumstances: an element represented only the last term which analysis had attained; it was therefore elementary for the chemist only phenomenologically, because he had not *shown* it to be complex; but it needed not have been elementary ontologically, because its proximity to the ultimate foundations of the structure of matter remained unknown.[1] As it appeared late in the century, this view was amply illustrated in the writings of Guyton de Morveau. Working, as pointed out below, under the influence of Buffon's Newtonian affirmation of the homogeneity of matter,[2] Guyton asserted,

[5] *Ibid.*, p. 252.

[6] *Ibid.*, pp. 261—262.

[7] *Ibid.*, pp. 217—219.

[8] *Ibid.*, pp. 215—216, 223—224.

[9] *Ibid.*, p. 262.

[1] On the "negative-empirical" definition, see Thackray, *Atoms and powers,* pp. 168—171, and cf. Metzger, *NSB,* pp. 128—129.

[2] Buffon's assertion that matter was fundamentally the same is in his "Second vue de la nature" (1765), reprinted in Jean Pivetau, ed., *Oeuvres philosophiques de Buffon,* Corpus général des philosophes français, Vol. XLI,i, Raymond Bayer, ed. (Paris: 1954), pp. 35—41. This text of the "Second vue" is cited henceforth as Buffon, "Seconde vue." For homogeneity of matter, see "Seconde vue," pp. 39—40, and cf. Metzger, *NSB,* pp. 57—59.

Il est très-probable qu'il n'y a qu'une matière; il est du moins très-certain qu'il n'y a pas autant de matières que de corps pourvus de différentes propriétés: c'est donc la modification de la matière homogène, qui constitute tous les différens corps, même les élémens, & cette modification est la densité, la porosité & la figure. Quatre modifications primitives ont formé dans ce systême les quatres élémens naturels, qui sont le feu, l'air, l'eau & la terre. Ces élémens sont-ils altérables ou convertibles les uns dans les autres? Se detruisent-ils, se reproduisent-ils par l'action continuelle des forces de la nature, ou s'ils conservent inviciblement le caractère que chacun d'eux a reçu dans la première formation? Ce sont des questions sur lesquelles les plus grands Physiciens ne sont pas d'accord, & dont l'examen, plus curieux qu'utile à la théorie chymique, nous meneroit trop loin: nous nous contenterons d'observer que la décomposition des rayons de la lumière par les prisms, annonce que l'élément du feu, qui est le même que celui de la lumière [a favorite Buffonian identification], le plus subtil de tous les élémens, est pourtant composé de principes qui manifestent encore des propriétés différentes.

Les élémens chymiques [stable chemicals apart from the four elements, or "élémens naturels"] sont donc à bien plus forte raison des corps déjà composés, peut-être même dans un ordre très-avancé, mais qui sont simple pour l'art, puisqu'il ne peut désunir leurs principes, ni faire cesser, par aucun moyen, la dernière surcomposition qui constitue leur caractère actuel. De ce nombre ont été jusqu'à présent les acides, les terres métalliques, vitrifiables, calcaires, les alkalis, les huiles, &c. Ainsi le soufre est un composé pour nous, parce que nous le formons de deux principes, qui sont le feu [phlogiston] & l'acide [vitriolique], & que nous les en retirons à volonté séparément. Ainsi tout métal est un composé d'une terre métallique particulière, du feu fixe ou phlogistique, mais chacune de ces parties composantes résistant à toute analyse ultérieure, nous sommes fondés à l'appeller corps simple ou élément chymique.[3]

Arguing in this passage that there was ultimately but one kind of matter, Guyton differentiated its manifestations into two types. First were the "natural elements," or the traditional four. They were ontologically closer to the prime matter than the diverse collection of stable, specific chemicals or "chemical elements" that he enumerated here. Despite their greater distance from the prime matter, Guyton and Bergman both treated them as the effective elements or "proximate constituents" of bodies, as shown in the next chapter. The most important example of this approach was the acids, which Guyton included in his list in this passage, and which Bergman considered proximate constituents of other bodies, notably metals.

[3] L. B. Guyton de Morveau, Hughes Maret, and J. F. Durande, *Elémens de chymie, théorique et pratique, rédigés dans un nouvel ordre, d'après les découvertes modernes, pour servir aux cours publics de l'académie de Dijon,* 3 vols. (1777—1778), I, pp. 11—13. For this work, henceforth cited as Guyton, *Chymie de Dijon,* and for the public courses on which it was based, see W. A. Smeaton, "Guyton de Morveau's course of chemistry in the Dijon academy," *Ambix,* 9 (1961), 53—69. Smeaton points out (p. 58): "The authors were not named on the title-page, but it was stated (vol. ii, p. xxi) that Guyton drew up the plan of the work, reviewed the whole manuscript before it was printed in order to ensure consistency in all its parts, and himself wrote thirteen of the twenty chapters; that Maret wrote the sections on materia medica and the chapter on volatile alkali (ammonia); and that Durande wrote the chapters on the vegetable acids and the oily solvents (spirits of wine, ether, essential and fatty oils)." Smeaton's paper is cited henceforth as Smeaton, "Guyton's course."

Of the four elements in Boerhaave's chemistry, only three, fire, earth, and water, entered into chemical combinations. The first of these, fire, was actually a binary concept in Boerhaave's mind,[4] for he distinguished two kinds of fire. The first was the *pabulum ignis* or aliment of fire. It was roughly similar to Stahl's phlogiston, in that its departure from chemical combination constituted combustion and its presence rendered combustible the substance of which it was an ingredient.[5] The second was the *materia ignis,* or element of fire, which was uniformly distributed throughout the universe. It consisted of fire particles, the motion of which was responsible for the phenomena of heat.[6] The *materia ignis,* therefore, like air, was an instrument of material change. None of the chemical elements was ever found completely pure (cf. the passage quoted above, p. 33, accompanied by n. 2) but all were necessarily mixed with the *materia ignis* and were laden with numerous other impurities.[7] Indeed, for Boerhaave, air resolved itself almost completely into its impurities. Only the impurities in the air, not the airy element, participated in chemical reaction.[8] Both water and earth entered into chemical combinations, but Boerhaave said little about the participation of water in mixtion, treating it chiefly in its instrumental role; and of earth, he said practically nothing.[9] Nevertheless, in his views about earth lay the chief similarity of Boerhaave's theory of the elements with that of Stahl. More so than in his discussion of the other elements, in his treatment of earth Boerhaave emphasized its role as a property-bearing principle. It conferred fixity and density on its mixts, and it was discerned, like Stahl's principles, only by reflection.[1] In regard to the details of its property-bearing role, however, Boerhaave expressed doubts and uncertainties.[2]

In general, then, Boerhaave's elements were very much less principles than were Stahl's. Though outside the element theory Boerhaave invoked a mercurial principle to account for the common properties of the metals[3]—above all, it conferred on them their high specific weight—this sort of argument did not pervade Boerhaave's chemistry.

3. Specificity in chemistry

A chief desideratum of a reconstructed chemical science was the ability to explain the phenomenological particulars of chemistry. Like the corpuscularians, Stahl

[4] For Boerhaave's fire concepts, see Metzger, *NSB,* pp. 209, 245 and these two papers by Rosaleen Love: "Some sources of Herman Boerhaave's concept of fire," *Ambix,* 19 (1972), 157—174; and "Herman Boerhaave and the element-instrument concept of fire," *Ann. Sci.,* 31 (1974), 547—559.

[5] Metzger, *NSB,* pp. 234—236.

[6] *Ibid.,* pp. 219—223.

[7] E.g., for the impurity of the air, see *ibid.,* pp. 252—253; of water, p. 260.

[8] *Ibid.,* p. 249.

[9] For Boerhaave's views on water, see *ibid.,* pp. 259—271; on earth, pp. 271—280.

[1] *Ibid.,* pp. 271, 275—276.

[2] *Ibid.,* p. 275.

[3] *Ibid.,* pp. 202—203, 276—280; Hélène Metzger, *Les concepts scientifiques* (Paris: 1926), pp. 56—58. The latter work is cited henceforth as Metzger, *Concepts scientifiques.*

and Boerhaave invoked mechanism as the general instrumental cause of all material changes.[4] However, unlike the corpuscularians, they insisted that mere citation of mechanical causes sufficed to explain only the general course of certain events, like solution, and was inadequate to account for the specific properties of bodies or the generic properties of groups of bodies, like combustibles. Moreover, they considered inadmissible the corpuscularians' technique of explaining particular properties by imagining suitable details of particulate size and shape.[5]

Stahl and Boerhaave divided into two categories the phenomena that corpuscularians had attempted to explain by particulate descriptions, and they invoked separate explanatory techniques for each category. The first kind of property was the specific reactivities of chemicals with one another, e.g., solubilities in water, oils, or acids. The second category was the properties that chemicals possessed independently of other bodies and included, for example, color, hardness, lustre, combustibility. This distinction between properties was a purely phenomenological one. To each category belonged its own explanatory technique. The properties (apart from reactivities) of individuals or genera were considered to derive from the properties (apart from reactivities) of their principles. As bearers of qualities, the principles conferred on the mixt its particular attributes—combustibility, friability, and the like. Reactivity, however, was explained by affinities, that is, by attractive powers operating between particles of chemicals. These two kinds of explanation are considered here in turn.

a. Generic and specific properties

1) *Two kinds of theories*

Similarities and differences among the particular properties of chemicals were treated on two levels in the eighteenth century, expressed in these two questions: what renders the members of a group, such as combustibles or metals, similar to one another, that is, what accounts for their generic properties? and what produces the distinctions among the members of a group, such as that between vitriolic and muriatic acids or that between gold and silver, that is, how are specific properties to be explained? Two conceptual techniques, Metzger argues, were commonly used in the eighteenth century to answer these questions. The first technique was to invoke two-component theories, which regarded members of generic groups as composed of two ingredients. One ingredient was common to all members of the genus and served to define it. The other ingredient was specific to each member of the genus and served to distinguish each from the others. The second technique was to regard each member of a generic group as composed of the same substances as were all the others. The common presence of all the ingredients served to define the genus and the differing proportions

[4] Metzger, *NSB*, p. 104 for Stahl; pp. 204—206, 289 for Boerhaave.

[5] See above, pp. 33—34 nn. 4—6 and accompanying text.

of the ingredients served to distinguish the species. This second kind of theory is here called a "diverse-proportion theory." Both two-component theories and diverse-proportion theories were often used when the ingredients were not directly discernable but were revealed only by reflection on experience—as was the case with the principles of the eighteenth-century chemists.[6]

Diverse-proportion theories, Metzger notes, characterized the chemistry of Stahl. Both his saline theory and his metallic theory accounted for the generic and specific properties in terms of differing proportions of the principles shared by all members of the genus. One or more of the three kinds of earth, mixed with water, produced everything.[7] Salts, a category that included both acids and alkalies (cf. below, section 4), were all compounds of earth and water.[8] The first or vitrifiable earth was necessary for the production of all salts. The proportions subsisting among the vitrifiable earth and the other ingredients determined the specificity of salts. For example, the vitrifiable earth and water formed vitriolic acid, the universal acid or universal salt. The latter, in combination with other components, produced all the other salts of all three realms of nature; for example, vitriolic acid, in combination with the second earth, phlogiston, produced nitric acid; and the addition of the third or mercurial earth gave muriatic acid.

Stahl's metallic theory was also a diverse-proportion theory, but it could be construed in two ways as a two-component theory.[9] Fundamentally, it was a diverse-proportion theory, because Stahl assumed that all three earths were present in each metal and that the metals were distinguished from one another by the differing proportions among the three earths.[1] The theory amounted to a two-component theory, however, when viewed with reference to the mercurial earth. Designed especially to account for the generic properties of metals, the mercurial earth constituted the generic ingredient, while the remaining two earths together formed the specific ingredient.[2] In this sense, Stahl's metallic theory was quite similar to Boerhaave's two-component theory of metals, which invoked a mercurial principle to account for the generic property of high specific gravity.[3] Moreover, in both theories the mercurial earth remained in the calx after calcination, a circumstance indicating that the mercurial earth was also the generic component of the calces. The difference between Boerhaave's and

[6] For the two explanations of generic and specific and specific properties, see Metzger, *Concepts scientifiques*, pp. 51—73. She analyzes two-component theories on pp. 53—64 and diverse-proportion theories on pp. 65—73. She notes application of these theories to constituents discerned only by reflection on pp. 52 and 54.

[7] Metzger, *NSB*, p. 130.

[8] For Stahl's saline theory, *ibid.*, pp. 148—159; cf. also Metzger, *Concepts scientifiques*, p. 71.

[9] For Stahl's metallic theory, see Metzger, *NSB*, pp. 132—134, 136—137, 161, 163—164, 169—171.

[1] *Ibid.*, pp. 132—133, 161.

[2] For the roles of the mercurial earth in Stahl's metallic theory, see *ibid.*, pp. 132, 134, 161.

[3] For Boerhaave's metallic theory, see *ibid.*, pp. 277—279, and Metzger, *Concepts scientifiques*, pp. 56—58.

Stahl's metallic theories, then, was more formal than real, as Metzger remarks.[4] Stahl's metallic theory was also a two-component theory when viewed with respect to phlogiston, which conferred the properties that distinguished the calces from the reguline metals. It was the generic ingredient of the reguline metals, and the two remaining earths, considered inseparable, together formed the specific ingredients.[5] As a two-component theory, Stahl's view of metals proved viable in the version that listed phlogiston as the generic ingredient, but not in the version that listed the mercurial earth as the generic component. The reasons for this difference are explored in the next subsection.

Stahl's explanation of combustibility, like Boerhaave's, was an unequivocal two-component theory.[6] The genus of combustibles was defined by the presemce of phlogiston in Stahl's chemistry and by that of the *pabulum ignis* in Boerhaave's. For both Stahl and Boerhaave, combustion was the process in which the generic principle was lost, leaving behind the incombustible component specific to each combustible. Metals, for Stahl, but not for Boerhaave, belonged to the genus of combustibles because they contained phlogiston. Calcination was assimilated by the Stahlians to combustion, because like combustion it consisted of the loss of the generic principle of combustibility, phlogiston.[7] The metallic calces remaining behind were the specific components of metals.

2) *Foundations of two-component theories*

The viability of two-component theories rested on two conditions. The first was the ability to point to the loss of the generic property with loss of the generic principle, such as loss of combustibility with loss of phlogiston. That is, once it had been assumed that combustibility indeed resulted from loss of a non-isolable

[4] Metzger, *NSB*, pp. 378—379.

[5] For the generic role of phlogiston in Stahl's metallic theory, see *ibid.*, pp. 132, 134, 161. It is possible that Stahl and his partisans would have denied that the metallic lustre and malleability of the reguline metals resulted solely from phlogiston, or that the high specific gravity of metals resulted solely from the mercurial earth. The passages quoted by Metzger in her analysis of the Stahlians' metallic theory provide evidence that the Stahlians believed all of these properties were somehow produced jointly by the mercurial and the phlogistic earth. A definitive interpretation is not possible, however, because, as Metzger emphasizes (e.g., *NSB*, p. 137), the theory was vague and the Stahlians had difficulty in assigning a clear-cut role to the mercurial earth. Whatever the difficulties of interpretation, they do not bear on the point here under discussion, that the difference between a two-component theory and a diverse-proportion theory could be slight, especially if one of the components of a diverse-proportion theory was considered essential to the genus in a stronger sense than were the others, that is, insofar as it was more thoroughly implicated than the others in production of the generic properties. This was indeed the case with the mercurial earth, and, insofar as the reguline metals were generically distinct from the calces, with phlogiston.

[6] For Stahl's theory of combustion, see Metzger, *NSB*, pp. 158—188; for Boerhaave's, pp. 228—245.

[7] For the Stahlians' assimilation of metals to combustibles and calcination to combustion, see *ibid.*, pp. 163—164, 169—171. Treatment of metals and combustibles in identical terms among the later Newtonian Stahlians was an important theme of the two-component chemistry of the late eighteenth century, as shown in Chapter II.

principle of combustibility, the theory could be used in new cases only if the circumtances of combustion were subject to interpretation as cases in which the principle was lost. (In some cases, a still stronger claim could be made for a two-component theory by citing examples of synthesis, that is, by pointing out the restoration of the generic property upon reintroduction of the generic principle into the mixt. For example, the revivification of metals, considered a synthesis from calx and phlogiston, added significantly to the prestige of Stahl's theory of calcination as a kind of combustion.[8]) The second condition for the viability of two-component theories was the ability to demonstrate directly the existence of the specific component after the generic component had left it, and to show that the specific component of each member of the genus indeed differed from those of all the other members. For combustibles and for metals when considered species of phlogiston compounds, Stahl could meet both these conditions. Each metal left a calx, and each calx differed from all the others. Each combustible left a residue, and each residue differed with the combustible. The existence of the specific component, unlike that of the generic component or indeed that of any principle that was non-isolable, was not merely inferred, but directly demonstrated. Moreover, the specificity of the specific components was likewise directly demonstrated.

By contrast, Stahl could not meet these conditions in either the metallic theory that treated metals as species of compounds with the mercurial earth or the saline theory. Unable to point to examples of the loss of the mercurial principle or to the specific components left behind upon loss of the mercurial principle, Stahl and Boerhaave could provide only weak foundations for the mercurial theory. The appeal of the theory derived primarily from its ability to account for the high specific gravity of metals and calces, but in the absence of experimental material that could be construed as direct support for the theory, it was abandoned. As this occurred, the phlogiston theory of metals was acquiring additional support from the failure of analysis to decompose the calces, which increasingly looked like a collection of specific, elementary chemicals, as noted shortly below.

Stahl also could not devise a saline theory that met the two conditions for the viability of two-component theories. As noted below in section 4, salts were not formally defined in Stahl's day. Accordingly, there was no unequivocal generic property or collection of generic properties that could be attributed to a saline principle. Moreover, even assuming such a principle, it was impossible to find examples of the loss of salinity with loss of a saline principle or to point out the specific components left behind. The post-Stahlian theory of neutral salts, described below in the next subsection, did meet the conditions, permitting

[8] Metzger notes (*NSB*, p. 171) that because metals, unlike other combustibles, could be reconstituted from their decomposition products by addition of phlogiston, metallic calcination became the archetype of combustion. She also notes that the supposed synthesis of sulfur added to the prestige of the theory.

a theoretical accommodation of neutral salts different from that of other saline bodies.

Supporting the claims made in the preceding paragraphs is that new two-component theories emerged in the late eighteenth century at the same time as advances of analysis made available new collections of specific chemicals or showed that known isolable chemicals could not be decomposed. Advances in analysis are considered in section 5 and several leading two-component theories prevailing prior to Berzelius are explored in depth in Chapter II. In preliminary illustration of the appearance of new two-component theories, two examples are noted here. First, Metzger points out that Lavoisier, working against the background of eighteenth-century failures to achieve the transmutation of salts that Stahl's saline theory had given cause to expect, had accomplished much by his ability to distinguish the acids sharply from one another.[9] He did so by invoking a two-component theory. For Lavoisier, acids consisted of a generic component, oxygen, common to each acid and conferring the generic property of acidity; and a specific component, the acidifiable base, different for each acid and determining specific properties.[1] The oxygen theory of acidity was clearly marked as a two-component theory of generic and specific properties from its earliest statement by Lavoisier in 1776. Referring to earlier experiments on the combustion of phosphorus and sulfur, Lavoisier asserted,

> Ces premiers m'ont fait réfléchir sur la nature des acides en général, et, en examinant les circonstances de leur formation et de leur destruction, j'ai cru entrevoir que tous étaient composés en grande partie d'air; que cette substance était commune à tous et qu'ils étaient ensuite différenciés les uns des autres par l'addition de différents principes particuliers pour chaque acide.[2]

This innovation rested on the two conditions above noted as the foundations of two-component theories. Lavoisier could synthesize acids by oxygenating bodies and could deprive bodies of their acidity by depriving them of their oxygen. Moreover, he had a collection of specific chemicals, the acidifiable bases, that served as the specific components. The latter were isolable (the

[9] Metzger, *NSB,* pp. 155—157; *idem, Concepts scientifiques,* p. 71.

[1] Metzger, *NSB,* pp. 155—156; *idem, Concepts scientifiques,* pp. 58—59; *idem,* "La philosophie de la matière chez Lavoisier," *Actualités scientifiques et industrielles,* no. 218 (1935), 1—48 (separate pagination), esp. 28—29.

[2] A.L. Lavosier, "Mémoire sur l'existence de l'air dans l'acide nitreux, & sur les moyens de décomposer et de recomposer cet acide" (1776; rev., 1779), in Lavoisier, *Oeuvres,* II, pp. 129—138, esp. p. 129. For bibliographic details about this memoir, including proof that it dated from 1776, see Partington, *History,* III, pp. 410—411. In this study, Lavoisier's memoirs are cited from his *Oeuvres* and, unless otherwise noted, are dated by the year of the *Mémoires de l'Académie* in which each appeared. These dates were often, but not always identical with the years in which the memoirs were read (though the date of reading is not always known), and differed from the date of publication, often appearing in print three or four years after having been read. Dates of the *Mémoires de l'Académie* for each of Lavoisier's memoirs, dates of reading, when known, and dates of publication are listed by Partington, *History,* III, pp. 368—371.

organic radicals excepted), and they all differed from one another. Indeed, Lavoisier's oxygen theory was the culmination of the tradition of two-component theories. As argued in Chapter II, his list of elements was nothing other than a collection of isolable, specific chemicals, each capable of producing a specific acid, or several specific acids, with oxygen.

A less dramatic but equally illustrative case is the transformation of Stahl's metallic theory. Among the later French Stahlians, a group discussed below in section 6, Stahl's metallic theory was changed from a diverse-proportion theory into a full-fledged two-component theory.[3] The change depended on the acceptance of the metallic calces as a set of specific components that analysis suggested were not compound, at least, not in the sense that Stahl believed them compound (containing the other two earths). Under consideration in the work of Macquer, this step was taken by Guyton.[4] He accepted a collection of specific metallic earths, the calces revealed by loss of phlogiston, as is shown by the passage quoted above (p. 35, accompanied by n. 3), in which he listed the "terres métalliques" among the "élémens chymiques." He eliminated the mercurial earth of Stahl and did not trouble himself with the problem that disturbed Macquer, that without the mercurial earth the chemist could not explain the high specific weight of metals and calces. Guyton accepted, but did not explain, this generic property. Henceforth, phlogiston, not the mercurial earth, was called the metallizing principle, because it conferred on *these* earths, but not on others, the properties that distinguished the reguline metals from their earths.[5] The shift from a diverse-proportion theory to a two-component theory of metals amounted to an exchange of a non-isolable generic principle for a collection of isolable, elementary specific chemicals.

3) *Generic bias of two-component theories*

The growing acceptance of two-component theories in reflection of the advances of analysis was ironic. Despite their foundation in the availability of collections

[3]Three of the most important sources for the views of the later French Stahlians, in addition to Guyton's *Chymie de Dijon*, are P.J. Macquer, *Elémens de chymie théorique* (Paris: 1749), A. Baumé, *Manuel de chymie ou exposé des opérations de la chymie et de leurs produits* (2nd ed.; Paris: 1766), and Macquer, *A dictionary of chemistry*, [tr. James Keir,] 2 vols. (London: 1771). Baumé's second edition scarcely differed from the first, as he himself stated (pp. x—xj). Macquer's 1766 *Dictionnaire* was unavailable to the present author. Though Keir's 1771 translation offered numerous additions to the original, it carefully distinguished them from Macquer's own text. The first of these three works is cited henceforth as Macquer, *Elémens;* the second, as Baumé, *Manuel;* and the third, as Macquer, *Dictionary*.

[4]For Macquer's hesitant consideration of the case for eliminating the mercurial earth see his *Elémens*, pp. 74—75 and his *Dictionary*, I, pp. 420—422. Like Guyton, Baumé readily rejected the mercurial earth, claiming that Stahl himself had already done so. See his *Manuel*, p. 23 and esp. p. 67. Relevant passages from Guyton are cited in the next note.

[5]For Guyton's acceptance of a collection of metallic earths and his treatment of phlogiston as the metallizing principle, see his *Chymie de Dijon*, I, pp. 108, 207, 215, 223, 224—225.

of specific components, the theories characteristically placed great emphasis on the generic component and on generic properties, while neglecting the specific. Following Metzger's suggestive example of employing Aristotelian language to describe two-component theories of generic and specific properties, one may say that, viewed from the standpoint of the genus, the specific component was a "primitive matter" and the generic components were "forms."[6] By combining with the specific component, the generic ingredient informed it, or conferred on it certain generic properties. Just as a primitive matter was devoid of properties and therefore also of interest, so the specific ingredients of two-component theories, in comparison to the generic, seemed relatively undistinguished by their properties and accordingly received little attention. Their ability to receive forms by union with the property-bearing generic components was the main concern. The generic principles therefore were studied intensely and characterized fully, even if non-isolable. Boerhaave and Stahl spoke of their mercurial principles, for example, but said almost nothing of the specific components of metals; insofar as metallicity was concerned, only the generic ingredient mattered. Similarly, membership in the genus of combustibles depended upon the generic ingredient, phlogiston, but the specific ingredients were diverse materials of little importance to the theory of combustion.

Other practitioners in the eighteenth century invoked two-component theories, exhibiting the same emphasis on the generic, to account for other kinds of substances. One prominent example was Rouelle's saline theory, which, as Partington points out, was an import from Germany.[7] In his discussion of neutral salts, Rouelle classified them first, according to their phenomena of crystallization, and second, according to their composition. As he maintained in 1744, they consisted of two ingredients:

> Je donne à la famille des sels neutres toute l'extension qu'elle peut avoir; j'appelle sel neutre[,] moyen ou salé, tout sel formé par l'union de quelque acide que ce soit, ou minéral ou végétal, avec un alkali fixe, un alkali volatil, une terre absorbante, une substance métallique, ou une huile.[8]

Ten years later, he expressed their binary composition in still more general terms, saying that a neutral salt

> [est] formé par l'union d'un acide avec une substance quelconque qui lui sert de base, et lui donne une forme concrète ou solide.[9]

[6] Metzger, *Concepts scientifiques*, pp. 57—64.

[7] Partington, *History*, III, p. 74.

[8] Quoted by Partington, *ibid.*, from G.F. Rouelle, "Mémoire sur les sels neutres, dans lequel on propose une division méthodique de ces sels, qui facilite les moyens pour parvenir à la théorie de leur cristallisation," *Mém. de l'Acad.* (1744), 353—364. For this study, the Amsterdam-reprint of the memoirs is consulted for Rouelle's paper: *Mém. de l'Acad.* (1744; rpt., Amsterdam: 1750), pp. 480—495. Rouelle's paper is cited henceforth from this edition as Rouelle, "Sels neutres."

[9] Quoted by Partington, *History*, III, p. 74. Cf. Lavoisier's 1778 paper, "Considérations générales sur la nature des acides, & sur les principes

Neutral salts, then, were two-component substances and could be so treated theoretically. The generic component was the acid; the specific, any other substance serving to coagulate the acid into a typically crystalline solid. Accordingly, within groups first defined crystallographically, Rouelle classified neutral salts generically by acid and specifically by base:

je tire le caractère générique de l'acide qui est essentielle à plusieurs sels, par exemple, tous les sels formés par l'acide vitriolique qui sont dans une même section [defined crystallographically], forment un genre, & je tire le caractère spécifique de la base de chaque sel: ainsi l'acide vitriolique uni à un alkali fixe, à un alkali volatil, à une terre absorbante & à une substance métallique, fait autant d'espèces.[1]

Metzger points out that modern theories of neutral salts need be further distinguished as "double-entry two-component theories," because either component may be treated generically and the other specifically. Classification therefore is never entirely satisfactory; an exposition of the subject of neutral salts according to their acids must make provisional references to the properties of their bases, and *vice versa.*[2] Underlying this interchangeability, it may be added, is the ontological equivalence of the two saline components. In Rouelle's saline theory the components were not interchangeable, however, because they were not ontologically equivalent: essential to the genus of neutral salts was an *acid*—a member of another category of chemicals of lesser complexity than neutral salts—but the specific components, like those of the genus of combustibles, were diverse materials of little importance to the generic properties of neutral salts. There was no genus of bases exhibiting positively defined "basic" properties opposed to those of the genus of acids. A base was just what its name suggested: a passive entity, a substrate for the coagulation of the generic component into a solid. Rouelle's theory of neutral salts was therefore not an example of a double-entry two-component theory. As noted in the next subsection, Guyton employed the term "base" very generally to describe the specific components of two-component theories as passive substrates suitable for receiving generic properties. Lavoisier acquired the term from Guyton and used it in precisely the same way, as pointed out in Chapter II.

If Lavoisier's oxygen theory of acidity, with its list of elements as the collection of available specific components, was the culmination of the tradition of two-component theories, it was also the culmination of the ironic preoccupation of these theories with the generic level. As argued in Chapter II, Lavoisier and his

dont ils sont composés," in *Oeuvres,* II, pp. 248—260, esp. 248: "A mesure que l'art fit de nouveaux progrès [beyond the iatrochemists and their principles], les chimistes qui leur succédèrent s'aperçurent que les substances qu'ils avaient regardées comme principes étaient encore suspectibles de décomposition, et ils reconnurent successivement que tous les sels neutres, par exemple, étaient formés par la réunion de deux substances, d'un acide quelconque et d'une base saline, terreuse et métallique."

[1] Rouelle, "Sels neutres," 489; cf. also 488.

[2] Metzger, *Concepts scientifiques,* pp. 62—64.

contemporaries, Guyton and Bergman, all needed specific components to make their theories work, but none of them developed a chemistry of specific components or specific properties. What mattered to them on the specific level was that specific components existed *at all* and that they differed from one another *in any way*. What the specific properties were, and what their differences, remained of scant importance. A significant illustration of this point is provided by the analysis in Chapter II of the competition between Guyton's radical theory and Lavoisier's oxygen theory of acidity. Guyton conceded the superiority of Lavoisier's theory, because the specific components it invoked were known, isolable substances, while those of his own theory were merely inferred to exist by reflection upon experience (just as the existence of Stahl's principles had been inferred). The advantage that Guyton acknowledged in Lavoisier's theory was not its greater ability to treat specific properties, but its concrete evidence for the existence of specific components. Lavoisier was more interested in oxygen than in acidifiable bases; in acids than in salifiable bases; and in the caloric fluid than in the bases it rendered elastic; generally, in generic principles than in specific chemicals. Lavoisier's list of specific components, his list of elements, remained unexploited in his own hands.

It was Berzelius who first developed a chemistry of specific components and specific properties. By making acidity a specific and not a generic property; by elevating bases to a status equal to that of acids and defining basicity as a positive, specific property; by affirming that a collection of diverse specific elements was the primary foundation of chemistry, Berzelius transformed Lavoisier's taxonomic two-component chemistry, which subordinated species to genus, into a double-entry two-component chemistry, which treated both components equally.

b. Affinities

Two aspects of eighteenth-century affinity theory that bear on later parts of this study are examined here. First, affinity theories provided an explanation for the phenomena of mixtion and aggregation; second, they provided the definition of a reagent as any substance endowed with affinities. These functions of affinity theory became highly prominent in the work of the later French Stahlians and informed the two-component chemistry prevailing before Berzelius.

Just as two-component theories explained the specificity of the properties of chemical individuals or groups, so affinity theories explained the specificity of reactions. Accordingly, they occupied a central position in the restored chemical science of the eighteenth century. Like two-component theories, affinity theories were developed in reaction to the failure of corpuscularism to account for the particulars of chemistry, in this case, for specific reactivities. In the kinematic corpuscularisms of Boyle and Lémery, specificity of reactivity had been explained in the same way as other specific properties, by referring them to the mechanical "affections" of particles—size, shape, porosity, texture, structure. The advocates of dynamic corpuscularisms also had acknowledged the role of these factors in determining specific reactivities, but they had hoped to dispense with particulate descriptions by invoking another technique: force measurement. Assuming that, at microscopic distances, particulate shape as well as other factors not significant on the astronomical scale were important determinants of interparticulate forces, the exponents of dynamical theories had sought to reduce descriptions of specifity in reaction to measurement of forces.[3] Reactivities owed their specificity to particular values of the forces subsisting between particles of the reactants. Chemical practitioners of the early eighteenth century rejected both these techniques. Their belief in the inadmissibility of particulate descriptions was noted above. They resisted force measurement as well because in their eyes it failed adequately to meet their need for an explanation of specific reactivities. Boerhaave, for example, argued that the specification of Newtonian attractions could be accomplished only by conceding in advance that particular chemicals possessed particular properties. Dynamic corpuscularism could explain properties common to all chemicals, but not the reactivities of particular chemicals.[4]

Instead of either particulate description or force measurement, Boerhaave and Stahl invoked affinities, or specific attractive powers between the particles of particular chemicals.[5] For both Stahl and Boerhaave, as for most chemical practitioners of the eighteenth century, chemical reaction was typically a process of attack by dissolution.[6] Stahl's approach, Metzger points out, was a virtually animistic reaction against kinematic corpuscularism and later accorded well with Newtonian dynamics.[7] Stahl's reagents, water, the three kinds of earth, mixts, even aggregates, reacted according to a rule of affinities that stated simply "like attracts like."[8] Though the rule failed to account for the union of heterogeneous principles into secondary principles, it could account for so large a number of reactions that the Stahlians were not prepared to let this flaw, serious as it was, compel their abandonment of the doctrine.[9] Reactions usually were double displacements, and it was by reflecting upon them that the chemist discerned the existance of the principles, themselves always combined.[1] The rule could be employed, moreover, as a general technique of analysis: the composition of a substance was revealed by that of the substances with which it entered into

[3] For the appeal of force measurement and calculation, see Metzger, *NSB*, pp. 40—45, 48—50, and Thackray, *Atoms and powers*, pp. 67—73.

[4] For resistance to force calculation and concern for specific reactivities, see Metzger, *NSB*, pp. 48—51, 101—108, 116, 205—206, 289.

[5] *Ibid.*, pp. 116, 130—131, 205—206, 289.

[6] For this point as it applies to Stahl, *ibid.*, pp. 104 and 105; to Boerhaave, pp. 280—293.

[7] *Ibid.*, p. 116.

[8] *Ibid.*, pp. 139—148; Metzger, *Concepts scientifiques*, pp. 42—43.

[9] Metzger, *NSB*, pp. 145—146.

[1] *Ibid.*, pp. 119—120, 123.

composition.[2] Accordingly, acids and alkalies were not the opposing antagonists that a prominent seventeenth-century theory had made of them;[3] on the contrary, they were similar. With the intensification of interest in affinities notable from mid-century, the notion that interacting substances were similar gave way gradually, under the impact of Newtonian chemistry of a more Boerhaavean bent, to the notion that they were, if not quite opposites, then likely very dissimilar.[4]

Boerhaave's emphasis on dissolution as the chief process of reaction was greater than Stahl's and his account of dissolution by particular reagents or menstrua was correspondingly more elaborate.[5] Reaction was the mutual dissolution of the solvent and the solute, and the forces responsible were accordingly particular to the pair of reagents in question.[6] Boerhaave was unprepared, however, to countenance the physicalist attempt to determine variations in the universal law of attraction that were supposed to be responsible for these particular effects. Rather, he argued that experimental investigation should be pursued to develop a rational classification of reaction phenomena. This attitude, increasingly prominent from the mid-eighteenth century, sanctioned the revival of affinity tables originally pioneered by Geoffroy.[7]

The operation of menstrua required that they be fluid. Their particles interacted with those of the solute, rendering it also fluid. The particles of the menstruum and those of the solute each occupied the spaces between those of the other and combined with one another by virtue of the operant attraction. A solution was therefore a compound, and the circumstances of its formation required that both of its constituents be in the same state of aggregation, initially a fluid, but perhaps, for example on cooling of a molten metallic solution, a solid.[8] These notions, Metzger remarks, were entirely reconcilable with Newtonian thought;[9] they enjoyed a considerable vogue from mid-century, particularly among the Newtonians inspired by Buffon, as described below in section 6. Even though Buffon asserted the absolute universality of the inverse square law, he recognized that mechanical and geometrical considerations modified its microscale effects. Though he encouraged renewed attempts to determine those modifications by calculation, he also encouraged a more qualitative taxonomic chemistry of affinities that was similar to Boerhaave's, spurring renewed interest in affinity tables. The same general notions of the mechanism

[2] *Ibid.*, pp. 145—146.
[3] *Ibid.*, pp. 142—143, and Metzger, *Concepts scientifiques*, pp. 46—47.
[4] For this change see A.M. Duncan, "Some theoretical aspects of eighteenth-century tables of affinity," *Ann. Sci.*, 18 (1962), 177—194, 217—232, plates VI--VII, VIII—IX, esp. 182—185, 191—192, 218. This paper henceforth is cited as Duncan, "Affinity tables."
[5] For Boerhaave on menstrua, see Metzger, *NSB*, pp. 280—305.
[6] *Ibid.*, pp. 281—282.
[7] *Ibid.*, p. 50.
[8] *Ibid.*, pp. 282, 288—289.
[9] *Ibid.*, p. 285.

of solution; of the requirement that both reagents be fluid, or, as it was said later, "equiponderant"; of the specificity of forces; and of the coincidence of solution and composition; all received regular enunciation in the two-component chemistry in the late-eighteenth century.

In Boerhaave's chemistry, the action of fire on all bodies was similar to the action of menstrua on solutes. The complexities of Boerhaave's fire concept, however, involving two entities and their differing roles, as well as Boerhaave's failure to maintain this distinction consistently, prevented him from elaborating the similarity. Nevertheless, the comparison became quite explicit later in the century among the French Newtonian disciples of Stahl, as shown below. A solvent, like fire, was considered to oppose the mutual attractions among the particles of a homogeneous body, to separate, and then combine with them. These Boerhaavean ideas were described in a Stahlian vocabulary, however. The tendency of the homogeneous or integrant parts of a body to form aggregates was called affinity of aggregation; and the mutual affinity between particles of different bodies, between a menstruum and a solute—the constituent parts of a new, compound body—was called affinity of composition.

Having rendered the action of fire and solvents analogous, the French Stahlians tended to reduce the former from its role as a universal, general instrument, to a menstruum; that is, to make fire a chemical.[1] This change was accompanied by another. Boerhaave had emphasized that the interaction between menstruum and solvent was mutual in that each was the solvent of the other, and each was solute for the other.[2] The later French Stahlians, on the contrary, adopted the two-component view, in which the menstruum was the generic agent, and the solute the passive substrate, or base. The very name "base," especially in Guyton's theory, served to describe the role of the solute as patient, as passive primitive matter, which the agent, the menstruum, informed with generic properties. Since the French Stahlians made Boerhaave's fire a chemical instead of a universal instrument, they thereby transformed it into a generic component in a two-component theory, aptly naming it "phlogiston."

By calling for the construction of affinity tables, the French Stahlians also

[1]Though this change is not explored in this study, it need be noted in anticipation of the analysis of Guyton's two-component theories in Chapter II. Guyton consummated the transformation of fire into a chemical in his *Chymie de Dijon* (I, pp. 158—166). The model for this change was air, a subtle fluid that proved itself to be a chemical by becoming fixed in bodies. Emphasizing the similarity that he saw between phlogiston and air, Guyton introduced the synonym "fixed fire" for phlogiston. For the synonym, see Guyton, *Chymie de Dijon,* I, p. 164; for the express parallelism of fire and air, see Guyton's "Conciliation des principles de Sthaal [*sic*] avec les expériences modernes sur l'air fixe," *Obs. sur la physique,* 7 (1776), 389—295, esp. 391—392. Aspects of this paper are analyzed in detail in Ch. II. For further development of this theme, see Melhado, "Phlogiston, oxygen, and caloric: an interpretation," forthcoming.

[2]Metzger, *NSB,* p. 281.

realized Boerhaave's view that specific reactivities should be used to develop a rational classification of matter.[3] The Stahlians began the vogue for affinity tables after mid-century by calling attention to the earliest table, compiled by E. F. Geoffroy (1672—1731) in 1718.[4] Under the symbol for a given chemical, Geoffroy's table listed those of other chemicals that reacted with it, in order of the strength of reaction, which was identified with the strength of affinity. The latter was most often determined by examining the displacements of one substance from combination with another. Any substance that displaced another from combination with a third was considered to have a greater mutual affinity with the third. Geoffroy's table had few entries and some of the chemicals heading the lists were not specific substances, but genera. A significant example was provided by the column headed by the notation "SM," for "substances mĕtalliques." Beneath the abbreviation were listed several acids, and their order in the list exhibited the strengths of their affinities for metals. With the revival of interest in affinity tables among the French Stahlians after mid-century, the reactivities of known chemicals were sharply distinguished and ordered, and those of newly discovered specific reagents, such as new earths and metals, were determined. Much experimental effort went into constructing the tables, which recorded the progress achieved in acquiring knowledge of specific substances and their reactivities. A principal preoccupation of the late-century Stahlians, the tables reached their culmination in the work of Bergman.

4. Acids, alkalies, and salts

The two-component theories of Bergman, Guyton, and Lavoisier explored in Chapter II were developed chiefly to accommodate acids and neutral salts. The theories contained articulate definitions of these substances that replaced the older, vague, and broadly inclusive notion of salts that is the subject of this section. None of the two-component theories of the late eighteenth century accommodated bases in the positive sense of possessing properties having status equivalent with those of acids. Only in Berzelius's chemistry did the concept "base" receive a positive theoretical treatment.

In the seventeenth and eighteenth centuries, Metzger points out, the concept of salt was a notion of common sense and therefore received no formal definition.[5] However, various criteria typically were cited as the characters of salts. The principal ones were taste, said to be exemplified above all by the mineral acids; solubility in water; crystallization (with reference to neutral salts); and last, corrosive properties. These characters were never cited systematically and

[3] On the affinity theory of the French Stahlians, see below, subsection 6.*b*.

[4] On Geoffroy's table, see Duncan, "Affinity tables," 181—187, and pl. VI.

[5] Metzger, *NSB*, p. 150.

were not assembled into a precise definition. Rather, practitioners invoked one or more of these or other characters in accordance with their needs and predilections. Consequently, the prevailing notion of salts was vague, including in its purview acids, alkalies, and neutral salts.[6] Stahl's beliefs about the composition of acids, described in the last section, were central tenets of his saline theory. The removal of acids and alkalies from the category of salts was effected by Lavoisier, who argued, as is well known, that the term "salt" should be reserved for neutral salts, that alkalies should be grouped among the acidifiable bases, and that the acids should be considered a class apart, their properties explained by the oxygen theory of acidity. Despite this innovation, he occasionally resorted, when pressed, to the vague, traditional notion, as shown in the next chapter.

The example of Rouelle reported in the last section indicates that neutral salts were becoming recognized in the eighteenth century as a separate category of salts, composed of acids in combination with one or another sort of base. Neutral salts acquired increasing importance as the eighteenth century neared its end. After Rouelle, several thinkers made elaborate efforts to distinguish various kinds of neutral salts, providing precedents for Berzelius's elaborate saline theory. Bergman in particular was aware of numerous categories of neutral salts and classified them according to the number of their constituent acids and bases.[7] What Lavoisier called simply "neutral salts," Bergman labelled "double salts," because they contained two proximate constituents, an acid and a base. Lavoisier was little concerned with the other kinds of neutral salts that Bergman considered and that, as noted in Part II, Fourcroy explored in the 1790's. These included basic and acid salts, as well as what Berzelius, in a departure from Bergman's nomenclature, named "double salts," that is, those salts composed of two bases and an acid, and less frequently, two acids and one base.

In the theory of Stahl, acids were considered mixts of water and one or more of the various earths, as shown in the last section. Acidity could be quenched, however, by addition of a sufficient dose of phlogiston.[8] The product of quenching was often a combustible. Sulfur, for example, resulted from the combination of phlogiston and the universal acid, as Stahl showed in a famous experiment. Similarly, phosphorus could be obtained by phlogisticating its acid. Viewed from this standpoint, the acid took on the character of a principle: though itself compound, it was the proximate constituent of a combustible. Indeed, the universal acid was a secondary principle, and as such was considered

[6] For the traditional notion of salts, see Metzger, *ibid.*; Maurice P. Crosland, *Historical studies in the language of chemistry* (London: 1962), pp. 77-78; and *idem*, "Lavoisier's theory of acidity," *Isis*, 64 (1973), 306—317, esp. 309.

[7] For Bergman on salts, see his dissertation, *Meditationes de systemate fossilium naturale* (1784), in his *Opuscula physica et chymica*, 6 vols. (Leipzig: 1779—1790), IV, pp. 180—278, esp. pp. 225—230. (For details about Bergman's dissertations and his *Opuscula*, see below, p. 65, n. 3). The dissertation here in question is henceforth cited as Bergman, *Meditationes*, with page references to the *Opuscula*.

[8] For the quenching of acidity by phlogiston, see Partington, *History*, III, pp. 671—672.

the first term of synthesis and the last of analysis, as noted earlier. For this reason, Stahl's experiment demonstrating the production of sulfur from sulfuric acid was considered a proof that sulfuric acid was an element. With this precedent, the later French Stahlians found it an easy step to consider the acids as proximate constituents or primitive chemicals like elements. The passage from Guyton quoted above (p. 35, accompanied by n. 3) listed the acids among the "chemical elements," and Bergman developed a theory of metals and combustibles in which acids, along with phlogiston, were the proximate constituents. To emphasize their elementary nature, Bergman called these acidic components "radical acids." From this derived the concept of radicals. These developments are explored in the next chapter.

5. Analysis

a. Introduction

During the seventeenth and eighteenth centuries, the issues that the corpuscularians raised about chemical analysis remained prominent and vexatious. The main problem was whether, as Boyle had argued, the products of analysis were creations of fire, the almost universally employed instrument of analysis, or whether, as he denied, they were present as ingredients in the mixt. A second problem was whether the products, if they were indeed ingredients, were subject to further analysis. A third was the overall uniformity in the products obtained by analysis from varieties of vegetable matter. If the products of vegetable distillations differed little from one plant to the next, what significance did the products possess? Questions of this kind are examined repeatedly in Metzger's *Doctrines chimiques en France,* which offers among the best discussions of analysis in the seventeenth and eighteenth centuries.[9] An immense historical problem, the development of conceptual issues of chemical analysis awaits further scholarship. Only a few remarks of a general nature are offered here.

The impact of corpuscularism on the chemistry of the mineral realm was more immediate than on the chemistry of the vegerable realm. All five Paracelsan principles were expected to be present in every mixt, vegetable or mineral. Though the five principles were readily identified with the residue and with the four typical distillate fractions customarily obtained in vegetable analysis, mineral matter, especially metals, could not always be made to yield the same numbers

[9] Cited above, p. 27, n. 1. Other useful secondary literature addressed to the conceptual issues of chemical analysis in the sixteenth and seventeenth centuries include Kuhn, "Structural chemistry"; Allen G. Debus, "Solution analysis prior to Boyle," *Chymia,* 8 (1962), 41—61; *idem,* "Fire analysis and the elements in the sixteenth and seventeenth centuries," *Ann. Sci.,* 23 (1967), 127—147; and Frederick L. Holmes, "Analysis by fire and solvent extractions: the metamorphosis of a tradition," *Isis,* 62 (1971), 129—148. Holmes's paper is cited henceforth as Holmes, "Solvent extraction."

and the same sorts of ingredients. The vulnerability of mineral chemistry to corpuscularian attack was accordingly greater. Moreover, because in the chemical science undergoing reconstruction in the eighteenth century mineral chemistry increasingly eclipsed the formerly dominant vegetable chemistry, the evolution of analytical issues in mineral chemistry was correspondingly faster and more pronounced than in the other realms. This change in orientation was hastened, especially in Germany and Sweden, by the growth of mining, metallurgy, assaying, other aspects of mineral chemistry, and mineralogy. The appeal of Stahlian chemistry to metallurgists is well known. To the extent that eighteenth-century chemistry was primarily that of the mineral realm, issues of analysis were focussed primarily there.

b. Mineral analysis

Several developments, both theoretical and practical, contributed to the clarification of issues of mineral analysis. One was that the attribution of secure ontological status to certain chemicals, especially those recoverable after having undergone a sequence of reactions, permitted their use both conceptually and practically as "proximate constituents." Such chemicals were substances that, though not necessarily elementary, did serve as ingredients of mixts. Among the earliest examples was that of the metals. Lémery had suggested the utility of treating them as elements,[1] and their use as proximate constituents persisted throughout the eighteenth century—even, as seen in Chapter II, among later Stahlians such as Bergman, who clearly believed them to be compounds containing phlogiston. Developments of this sort were sustained by the Stahlian notion that mixts or secondary principles were considered the last term of analysis and the first of synthesis. Stahl himself believed that at least two of the metals, gold and silver, were secondary mixts. For later Stahlians it was an easy step to subsume the remaining metals under this view by considering them as compounds of phlogiston and earths.

Another factor contributing to clarifying problems of mineral analysis was the increasing resolution of each of the hypostatical principles and elements into several specific, isolable chemicals. The rise of pneumatic chemistry exemplified this development in regard to air. Various theoretical developments that cannot be pursued here had contributed to establishing a conceptual foundation for a single airy element. The growth of British pneumatic chemistry forced the resolution of this element into several airy chemicals. Less dramatic, but scarcely less important, was the resolution of the hypostatical earth into a number of particular, earthy chemicals. The dynamics of this process are visible in several late-eighteenth-century contexts. Two of these, the elimination of the mercurial earth in favor of a collection of metallic earths and Guyton's juxtaposition of the

[1] Metzger, *Doctrines chimiques,* pp. 315—316.

four "élémens naturels" with a diverse collection of "élémens chymiques" have already been noted (cf. the passage quoted above, p. 35, accompanied by n. 3). Baumé's discussion of earth in his 1765 *Manuel* provides another example. His treatment of the subject opened with the remark, "La terre élémentaire n'est pas aussi facile à reconnoître que les autres élémens . . . Cette difficulté vient de la prodigieuse variété de pierres et de terres que nous offre la nature."[2] Though Baumé felt he could reduce this variety to two types, which he believed to correspond with those recognized by Stahl, his discussion clearly exhibited the pressure exerted by the studies of mineralogists on belief among chemists in a restricted number of elementary principles.

In the German and Swedish mineralogical traditions this effect was still more pronounced (indeed, Baumé's citations reveal his familiarity with this literature). The mineralogist, mining engineer, assayer, and mineral chemist all discovered and analyzed a variety of naturally occurring minerals containing numerous distinguishable earths. Cronstedt's mineralogy listed nine particular earths that Marggraf's analytical work later showed to be compounds and mixtures of only three "primitive earths," that is, ones serving as proximate constituents: alumina, silica, and chalk. In Bergman's mineralogy, the number of primitive earths rose to five, two recently discovered ones, barytes and magnesia, having been added to the list.[3] Bergman's work culminated the chemical approach to mineralogy taken by eighteenth-century Swedish practitioners. These figures enumerated an abundance of specific substances that earlier in the century had been scarcely surmised to exist. They thus contributed to chemistry a collection of particular and sometimes "primitive" chemicals that could be deployed as specific components in two-component theories or required explanation in terms of such theories.

Bergman's achievements (and Marggraf's) rested on the rise of wet chemical analysis, a third factor leading to clarification of issues in mineral analysis. During the eighteenth century, wet chemical procedures occupied an increasingly prominent place beside the formerly predominant fire analysis. Typically milder and potentially more discriminating, wet chemistry permitted examination of a broad range of reactions, especially displacements, in which the presence of proximate constituents could be discerned, leading, eventually, to their isolation and identification. Wet chemical phenomena received theoretical articulation in the notions of dissolution and menstrua, so prominent in Boerhaave's chemistry,

[2] Baumé, *Manuel*, pp. 65—68.

[3] For aspects of the history of the earths, see D.R. Oldroyd, "A note on the status of A.F. Cronstedt's simple earths and his analytical methods," *Isis*, 65 (1974), 506—512; *idem*, "Some phlogistic mineralogical schemes illustrative of the evolution of the concept of 'earth' in the seventeenth and eighteenth centuries," *Ann. Sci.*, 31 (1974), 269—305; Olsson, *Kemiens historia*, pp. 188—192, 318—330; and R. Siegfried and B.J. Dobbs, "Composition, a neglected aspect of the chemical revolution," *Ann. Sci.*, 24 (1968), 275—293. Cf. also below, p. 69, n. 1 and accompanying text.

and thereby became connected with doctrines of affinity. Among the Newtonian Stahlians of the latter part of the century, these ideas emerged in Baumé's call for a careful discrimination of wet and dry chemical affinities, for example, and in the very organization of Guyton's *Chymie de Dijon*. They came to fruition in Bergman's construction of the most elaborate set of affinity tables to appear in the eighteenth century, one group being for "the wet way" and the other for the "dry." The former represents the zenith of the labors of a pioneer of wet chemistry and wet chemical analysis.[4]

c. Vegetable analysis

The sixteenth-century herbal distillers, desiring to improve herbal pharmacy by separating the medically efficacious ingredients of herbs, were among the earliest figures interested in separating vegetable matter into its supposed constituents. DuChesne's identification of the products of distillation with the five principles (two Aristotelian and the Paracelsan *tria prima*) amounted to the subsumption of vegetables, that is, entire individual plants, under the category of mixts and entailed the equation of herbal distillation with fire analysis.[5] Many other practititioners followed his lead, and this widespread view was a major target of Robert Boyle.

Equation of vegetable distillation with fire analysis of mixts entailed the belief that each vegetable should yield all five Paracelsan principles. However, differences in the corresponding distillation products obtained from one plant to another raised questions about the specificity of vegetable distillates. Was there just one oil, for example; or was the principle "oil" a genus of substances, of which different species could be obtained from different plants; or was each oily

[4]Though Holmes, in his "Solvent extraction," explores the increasing use of extractive techniques in vegetable analysis, there is no general account of the rise of wet methods in mineral chemistry during the eighteenth century. For the importance of wet methods to Bergman, see his *Meditationes*, where (pp. 193—196) he expressly stated that his achievements in mineral chemistry rested on the shift from fire analysis by the blowpipe to wet chemical analysis. As his principal predecessor in the use of wet methods for mineral analysis, he cited (p. 196) Andreas Sigismund Marggraf (1709—1782). In this connection, see Franz von Kobell, *Geschichte der Mineralogie von 1650—1860*, Geschichte der Wissenschaften in Deutschland. Neuere Zeit, Vol. II (Munich: 1864), pp. 133—136 and Partington, *History*, II, pp. 723—729. Marggraf's role in wet mineral analysis seems to parallel his role in the rise of extractive vegetable analysis to which Holmes points ("Solvent extraction," 142—143). For Bergman's affinity tables, see Duncan, "Affinity tables," 227—231; Frederick L. Holmes, "From elective affinities to chemical equilibria: Berthollet's law of mass action," *Chymia*, 8 (1962), 105—145, esp. 106—108; and Olsson, *Kemiens historia*, pp. 238—242. For Baumé's call, see below, p. 60, n. 8 and Duncan, "Affinity tables," 221. For the organization of Guyton's *Chymie de Dijon*, see Smeaton, "Guyton's course," 59—61. Guyton had a list of twenty solvents and thirty solutes. The solvents were menstrua that informed the solutes, or "bases," as Guyton called them. There was one chapter for each solvent, in which its action on each of the bases was described.

[5]On DuChesne's equation of distillation products with principles, see Hooykaas, "Elementenlehre." On the role of organic distillation as the model for inorganic element-theory, see Metzger, *Doctrines chimiques*, pp. 310—316.

distillate itself a generic mixture, differing in its mix of species from plant to plant? The history of this issue remains largely unanalyzed.[6] However, it seems clear that no sharp distinction between the generic and specific levels in vegetable chemistry had emerged by the beginning of the nineteenth century. Even in Lavoisier's *Traité,* substances that the modern mind recognizes as generic remained for the most part conceptually undifferentiated from the specific, as shown in Chapter III. Only in Berzelius's work was this conceptual distinction firmly established and the need fully recognized for analytical techniques that could resolve generic mixtures.

More prominent in the seventeenth and eighteenth centuries than the problem of the generic or specific character of the products of analysis was that of the possibly creative, rather than the analytical role of distillation. First raised among the herbalists themselves, the issue was pressed by Boyle, and it remained a constant theme throughout most of the eighteenth century. Hoping to avoid modification of the ingredients, vegetable analysts of the eighteenth century increasingly turned to extractive procedures. By mid-century, extractive analysis had gained nearly coequal status with distillation in the chemist's analytical repertoire. Holmes illustrates the shift to increasing use of extractive analysis by citing Rouelle's approach, which he advocated at the *Jardin du roi:*

> Chemistry derives from the decomposition of different parts of plants 1° an aromatic spirit . . . 2° an extractive material . . . soluble in water. 3° a resin which alcohol dissolves 4° a resinoextractive material, of which one part dissolves in water, and the other in alcohol, but the first is in greater proportion than the second. 5° a matter analogous to the preceding, but in which the part soluble in alcohol is predominant, . . . [named] extractoresinous. 6° Finally, a green matter which gives plants their color, and which is found at their surface. This matter is purely resinous and is not yet well known.
>
> It is by following that order of analysis that one attains a true idea of the composition of plants. The old chemists neither followed nor were aware of that kind of analysis: their practice of distilling plants over an open fire caused everything to be confounded together for them.[7]

Extractive analysis sustained the notion that the products of analysis were proximate constituents, even principles, rather than mere artifacts of the technique. Rouelle's resino-extractive matter, for example, was a vegetable principle. Still open to question, however, was the generic or specific nature of the products of extractive analysis, a problem that Berzelius, in his own analytical work in the early nineteenth century, considered to be a weakness of extractive analysis. Extractive analysis did avoid the problem of the creative role of fire; ultimately, in the nineteenth century, it did provide a central technique for the resolution of generic mixtures. Through the first three or four decades of the nineteenth century, however, it offered no sure access to the specific level.

[6] But see Holmes, "Solvent extraction," for eighteenth-century attitudes about the products of vegetablen distillation.

[7] Holmes, "Solvent extraction," 144.

Extractive analysis did not replace fire analysis but only complemented it. Despite the problem of the possibly creative role of fire, distillation of organic matter continued in use throughout the eighteenth century and its products were often accepted uncritically as the preexistent ingredients of the material analyzed. An interesting example of this uncritical acceptance is provided by the work of Stephen Hales (1677—1761). From the distillation and fermentation of vegetable matter, Hales obtained prodigious quantities of air. In his *Vegetable Staticks* of 1727, he claimed that the air so obtained was originally present in the vegetable, where it behaved like a chemical ingredient and served as a sort of cement holding the other ingredients together. As Guerlac shows, Hales's work on air lived an underground existence while the doctrines of Boerhaave and Stahl reigned, and it kept alive the notion that air could enter into mixtion.[8] Hales's views became known in late-eighteenth-century France, and there contributed to the acceptance of a chemical role for air.[9] As his ideas were assimilated, so too was their principal pillar, the assumption that the products of fire analysis were the ingredients of the material analyzed. This assumption was accepted until it was attacked explicitly by Lavoisier, as shown in Chapter III.

The attack resulted from Lavoisier's success, also analyzed in Chapter III, in solving the problem of the analytical and creative roles of fire. With this achievement, he provided the foundation for organic combustion analysis, though its promise could not be exploited until Berzelius had clearly formulated the other issue of organic analysis, the generic/specific distinction, and his successors had come to grips with it. Lavoisier showed that three stable and apparently elementary chemicals, carbon, hydrogen, and oxygen, were the constituents of all vegetables. He deployed his oxygen theory of combustion, particularly after the discovery of the composition of water, to analyze the processes of distillation. He showed that, though certain ingredients could be separated intact by distillation, as even Rouelle recognized, most distillation proceeded by reorganization of the elementary constituents. He also showed that reorganization occurred by addition of oxygen, by addition of the elements of water, or by rearrangement without addition. He articulated a notion of affinities for organic matter appropriate to his understanding of organic distillation, and he developed a means of combustion analysis to determine the quantities of the elementary constituents. On this foundation, Berzelius performed his own combustion analyses, with essential improvements in technique. However, even in the mid-1820's, as

[8] For Hales's ideas and their underground existence, see Henry Guerlac, "The continental reputation of Stephen Hales," *Arch. int. d'hist. sci.*, 4 (1951), 393—404; and *idem, Lavoisier—the crucial year: the background and origin of his first experiments on combustion in 1772* (Ithaca, N.Y.: 1961), p. 28n43. See also Milton Kerker, "Herman Boerhaave and the development of pneumatic chemistry," *Isis,* 46 (1955), 36—49, esp. 39—44. Guerlac's book is cited henceforth as Guerlac, *Lavoisier.*

[9] For the entry of Hales's ideas into France and their reception, see Guerlac, *Lavoisier.*

seen near the conclusion of this study, he warned the readers of his textbook that a combustion analysis indicated nothing if the material analyzed was not known to be a specific compound.

6. The French Stahlians

a. Introduction

During the two decades preceding the start of Lavoisier's famous train of experiments, a group of French Stahlian chemists emerged.[1] In a recent study it is argued that the Stahlian group was a major force in the establishment of a chemical role for air, thereby setting the stage for Lavoisier's work.[2] The later recruits to this group also served at least one other major function, which is the subject of the next chapter. Exponents of a resurgent Newtonianism, the later French Stahlians, and their fellow traveller, Bergman, elaborated the leading two-component theories of the late-eighteenth century. In Lavoisier's hands, their ideas were modified and broadened for transmission to others, notably Berzelius. The chemical ideas of the later French Stahlians were expressed in terms of novel concepts, which this section explores.

The appearance of the Newtonian group marked the second stage in the flourishing of French Stahlian chemistry. The first stage began with G. F. Rouelle (1703—1770), who expounded at the *Jardin du roi* a modified Stahlianism that was adopted and disseminated by his pupils and advocates, among whom were some of the prominent younger chemists of the day, including G. F. Venel (1723—1775), P. J. Macquer (1718—1784), and A. Baumé. Unlike those active in the second stage, Rouelle and Venel shared Stahl's antipathy to physicalism and resisted physicalist domination of chemistry. Macquer is the transition figure between them and the second stage. His role as an agent of dissemination of Rouelle's teaching is well documented, but from 1766 he participated fully in the second stage. The latter centered on Macquer and Baumé and culminated in L. B. Guyton de Morveau (1737—1816).[3]

b. The circle of Buffon: Newtonian Stahlianism

The novelties introduced by the later French Stahlians appeared under the banner

[1] The group is identified by Rhoda Rappaport, in her "Rouelle and Stahl—the phlogistic revolution in France," *Chymia,* 7 (1961), 73—102, and is explored by Martin Fichman, in his "French Stahlism and chemical studies of air, 1750—1770," *Ambix,* 18 (1971), 94—122. Rappaport's paper is cited henceforth as Rappaport, "Rouelle and Stahl"; Fichman's, as Fichman, "French Stahlism."

[2] This is the argument of Fichman, "French Stahlism."

[3] Rappaport, "Rouelle and Stahl," and Fichman, "French Stahlism," concentrate on the earlier French Stahlians. For Rouelle's role as a principal source of French Stahlian thought, and for Macquer and Venel as disseminators of Rouelle's teachings, see Rhoda Rappaport, "G.-F. Rouelle: an eighteenth-century chemist and teacher," *Chymia,* 6 (1960), 68—101. For the Newtonian group see the next note.

of Newtonianism. Its standard bearer was their patron, Georges Louis Leclerc, Comte de Buffon (1707—1788), among the most vociferous French exponents of the Newtonian creed. Buffons's orthodoxy lay chiefly in his insistance on the absolute universality of the inverse square law, governing all material phenomena, from the astronomical scale to the chemical. The difference between these extremes lay only in the importance to the latter of particulate size and shape, which, at the small distances in question, influenced the net attractive force. In principle, therefore, particulate shapes could be inferred from the attractive forces, if the latter could be measured, and Buffon's ideas amounted to an invitation to do so. The invitation was accepted by Guyton, whose efforts to measure affinities directly and calculate the attractive forces between particles of given shape are well known and set the tone for subsequent labors in what Thackray aptly calls a "force quantification program."[4]

The Newtonianism of two of Buffon's principal clients, Macquer and Baumé, reinforced their prior Boerhaavean commitments, which they passed on to Guyton. Though Boerhaave had expressly denied the universal applicability of the inverse square law and had treated affinities with circumspection as ontologically vague and mathematically ill-defined species of sympathy, Boerhaavean thought was readily integrated with more orthodox Newtonianism. Buffon accepted ontologically stable chemicals, which, precisely because their particles had specific sizes and shapes, had specific forces and therefore exhibited specificity of attractions. Stable chemicals with specific reactivities were the common fundamental ground of Boerhaavean and Buffonian thought; Boerhaavean ideas could only be confirmed by Buffonian ones and blessed by their orthodox label. Specific reactivities, their summary in tabular form, elaboration of solution chemistry, all were hallmarks of the Newtonian Stahlians and all were fully consonant with antecedent Boerhaavean tradition.

The full effects of Buffon's Newtonianism were not felt immediately, either sociologically or conceptually. Macquer, who was originally under the direct influence of Rouelle, and who therefore displayed reserve toward a self-conscious physicalism in chemistry, embraced Buffonian views slowly. The cautious Newtonianism of his 1749 *Elémens de chymie théorique* and its anonymous appearance testify to Macquer's original hesitancy. Later editions of the work were revised only marginally, Macquer's wholehearted embrace of Newtonianism coming just after he had accepted Buffon's patronage and appearing in the 1766 *Dictionnaire* and its later editions. Baumé, Macquer's younger associate from 1757, was perhaps less encumbered by antiphysicalist sentiments. His conceptual contributions to Newtonian chemistry invite further investigation. It was on the still

[4] Metzger, *NSB,* 57—66, briefly explores Buffon's views and signals the existence of his circle. For more about Buffon's ideas, see Thackray, *Atoms and powers,* pp. 155—160. Thackray explores the Buffonian group on pp. 215—218 as part of a general discussion of force quantification in his Ch. 7.

younger Guyton, however, that Buffon exercised a dominant influence. A lawyer, Guyton came to chemistry in the early '60's without prior commitments and directly under Buffon's tutelage. Buffon's Newtonianism, then, began to take hold primarily in the mid to late 1760's and flourished especially in Guyton's hands in the '70's and '80's. Some of his achievements, as well as of Bergman, whose connections with the French Stahlians were close, are consired in the next chapter.[5]

The conceptual thrust of a renewed Newtonianism likewise developed gradually. Chief among the changes introduced by the Newtonian Stahlians was a resurgent interest in affinities and affinity tables. In 1749 Macquer republished Geoffroy's original table unaltered, but with critical commentary. He refrained from constructing a new one because of the absence of sufficient experimental material.[6] He still tended to see affinities in a Stahlian way as attractions of likes for likes, though by 1766 he viewed them in a more Boerhaavean and Newtonian fashion as attractions between usually dissimilar chemicals[7]—a view henceforth typical of late-eighteenth-century affinity theorists. In addition, the Newtonians articulated an elaborate account of affinities, one expounded in terms of an altered Stahlian vocabulary. In the 1765 "Second Vue de la Nature," Buffon himself gave currency to the terms "integrant" and "constituent" parts, terms present, though scarcely prominent, in the school of Stahl.[8] Absent from Macquer's 1749 *Elémens,* these terms were conspicuous in his writings, as well as in those of Baumé and Guyton, after the mid-'60's.[9] In addition, Baumé replaced the term "mixtion," used almost exclusively by Macquer in 1749, with those of chemical "combination" or "composition." "Compound bodies" were formed by the union of heterogeneous substances, that is, their constituent parts, which combined to form integrant parts.[1] This terminology of composition coexisted for a time with the older Stahlian terminology of mixtion and slowly eclipsed it. The last linguistic innovation of the Newtonians, one that did not gain wide currency, was Macquer's replacement of the Stahlian designations for levels in the hierarchy

[5]For Bergman's Newtonianism and his ties with Buffon's circle, see Thackray, *Atoms and powers,* pp. 218—221. See also the correspondence of Guyton and Macquer with Bergman, in Bergman, *Torbern Bergman's foreign correspondence,* eds. Göte Carlid and Johan Nordström, Lychnos-Bibliotek: Studier och källskrifter utgivna av Lärdomshistoriska samfundet, Vol. XXIII,i (Stockholm: 1965), pp. 100—138, 229—255, 437—438.

[6]For Macquer on Geoffroy's table, see Partington, *History,* III, p. 86 and Duncan, "Affinity tables," 192.

[7]For this shift, see Duncan, "Affinity tables," 184, 185, 190—191; and L.J.M. Coleby, *The chemical studies of P.J. Macquer* (London: 1938), p. 39; henceforth cited as Coleby, *Macquer.*

[8]For the "Second vue," see above, p. 34 n. 2. For the early appearance of the terms "integrant" and "constituent" parts, see above, p. 32 n. 5 and accompanying text. For Buffon's use of these terms, see "Second vue," p. 39.

[9]See, for example, Macquer, *Dictionary,* I, pp. 28—29; Baumé, *Manuel,* pp. 2—3; Guyton, *Chymie de Dijon,* I, pp. 19—20.

[1]Baumé, *Manuel,* p. 3. Cf. Partington, *History,* III, pp. 91—92, and P. Walden, "Die Lösungstheorien in ihrer geschichtlichen Aufeinanderfolge," *Samml. chem. u. chem.-tech. Vorträge,* 15 (1910), 277—454, esp. 316—317.

of mixtion with a vocabulary of orders: first were principles, then the secondary principles, which became principles of the second order, next were principles of the third order, and so on (though Baumé preferred to speak of orders of compounds, not principles).[2] The Stahlian term "aggregate" was retained to indicate the aggregation of the integrant parts into massive quantities;[3] and the process of mechanical separation, which merely separated aggregated integrant parts from one another, was placed in contrasting juxtaposition with chemical analysis, which separated constituent parts[4]—a comparison reminiscent of Boerhaave's discussion of the action of menstrua.

In these terms Macquer's innovations in affinity theory became widely known. The tendency of the homogenous parts of all bodies to attract one another and thereby to oppose the expansive power of heat, first elaborated by Boerhaave and later by Buffon—another area of agreement between them—was called "affinity of aggregation" and was deemed a phenomenon operating among the integrant parts of an aggregate.[5] This was one kind of simple affinity, of which the other was "affinity of composition," operating between constituent parts and forming new substances from them.[6] Other, more complex sorts of affinities were complications of affinity of composition and included various cases of displacements.[7] These ideas, together with Baumé's call for simultaneous construction of two sets of affinity tables, one for wet and one for dry reactions,[8] were elaborated experimentally by Bergman.[9]

With these innovations, the Newtonian Stahlians established the intellectual context in which two-component chemistry late in the eighteenth century was to flourish. Two of the Stahlians, Guyton and Bergman, articulated the foundations of the radical theory that, in modified form, occupied a prominent position in the chemistry of Lavoisier. Though Lavoisier did not share the Stahlians' concern to account for specific reactivities and had little sympathy with affinity theory, he did share their desire to account for generic qualities and unhesitatingly embraced the sort of two-component explantions that they had elaborated. He deployed them both in his oxygen theory of acidity and in his radical theory of organic chemistry. These, in turn, constituted the immediate background for the innovations of Berzelius. The two-component chemistry of the late-eighteenth century is the subject of the next chapter.

[2] E.g., Macquer, *Dictionary*, I, p. 171, and II, p. 583; Baumé, *Manuel*, p. 20.

[3] Macquer, *Dictionary*, I, p. 29; Baumé, *Manuel*, p. 2.

[4] Macquer, *Dictionary*, I, p. 199; Baumé, *Manuel*, pp. 1—2. Cf. Metzger, *NSB*, pp. 281—282.

[5] Macquer, *Dictionary*, I, p. 23.

[6] *Ibid.*, pp. 23—24; cf. Baumé, *Manuel*, pp. 8—9.

[7] For kinds of affinity, see Coleby, *Macquer*, pp. 39—40, and Duncan, "Affinity tables," 192—193, 220—221.

[8] Baumé, *Manuel*, p. 7.

[9] For Bergman's affinity tables, see above, p. 54 n. 4.

CHAPTER II

Two-component Chemistry in the Late Eighteenth Century

1. Introduction

Lavoisier's oxygen theory of acidity, his theory of neutral salts, and his radical theory of organic chemistry were the immediate predecessors of Berzelius's novel ideas. As two-component theories, they are shown here to have drawn on concepts and techniques that had been evolving, partly under pressure exerted by Lavoisier's own innovations, in the antecedent labors of Guyton and Bergman, also analyzed in this chapter.

The central theme of the two-component theories of Guyton, Bergman, and Lavoisier was the effort to make a special place in chemical theory for acids. In Stahl's chemistry, the acids had been treated along with other salts in the theory of salts, and among the French Newtonians they were considered to be ontologically stable, primitive chemicals. Guyton and Bergman, however, ultimately felt the need for a particular treatment of acids. Guyton's work, in turn, was partly a response to Lavoisier's own studies of acidity, and both Guyton's and Bergman's accomplishments provided inputs into the later evolution of Lavoisier's oxygen theory.

Bergman first felt the need for a special treatment of acidity following Scheele's discovery of several acids that could be obtained from metals, notably arsenic, by extreme dephlogistication. He persisted in the view of the Newtonian Stahlians that the acids were primitive chemicals and believed that Scheele's metallic acids were the specific components of metals. To emphasize their character as stable, virtually elementary chemicals, Bergman called these specific components "radical acids." To preserve the Stahlian assimilation of calcination and combustion, he provided a parallel treatment of the acidification and the reduction of combustibles, and he went so far as to blur the distinction between metals and combustibles, claiming that both were extremes of a continuum.

Guyton experienced the need to account for acidity as a result of Lavoisier's early studies of acidification. He adopted Bergman's attempt to treat the acidification of combustibles and metals by a single theoretical apparatus and accepted the blurring of the distinction between them. However, he denied that the specific components were acids. Instead, he drew on the belief of the French Stahlians that the specific components of metals were earths, and he extended it to combustibles as well, claiming that both metals and combustibles contained earths as specific components. To explain the passage from metal to acid or from combustible to acid, Guyton invoked a second generic chemical, the air, which Lavoisier had shown to be a reagent that conferred the generic property of acidity. Guyton's was a two-fluid theory of calcination, combustion, and acidification. He also drew from Bergman the description of the specific components as "radical" components. By their combination with air, radical components were acidified, and by their combination with phlogiston, they were reduced to metals or combustibles.

With his conversion to Lavoisier's views, Guyton bequeathed to the new chemistry the term "radical," now used as a substantive as well as an adjective, to designate the combustibles, which, in the oxygen theory of acidity, were all considered at least potentially acidifiable. However, the oxygen theory was undermined by a serious flaw. Not all combustibles proved to be acidifiable, though Lavoisier had tried to show that they were. Some oxides, especially the higher oxides of the metals, thus lay outside the domain of the theory and constituted an embarrassment to it.

Among Berzelius's principal innovations was to reduce the anomaly of the metallic calces by abandoning two-component theories. Instead of emphasizing the generic, he focussed on the specific. He made acidity a specific property, a property of the radicals. He elevated basicity to a status equivalent with acidity. In Berzelius's theory, whether a radical was predominantly acidic or basic was a function of its position in his electrochemical series. Moreover, his theory emphasized that acidity and basicity were relative concepts, and that the labels "acid" and "base" could be applied to any pair of reacting substances according only to their relative positions in the series.

The late-eighteenth-century explanations of the properties of acidity, metallicity, and combustibility were not the only antecedents of Berzelius's innovations. Corresponding to Lavoisier's theory of acidity was a two-component theory of neutral salts, a direct descendant of Rouelle's ideas. In it Lavoisier articulated the notion of base as a passive substrate, leaving it to Berzelius to emphasize that basicity was a positive concept of standing equivalent to that of acidity. In addition, as shown in Chapter III, Lavoisier deployed the radical theory in organic chemistry, an innovation that rested on significant advances in the understanding of distillation analysis. Berzelius relied on Lavoisier's conceptual analysis of organic distillation, but argued that it could not be exploited without prior discrimination of specific ompounds from generic mixtures, a distinction scarcely emergent in Lavoisier's chemistry.

This chapter begins with an analysis of the two principal developments to which Guyton responded. First it explores Bergman's metallic theories, and it studies his parallel treatment of the acidification of metals and combustibles. Next it turns to aspects of Lavoisier's oxygen theory of acidity, where the generic role of air was elaborated. Because of its importance to the later analysis of the work of both Guyton and Berzelius, the anomaly to the oxygen theory presented by the metallic calces is treated in detail. The chapter then takes up Guyton's work, analyzing the theoretical elaborations he pursued in response to both Lavoisier and Bergman. Last, the chapter notes the fate of the concept of radicals in Lavoisier's hands.

2. Bergman's metallic theories

a. A Newtonian Stahlian in Sweden

That Torbern Bergman (1735—1784) was a fellow traveller of the Buffonians has already been noted.[1] Like the French group, Bergman advocated a modified Stahlianism and a Newtonian affinity theory. His affinity tables surpassed in quantity and quality anything previously seen. Fulfilling Baumé's call for separate wet and dry tables, Bergman summarized an immense body of wet-chemical analytical materials and of more traditional analyses by fire. Also, in the theoretical apparatus accompanying the tables he employed the different categories of affinities developed in France to account for both regularities and anomalies in the tables.[2]

Unquestionably one of the seminal figures in the chemistry of the late eighteenth century, Bergman pursued a broad range of interests.[3] In many instances, the orientation of his work was significantly determined by his concern for the well-developed mining and metallurgy industries in Sweden and for the strong Swedish tradition in mineralogy. Bergman's Newtonian studies of crystallogenesis and crystal structure, though failing to establish a rational science of crystallography, provided important precedents for Haüy's work, as Hooykaas shows.[4] His chemical interests, colored by his awareness of industrial needs and by the naturalist's desire to attain a depiction of the natural order, led him to produce mineralogical works that are landmarks in both mineralogy and chemistry.

In this chapter, Bergman's metallic theories are important. Because of his industrial interests, much of his chemical work was devoted to the study of metals, and, as Fourcroy remarks, much of Bergman's chemical theory evolved from this work.[5] The theoretical developments that impinged on the work of

[1] For Bergman's connections with Buffon's circle, see above, p. 59, n. 5.

[2] For Bergman's affinity tables, see above, p. 54, n. 4 and accompanying text.

[3] For an overview of Bergman's research concerns, see Olsson, *Kemiens historia*, pp. 122—136, esp. 133—136.

[4] See R. Hooykaas, "Torbern Bergman's crystal theory," *Lychnos*, 1952 (1952), 21—54, esp. 44—54; *idem, La naissance de la cristallographie en France au XVIIIe siècle* (Paris: 1953), pp. 14—17; and *idem*, "Les débuts de la théorie cristallographique de R.J. Haüy, d'après les documents originaux," *Rev. d'hist. sci.*, 8 (1955), 319—337.

[5] A.F. de Fourcroy, *Encyclopédie méthodique, chimie, pharmacie et métallurgie*, III (Paris: 1796), p. 507 (misnumbered as 515; pp. 505—512 are misnumbered as pp. 513—520). For bibliographic details about the *Encyclopédie méthodique*, see below, p. 90, n. 7. For full citation of Vol. III, see the bibliography.

Guyton and Lavoisier are considered in this section. Most important was his effort to make a special place for acidity in chemical theory after the discovery of the metallic acids.

b. The metallic acids

The archetype for Bergman's theory was arsenic acid, obtained by Scheele in 1775. The theory received further support from Scheele's discovery of molybdic acid in 1778 and tungstic acid in 1781. Bergman's own research had provided him with what he believed to be a fourth example of this class of substances, on which he conferred the name "metallic acids." By analysis of samples of cold-short iron, he obtained what appeared to be a metallic calx; reducing it, he produced a metallic regulus, which he dubbed "siderum"; and he believed the origin of the cold-shortness lay in the presence of this impurity. The putative metal was later shown, however, by Meyer, Klaproth, and Scheele, to be a compound of iron and phosphorus.[6]

Prior to his discovery of arsenic acid, Scheele had reported in a study of black magnesia (pyrolusite, manganese dioxide) that by the action of arsenic (white arsenic, the trioxide) on a green alkaline fusion of black magnesia (potassium manganate), the color changed to white, and, when the mass was dissolved in water, a "phlogisticated" manganese precipitated.[7] "This is certainly a strange occurrence," he remarked. "I have perceived that the [white] arsenic truly contained phlogiston in its composition."[8] The observation that phlogiston was a constituent of white arsenic led him to study this substance. As he reported:

> After I had found in my experiments on [black magnesia] that inflammable matter was not only to be found in arsenic but could actually be separated from it, the question occured to me: What might the remainder of the arsenic be? I then found that it was an acid.[9]

[6] For an overview of the history of the metallic acids, with emphasis on Scheele's work, see Olsson, *Kemiens historia*, pp. 303—312. In this study, Scheele's papers are consulted in Dobbin's English translation (of which Partington [*History*, III, p. 211] says, "Apart from the originals, this is the most accurate and complete form of Scheele's memoirs"): Carl Wilhelm Scheele, *The collected papers of Carl Wilhelm Scheele translated from the Swedish and German originals by Leonard Dobbin* (London: 1931); henceforth cited as Scheele, *Papers*. In this study, the individual memoirs are cited by their shortened titles with page references to Dobbin's collection. For the metallic acids, see these memoirs of Scheele: "On arsenic and its acid" (1775), pp. 53—77; "Experiments with lead ore: molybdaena" (1778), pp. 186—194; and "The constituents of tungsten" (1781), pp. 225—229; the first two of these are henceforth cited, respectively, as Scheele, "Arsenic," and Scheele, "Molybdaena." For "siderum," see Partington, *History*, III, p. 194. For the name, "metallic acids," see below, p. 65, n. 3.

[7] Scheele, "On manganese or magnesia, and its properties" (1774), in Scheele, *Papers*, pp. 17—49; henceforth cited as Scheele, "Magnesia."

[8] *Ibid.*, p. 39.

[9] Scheele, "Arsenic," p. 53.

Scheele presented two methods for the *decomposition* of white arsenic, that is, for its dephlogistication.[1] He showed that the decomposition proceeded through two stages, and Bergman incorporated this observation prominently in his theory. The first method of decomposition was by the action of nitric acid on a solution of white arsenic in muriatic acid. The muriatic acid dissolved the arsenic to facilitate the dephlogisticating action of the nitric acid, which alone could not completely dissolve the arsenic. The dephlogisticated arsenic was an acid, the second of the two ingredients of white arsenic; because it was insoluble in muriatic acid, it precipitated out. A second method involved the action on moistened white arsenic of dephlogisticated muriatic acid (that is, of chlorine, discovered by Scheele in 1773 as the product of the reaction of muriatic acid with black magnesia). The dephlogisticated muriatic acid, by virtue of its avidity for the phlogiston of which it had been deprived, dephlogisticated the white arsenic, leaving the acid. Rephlogistication of the arsenic acid by treatment with inflammable substances generally reproduced the white arsenic. Scheele also obtained the regulus from the acid by heating it strongly with dry charcoal. The reduction of arsenic acid therefore revealed two stages of phlogistication: from the acid to white arsenic, and from the latter to the regulus. Scheele had shown that a metal consisted of two components, "a specific acid and the ordinary inflammable principle [phlogiston]."[2] Removal of one portion of the phlogiston left a calx, which contained the same two components, but in different proportions. Removal of the phlogiston remaining in the calx revealed the isolated acid. Applied to this case and to those of the other metallic acids, largely wet chemical techniques had provided a collection of stable and apparently elementary chemicals, the metallic acids, that served as the specific components of metals and metallic calces. The way was prepared for Bergman's two-component metallic theory, which invoked the specific metallic acids and the generic principle, phlogiston.

c. Bergman's modified two-component theory

Bergman's two-component theory of metals resembled that of the French Stahlians, but it was modified to account for the metallic acids. The theory appeared fully articulated in the 1781 *De acidis metallicis.*[3] There, Bergman began by asserting that

[1] *Ibid.*, pp. 53—56.

[2] Scheele, "Magnesia," p. 39. Cf. also Scheele's *Chemische Abhandlung von der Luft und dem Feuer* (Stockholm and Uppsala: 1777), republished, in the original German, as the first volume of Scheele, *Sämmtliche physische und chemische Werke, nach dem Tode des Verfassers gesammlet, und in deutscher Sprache herausgegeben von Sigismund Friedrich Hermbstädt,* 2 vols. (Berlin: 1793), I, esp. pp. 105—107, 157—160. Hermbstädt's collection is henceforth cited as Scheele, *Werke.* Scheele's *Abhandlung* is henceforth cited from Hermbstädt's collection as Scheele, *Abhandlung.* An English translation of the *Abhandlung* is in Scheele, *Papers,* pp. 85—178.

[3] The *De acidis metallicis* was one of sixty dissertations that were published (many of them in

In the analysis and synthesis of individual metals, two principles are encountered, phlogiston and earth. Phlogiston, simple and elementary, appears the same in all; for gold dissolved in acid receives back its complete and genuine form when combined with the inflammable part liberated from iron, copper, lead, or other metals; and most of the remaining metals can be similarly reduced with the phlogiston of other metals. [Here, Bergman referred to his *De diversa phlogisti quantitate in metallis* (1782).[4]]

However, each metal appears to be assigned its own peculiar and *specific* [*propria*] earth.[5]

The discovery of arsenic acid suggested that the earths were acidic:

White arsenic, by its saline and acidic nature, offered an opening, which, when widened, gave hope of perceiving the composition of at least one metal. This result is not lacking, for by analysis in the wet way and by synthesis it is already known that [white] arsenic consists of an acid, known to be different from all others, which is coagulated into earth or white calx by a certain dose of phlogiston but presents a complete regulus when fully saturated [with phlogiston].[6]

Bergman here distinguished two stages of phlogistication: coagulation and saturation (the latter he elsewhere also called metallization or reduction). The calx, white arsenic, was a coagulum of the acid with phlogiston; the regulus was the calx saturated or metallized with a second dose of phlogiston. This approach permitted the application of the two-component approach to metallic calces as well as to metals, a precedent that proved signifant for Guyton and Lavoisier. In this connection, Bergman remarked:

The metallic calces in turn are generally so similar, in origin and composition, that they merit the expection of [great] conformity [in properties]. By dephlogistication, all metals evidently assume the character of earth, each earth, however, being peculiar to the metal, [but each being] devoid of metallic lustre, pulverulent, soluble in acids, fusible in fire, coloring glass, [and, upon addition of phlogiston,] recovering inflammability and again receiving its original metallic state. However, experience shows that each calx may be differentiated from the others by specific marks, thereby not so much destroying but strengthening our hypothesis of the composition of metals.[7]

The generic and specific properties of the metallic calces could be explained, no less than those of the metals themselves, in terms of the same two-component theory. Both the metal and the calx contained the same specific component, the

revised form) as Bergman's *Opuscula physica et chymica,* 6 vols. (I, Stockholm: 1779; II, Uppsala: 1780; III, Uppsala: 1783; I—III, rpt., Leipzig: 1786—1792; IV—VI, Leipzig: 1779—1790). In this study, the six-volume set published in Leipzig is consulted; it is henceforth cited as Bergman, *Opuscula.* Quotations from the *Opuscula* are given in English translations prepared by the present author (with the assistance of Gerald Day). For bibliographic details pertaining to Bergman's dissertations, see Partington, *History,* III, pp. 182—185, esp. the chronological list of dissertations on p. 184, from which dates have been drawn for this study; and Birgitta Moström, *Torbern Bergman: a bibliography of his works* (Stockholm: 1957), pp. 48—51. The dissertations are henceforth cited by their (sometimes shortened) Latin titles with page reference to the *Opuscula.*

[4] In Bergman, *Opuscula,* III, pp. 132—156.

[5] Bergman, *De acidis metallicis,* in his *Opuscula,* III, p. 124.

[6] *Ibid.,* p. 125.

[7] *Ibid.*

acid, and the same generic component. The generic distinction between metal and calx depended on the dose of the generic component, phlogiston. Yet it was striking that the specific component was now no mere uninformed matter. It was itself a species of a genus; it was an acid.

d. Primitive acids

Bergman shared with the French Stahlians their belief that the acids were primitive chemicals, and accordingly he felt no need to account for the generic and specific properties of the metallic acids themselves. Moreover, he had found in metallic chemistry the same quenching of acidity that Stahl had seen in the acids obtained from combustibles. The new parallelism of the chemistry of metals with that of combustibles put Bergman in a position to expand the Stahlian analogy of calcination and combustion. He took this step in the 1784 *Meditationes de systemate fossilium naturali,*[8] where he expressed his conviction that acids were primitive chemicals and explicitly asserted the similarity of the metallic acids with the acids obtained from combustibles.

In his discussion of salts, Bergman observed that the traditional category contained two primitive or simple kinds of salts, acids and alkalies (the various kinds of neutral salts being derivative because they were compounds of primitive chemicals with one another).

> Among salts we readily distinguish two genera, by no means ambiguous, acid and alkali. Their proximate principles [*principia proxima*] cannot yet be obtained by chemistry, but it is not doubtful that they are diverse.[9]

Similarly, Bergman described the genus of sulfur, that is, the combustibles that Stahl had considered virtually elementary, "as presenting a very simple compound of only two principles, acid saturated with phlogiston."[1] There were only two combustibles possessing this simple composition:

> The species of sulfur are distinguished by the diversity of their acids, but up to now only two have been observed, the common, formed with vitriolic acid, and plumbago [graphite], containing the aerial acid [carbon dioxide] saturated with phlogiston.[2]

That both combustibles and metals had been shown to be compounds of acids with phlogiston permitted Bergman to sustain the Stahlian assimilation of calcination to combustion and of metals to combustibles. Illustrating, in the *Meditationes,* the natural-historical doctrine of transitions, Bergman offered a paragraph on the "sulphureous nature of the metals" and in it called attention to this univocal character of the theory of the composition of combustible matter:

> Metals, in their complete state, are either metallic acids saturated by phlogiston or a species of metallic sulfur which sometimes can inflame distinctly, like zinc and arsenic.

[8] In Bergman, *Opuscula,* IV, pp. 180—278; henceforth cited as Bergman, *Meditationes.*

[9] *Ibid.*, p. 207. Cf. p. 223.

[1] *Ibid.*, p. 216.

[2] *Ibid.*, p. 242.

Traces of flame, though mostly obscure, are also offered by gold in fusion and by copper with a greenish appearance; incandescent iron throws off sparks and tin can also inflame by proper treatment.[3]

Metals were like sulfurs and sulfurs were like metals. Some metals, such as arsenic, a "semi-metal," stood between the two. Setting a precedent for Guyton, Bergman obscured the distinction between the specific components of metals and those of other combustibles. Unlike Bergman, however, Guyton regarded them not as acids, but as earths.

e. Radical acids

Bergman coined the term "radical acids" to describe the acids that were the specific components of metals. The word derived from the Latin noun "radix," meaning root, base, or source. Thus it was Bergman, and not Guyton, as commonly believed,[4] who introduced this term that was to have so rich and complex a history over the next century. Originally an adjective, not a substantive, the term appeared prominently in the *De acidis metallicis:*

If we establish as many radical acids [*acida radicalia*] as metals, they doubtless act differently with phlogiston, as much in regard to the quantity [of phlogiston] necessary for coagulation and reduction, as in regard to the intensity of attraction that strengthens the compound and resists separation of the principles.[5]

The radical acids displayed the specificity of properties both instrumentally and materially: they gave rise to the order of metallic affinities and to the specific quantitative phenomena of metallic chemistry, which Bergman himself had investigated in his dissertation on metallic precipitates.

The description of the acids of the metals as "radical" issued from Bergman's profound awareness that they were ontologically stable chemicals representing a newly attained level of analysis. Bergman developed this point of view in his preface to Scheele's 1777 book *Chemische Abhandlung von der Luft und dem Feuer.*[6] There, he pointed out that the deepest levels in the hierarchy of composition often presented the greatest resistance to decomposition and that therefore chemistry could only provide limited insights into the secrets of nature.[7] Nevertheless, "Könnten nur wenigstens mit Sicherheit die nächsten Grundmaterien der Körper (*principia proxima*) wie auch dieser ihre Verbindung und Proportion entdecken, so sollte die Scheidekunst etwas Grosses einrichten."[8] He went on to describe how, in his view, the progress of chemistry had pushed more deeply into the hierarchy of composition.[9] Though the four Aristotelian

[3] *Ibid.*, p. 203.

[4] For example, see Partington, *History*, III, pp. 469, 527.

[5] Bergman, *De acidis metallicis*, p. 125.

[6] Bergman, "Vorbericht zur Abhandlung über Luft und Feuer" (1777), in Scheele, *Werke*, I, pp. 3—27; henceforth cited as Bergman, "Vorbericht." The preface does not appear in Scheele, *Papers*.

[7] Bergman, "Vorbericht," pp. 17—18.

[8] *Ibid.*, p. 18.

[9] *Ibid.*, pp. 19—20.

elements had once been seen as simple, all of them had eventually been decomposed. Bergman considered each of them in turn, his discussion of earth being of particular interest here. There was no longer but one earth; instead there were six.[1] Provisionally at least, they were considered primitive. An experiment of Scheele's, however, in which silica apparently was produced from the acid of fluorspar and water,[2] raised the possibility that the earths were acids coagulated with water, just as the metallic calces were acids coagulated with phlogiston:

Da ... Flusspat-Säure und Wasser, wenn sie sich einander unter der Form von Dunsten begegnen, zu Kalk-Erde, und Arsenik-Säure mit Phlogiston sich in weisen Arsenik coagulirt: so ist einige Vermuthung vorhanden, dass, sowohl Erdarten als metallische Kalke, nach ihrem eigentlichen Grundstoffe, für Säuren zu halten sind, welche beym erstern Fall durch Wasser, und im letztern durch Phlogiston in eine feste Masse verwandelt worden.[3]

In this passage Bergman displayed the vocabulary he employed to describe the deepest level attained by chemical analysis. Until Scheele's experiment suggested another possibility, the earths were the last term of the decomposition of certain substances and the first term of synthesis. They were thus labelled "proximate principles" and, because they had resisted further analysis, "primitive principles." Now Scheele's work indicated that chemistry was soon to take another step. Acids appeared to Bergman increasingly to represent the deepest level of analysis. They were obtained from the decomposition of several metals and it appeared that they were also among the constituents of earths. The adjective "radical" was Bergman's addition to the vocabulary expressive of the latest state of the analytical art. Acids were being shown to be the proximate principles of the proximate principles, the *eigentliche Grundstoffe,* the radical principles.

f. The radical acid of iron

Besides the *De acidis metallicis,* Bergman published in 1781 two other dissertations on metallic chemistry in which he attempted to provide other examples of metallic acids. One of the dissertations was the *De causa fragilitatis ferri frigidi,* in which Bergman reported the discovery of siderum, the supposed metal that he believed responsible for cold-shortness in iron and that he thought gave rise to a metallic acid on dephlogistication.[4] The third dissertation was the *De analysi ferri,* where Bergman described his principal attempt to provide another example of the pattern discovered by Scheele.[5] This effort, the subject of this subsection,

[1]The six included chalk, barytes, magnesia, silica, alumina, and "ädeljord." This last Bergman soon dropped, as Hermbstädt notes (Scheele, *Werke,* I, p. 19n). Cf. Bergman, *Meditationes,* p. 211, where Bergman listed only the five.

[2]For Scheele's experiments of 1771 on the acid of fluorspar, see Partington, *History,* III, p. 214.

[3]Bergman, "Vorbericht," pp. 20—21.

[4]In Bergman, *Opuscula,* III, pp. 109—123.

[5]*Ibid.*, pp. 1—108

provided a precedent for Lavoisier's attempt, using the example of iron, to show that even metals were acidifiable by oxygen.

The orientation of the *De analysi ferri* was largely metallurgical, its object having been to investigate the chemical differences among varietes of iron encoutered in metallurgical processes, such as cast iron, steel, malleable iron, and hot- and cold-short iron. The treatment of this industrial subject was organized expressly as an application to iron of the conception of metals as compounds of radical acids with a double dose of phlogiston. Moreover, one section of the dissertation was devoted to the proximate principles of iron, which Bergman considered to be phlogiston and a specific earth, and another section was devoted to isolating the earth by wet chemical means. Bergman expected to find that the specific component was an acid.[6]

Bergman made two attempts, one direct, and one indirect, to reveal the existence of an acidic component of iron. The former consisted of a series of experiments designed to bring about a second stage of dephlogistication, principally by repeated treatment of samples of iron with nitric or dephlogisticated muriatic acid.[7] However, these experiments failed to accomplish the removal of the second dose of phlogiston, the coagulating portion. The indirect experiments seemed more promising to Bergman. Their appeal derived from two of his conceptual commitments: his adherence to the traditional notion of salts that included under that rubric both acids and alkalies and his Newtonian equation of reaction with the process of solution. The indirect experiments were all indications of the reactivity of iron with water.[8] If iron was water-soluble, then it contained a saline constituent, which Bergman clearly believed was the radical acid. To make his point, he cited several examples: first, that concentrated vitriolic acid did not dissolve iron, but needed first to be diluted with water; second, that water alone resolved the metal into the calx, *aethiops martialis;* and last, that the action of sulfur on iron filings was dramatically intensified by the presence of water. These indications of water-solubility suggested the presence of a saline, and, he anticipated, acidic component:

> Do not these phenomena proclaim in some measure a hidden, radical, especially acidic salt [*salinum delitescens & radicale, praesertim acidum*] that seeks water?[9]

He went on to describe a series of experiments in which he used water hoping

[6] For a general discussion of the *De analysi ferri* (1781), see Partington, *History,* III, pp. 193—194. For the organisation of the dissertation around the conception of metals as compounds of radical acids with a double dose of phlogiston, see Bergman, *De analysi ferri,* pp. 6—8. For the proximate principles of iron, see section 9 of the dissertation, pp. 81—93, esp. p. 81. For the effort to find an acidic component of iron, see section 10, pp. 93—101.

[7] For the direct experiments, see Bergman, *De analysi ferri,* pp. 100—101.

[8] For the indirect experiments, see *ibid.,* pp. 94—95. Scheele had observed previously that inflammable air was evolved by the action of water on iron filings, as Partington notes (*History,* III, p. 194).

[9] Bergman, *De analysi ferri,* pp. 93—94.

to achieve a further dephlogistication of iron beyond the stage represented by martial ethiops. The experiments produced an ochreous material that seemed to Bergman not the radical acid, but a stage of dephlogistication closer to it than the ethiops. In the action of water on iron, Bergman first demonstrated an elevation of temperature, the production of inflammable air, the appearance of the calx, martial ethiops, and he pointed out the violence of the combination of sulfur and iron in the presence of water.[1] In several experiments, small amounts of the ochre appeared. It could be produced in larger amounts, Bergman showed, from iron vitriol. The latter, a neutral salt, contained the metallic calx, that is, the metal deprived of its first or reducing dose of phlogiston. When exposed to the air in aqueous solution, this salt produced first ochreous particles and then a yellow solution. This was an example, Bergman asserted, of a further dephlogistication of the metal, from the stage represented by the ethiops to that represented by the ochre. He showed in another experiment that heating a solution of iron vitriol in the absence of air likewise produced the ochre, that is, that further dephlogistication could be achieved without the presence of air.[2] Striking as these phenomena were, Bergman did not believe that in the ochre he had found the radical acid of iron. With reference to the last described experiments, he held that

> The experiments described above all sufficiently signal that iron, though it often releases little more than reducing phlogiston ... nevertheless, by means thus far used, cannot be robbed of its reducing phlogiston so as to expose its radical acid [*ut quodammodo denudetur acidum radicale*]. But this tenacity should not discourage us: persistent labor conquers all. At some time we shall seek stronger instruments to try to expose the radical acids not only of iron but also of the rest of the metals.[3]

Lavoisier, seeking to uphold the generality of the acidifying role of oxygen, picked up where Bergman left off and expended much energy to produce an acidic oxide of iron. He got no farther than Bergman. Having produced the ochre, he leaned on the traditional notion of salt to claim that this saline product was testimony of the passage of iron toward the acidic state.

3. Lavoisier's oxygen theory

a. Introduction

From the first, Lavoisier's oxygen theory of combustion was also a theory of the generic and specific properties of acids. The specific properties of the acids resulted from the combustion of specific substances, not from the proportions of their ingredients, as Stahl had believed. The generic properties of acids were conferred on the specific components by oxygen, the generic principle of acidity.

[1] *Ibid.*, pp. 94—95. The experiments were numbered 255—257.

[2] *Ibid.*, pp. 97—98 (experiment 263).

[3] *Ibid.*, p. 101.

As noted in the last chapter, the theory was a decided advance over that of Stahl, because it eliminated the problem posed by the failure of chemists to achieve the transmutation of acids that Stahl's theory had led them to expect. It was also an advance over that of the later Stahlians, who had simply labelled acids primitive chemicals, because Lavoisier had succeeded in explaining the properties of an entire genus, of which many new members were discovered late in the century. Accordingly, the theory evoked a response from Guyton, who tried to accommodate its novelties, while preserving a role for phlogiston. That Lavoisier's novel theory, like Bergman's, served as a stimulus to Guyton, is one motive for analyzing it here.

Another motive is the importance of the oxygen theory to understanding the novelties of Berzelius's chemistry. As shown below, Lavoisier's oxygen theory claimed that oxygen was both necessary and sufficient for the production of acidity. The belief that oxygen was necessary was later undermined by the observation of acids that contained no oxygen. However, this blow to the theory was not at first significant. More telling was the recognition that the presence of oxygen was not sufficient to give rise to acidity. Lavoisier himself was aware from the first that oxygen did not consistently engender the same set of generic properties. The metallic calces, in particular, were anomalies. The later decomposition of the alkalies and earths by means of the Voltaic pile rendered the exception of the calces an acute embarrassment to the theory, precipitating its downfall. On its ruins, Berzelius constructed his own theory that described acidity as a specific property. Lavoisier, however, spared no effort to reduce the anomaly, and this endeavor is also of importance to the analysis of Berzelius. To circumvent the problem Lavoisier devised a correlation of properties with degree of oxidation, and he claimed, much as Bergman had about the iron ochre, that oxides were saline substances, intermediate between metals and acids. He also asserted that the higher stages of oxidation exhibited fully acidic properties and that the strength of acidity depended on the degree of oxidation. Though Berzelius no longer considered acidity a generic property, he used his stoichiometric innovations to reconstruct this correlation.

Lavoisier's elaboration of the oxygen theory of acidity all but coincided with his efforts to reduce the anomaly of the metallic calces. The latter theme and its consequences for the chemistry of metals are emphasized in this section, the detailed history of the theory having been treated in other studies.[4] In the next section, Guyton's theoretical work is analyzed, the fate of Guyton's idea of radicals in Lavoisier's hands is noted, and Lavoisier's successes in providing qualitative explanations are estimated.

[4] See H.E. Le Grand, "Lavoisier's oxygen theory of acidity," *Ann. Sci.*, 29 (1972), 1—18; and Maurice P. Crosland, "Lavoisier's theory of acidity," *Isis*, 64 (1973), 306—317.

b. The theory and its counterinstance

The first full statement of the generic role of oxygen appeared in Lavoisier's 1776 paper on nitric acid.[5] Having come forth in favor of a two-component theory (in the passage quoted above, p. 41, accompanied by n. 2), Lavoisier added:

> Ce qui, d'abord, n'était qu'une conjecture assez vraisemblable, s'est bientôt converti en certitude, quand j'ai appliqué l'expérience à la théorie; et je suis en état d'avancer affirmativement aujourd'hui que non-seulement l'air, mais encore la portion la plus pure de l'air, entre dans la composition de tous les acides sans exception; que c'est cette substance qui constitue leur acidité, au point qu'on peut à volonté leur ôter ou leur rendre la qualité acide, suivant qu'on les dépouille ou qu'on leur donne la portion d'air essentielle à leur composition.[6]

The case reported in this paper, the production of nitric acid by combining nitrous gas with oxygen, appeared to have convinced him. He had shown, he asserted,

> que, lorsqu'on dissout du mercure dans l'acide nitreux, cette substance métallique s'empare de la portion de l'air pur dans l'acide nitreux et qui constitue son acidité; d'une part, ce metal combiné avec l'air le plus pur se réduit en chaux; de l'autre, l'acide dépouillé de ce même air entre en expansion et forme l'air nitreux; et les preuves que les choses se passent ainsi dans cette opération, c'est que, si, après avoir ainsi séparé les deux airs qui entraient dans la composition de l'acide nitreux, on les recombine de nouveau, on refait de l'acide nitreux pur, tel qu'on l'avait auparavant, avec cette différence seulement qu'il est fumant.[7]

This decomposition of nitric acid and the recomposition of the acid from the products of analysis seemed definitive to Lavoisier. The addition of vital air to nitrous gas was tantamount to the acidification of the gas. The conclusion that phlogiston was an unnecessary concept was all but inevitable. Lavoisier went on to broach that conclusion, without at this date urging it too forcefully:

> On ne manquera pas sans doute de demander ici si le phlogistique du metal ne joue pas quelque role dans cette opération; sans oser décider une question d'aussi grande conséquence, je repondrai que, puisque le mercure sort de cette opération précisement tel qu'il y était entré, il n'y a pas d'apparence qu'il ait perdu ni repris du phlogistique . . .[8]

The earliest summary account of the theory appeared in the 1777 "Mémoire sur la combustion en générale."[9] There Lavoisier identified in combustion

[5] Lavoisier, "Mémoire sur l'existence de l'air dans l'acide nitreux et sur les moyens de décomposer et de recomposer cet acide" (1776; rev., 1779), in Lavoisier, *Oeuvres*, II, pp. 129—138; henceforth cited as Lavoisier, "Acide nitreux." For a general discussion of Lavoisier's studies of nitric acid, see Partington, *History*, III, pp. 410—416; cf. also below, p. 77, n. 1. For the technique used in this study for citation of Lavoisier's memoirs, see above, p. 41, n. 2.

[6] Lavoisier, "Acide nitreux," pp. 129—130.

[7] *Ibid.*, p. 137.

[8] *Ibid.*, pp. 137—138.

[9] Lavoisier, "Mémoire sur la combustion en général" (1777), in his *Oeuvres*, II, pp. 225—233.

"quatre phénomènes constants, qui paraissent être des lois dont la nature ne s'écarte jamais . . ."[1] Of these only the fourth is important here:

> Dans toute combustion, le corps brûlé se change en un acide par l'addition de la substance qui a augmenté son poids: ainsi par example, si l'on brûle du soufre sous une cloche, le produit de la combustion est de l'acide vitriolique; si l'on brûle du phosphore, le produit de la combustion est de l'acide phosphorique; si l'on brûle une substance charbonneuse, le produit de la combustion est de l'air fixe, autrement dit, de l'acide crayeux [carbon dioxide], etc.
>
> La calcination des métaux est soumise exactement à ces même lois, et c'est avec très-grande raison que M. Macquer l'a considéré comme une combustion lente: ainsi, [as in all combustion,] 1° dans toute calcination métallique il y a dégagement de la matière du feu; 2° il ne peut y avoir de véritable calcination que dans l'air pur; 3° il y a combinaison de l'air avec le corps calciné, mais avec cette différence, qu'au lieu de former un acide avec lui il en résulte une combinaison particulière, connue sous le nom de *chaux métallique*.[2]

This last paragraph displayed the dissimilarity of the properties imposed by oxygen on metals from those imposed on combustibles. In this memoir, however, Lavoisier did not emphasize the role of oxygen as generic component, but reserved treatment of it for another one, apparently composed at the same time (1778), offering "Considérations générales sur la nature des acides, & sur les principes dont ils sont composés."[3] There he called attention to contemporary usage that labelled as "principles" the proximate constituents of substances as revealed by analysis and synthesis. Chemists, he noted, having now generally recognized that neutral salts were compounds of acids with saline, earthy, and metallic bases, were faced with finding the principles of the acids and bases themselves.[4] In the introduction to his memoir, he took the opportunity to display his ideas on the principles of the acids:

> J'ai déjà fait part à l'Académie de mes premiers essais sur ce sujet: je lui ai démontré, dans de précédents mémoirs . . . que l'air le plus pur, auquel M. Priestley a donné le nom *d'air déphlogistiqué,* entrait, comme partie constituante, dans la composition de plusieurs acides, et notamment de l'acide phosphorique, de l'acide vitriolique et de l'acide nitreux.[5]

In this passage he revealed again the primary source of the oxygen theory: three archetypical acids (notably two of the trio labelled "mineral acids") were produced by direct combination of vital air with another substance. He went on to claim himself now authorized to extend these ideas.

[1] *Ibid.*, p. 226.
[2] *Ibid.*, p. 227.
[3] Lavoisier, "Considérations générales sur la nature des acides et sur les principes dont ils sont composés" (1777), in his *Oeuvres,* II, pp. 248–260; henceforth cited as Lavoisier, "Nature des acides."
[4] For remarks on principles and on composition of acids and bases, see *ibid.*, p. 248. Though Lavoisier claimed that it was time for chemists "à attaquer les acides et les bases," he concerned himself only with the acids.
[5] *Ibid.*, p. 249.

Des expériences plus multipliées me mettent aujourd'hui dans le cas de généraliser ces conséquences, et d'avancer que l'air le plus pur, l'air éminemment respirable, est le principe constitutif de l'acidité; que ce principe est commun à tous les acides, et qu'il entre ensuite dans la composition de chacun d'eux un ou plusieurs autres principes qui les différencient et qui les constituent plutôt tel acide que tel autre.[6]

Here was the oxygen theory in its full form as a two-component theory, and here too Lavoisier bestowed upon the base of vital air its name:

D'après ces vérités, que je regard déjà comme très-solidement établies, je désignerai dorénavant l'air déphlogiostiqué ou air éminemment respirable dans l'état de combination et de fixité, par le nom *principe acidifiant,* ou, si l'on aime mieux la signification sous un mot grec, par celui de principe oxygine . . .[7]

Lavoisier chose his words with great care. A participant in the Condillacian tradition, Lavoisier saw in language as much an instrument of analysis as any employed in the laboratory. It was doubtless no accident that in this statement of the oxygen theory he refrained from saying that the presence of oxygen *must* confer acidity on the compound in which it was an ingredient. His statement amounted only to asserting that the presence of acidity presupposed the presence of oxygen, because oxygen was constititive of acidity. In this form the theory appeared as an induction from the facts: since acids arose from oxygenations, and since no acids appeared to be without oxygen, then all acids, to be such, doubtless required oxygen. Oxygen, in brief, was necessary to acidity, but not sufficient. In this formulation of the theory the absence of acidity in substances containing oxygen was not, of logical necessity, problematic. In the succeeding paragraphs Lavoisier presented the results of his well-known train of experiments in which oxygenation produced either acids or metallic calces, and he appeared quite untroubled by the latter category.

Having taken this linguistic precaution, Lavoisier tended nevertheless to ignore it in practice. He would claim in effect that oxygen was not only necessary to the appearance of acidity, but also sufficient. There appear to have been at least two factors making it easy for him to do so. First, it was natural to expect that if acids had been obtained before by oxygenating various substances, more acids, new ones, could be obtained by oxygenating new substances. There was little to suggest that this approach might fail, except for the anomaly of the metallic calces. There were significant indications, however, that this exception was only apparent, the most important indication being the case of arsenic acid, an acid that resulted from the oxidation of a metal. The evidence provided by the metallic acids that metals, too, were acidified by oxygen was the second factor enabling Lavoisier to violate his own cautions inductive statement. Both factors appeared in the example that Lavoisier chose to present in this memoir

[6] *Ibid.*

[7] *Ibid.*

on acids as the first illustration of the power of his two-component theory: the case of the acid of sugar (oxalic acid), first obtained by Bergman:

Cet acide, d'après les expériences dont je vais rendre compte, ne me paraît être autre chose que le sucre en entier combiné avec le principe acidifiant ou oxygine, et je me propose de faire voir successivement, dans différents mémoires, qu'on peut, par des procédés analogues, unir ce même principe à la corne des animaux, à la soie, à la lymphe animale, à la cire, aux huiles essentielles, aux huiles par expression, à la manne, à l'amidou, à l'arsenic,au fer, et probablement à un grand numbre d'autres substances des trois règnes, et les convertir en de véritables acides.[8]

Animal, vegetable, and mineral substances, even the metals, could be converted into acids. Having solved the problem of the specificity of the acids and having opened the prospect of creating an essentially unlimited number of other acids, Lavoisier was too taken with the potential of the oxygen theory to persevere in his initial caution about the sufficiency of oxygen for the generation of acidity. The thrust of his thought was all but inevitably directed to rendering the anomaly of the calces one of mere appearance.

It is not surprising, then, that the first two conclusions offered by the memoir on acids were reaffirmations of the generic role of oxygen as a property-bearing principle and of the possibility of assimilating the apparently exceptional metals to the category of acidifiables. From his reflections and experiments, Lavoisier claimed, it followed:

1° Qu'en général, lorsque le principe acidifiant ou oxygine se combine avec un corps quelconque, sans le décomposer (si on en excepte le plus grande nombre des substances métalliques), il le convertit en un acide particulier, qui, indépendamment des caractères généraux, communs à tous les acides, en a qui lui sont propres;

2° Qu'à l'égard des substances métalliques, il forme, avec la plupart, des composés connus sous le nom de *chaux métalliques;* j'ajouterai cependant ici, que, même dans cette classe des substances, il en est quelques-unes, comme l'arsenic, le fer et peut-être plusieurs autres, qui, combinées avec le principe acidifiant ou oxygine, jusqu'à un certain degré de surabondance, prennent non-seulement un caractère salin, mais encore acquièrent les propriétés communes aux acides, et deviennent comme eux de véritables dissolvants.[9]

The first of these conclusions was a restatement of the oxygen theory that went beyond the cautious induction in the introductory section of the memoir. The second statement revealed the technique by which Lavoisier sought to eliminate the anomaly of the metals: the attempt to correlate the intensity of acidic properties with the proportion of oxygen in the compound, coupled with the idea of a rapprochement of the calces with the acids by the traditional category of salts.

c. The correlation of properties and degree of oxidation

Lavoisier already had made successful use of this correlation technique in his studies of the typical acids. One of its most striking examples was in his studies

[8] *Ibid.*, p. 251.

[9] *Ibid.*, p. 259.

of nitric acid. In the original 1776 version of this paper, the one in which the oxygen theory was first suggested clearly, Lavoisier had no explanation, as Partington points out, for the disparity in the quantities of the constituents obtained from analysis in comparison with those necessary for synthesis. In the version of 1779, however, this discrepancy was explained by noting that the acid produced from the decomposition of nitre with clay contained a relatively greater quantity of pure air than the acid obtained by synthesis from nitrous air and pure air, a point that Lavoisier also stated in the 1778 paper on the nature of acids.[1] The distinction was later embodied, as is known, in the linguistic formulations of the new nomenclature. Nitrous and nitric acids were two oxidation stages of the same acidifiable base; the latter, containing more oxygen, was the stronger. Lavoisier had also distinguished sulfurous and sulfuric acids in another memoir, the former having been obtained by the removal of oxygen from the latter. In the *Traité* he made an express correlation of their relative acid strengths with the degrees of saturation with oxygen.[2]

Apparently it was the case of arsenic, to which Lavoisier had called attention in his memoir on acids, that seemed to hold out the possibility of applying similar considerations to the metals.[3] White arsenic was clearly a calx and displayed the properties of that genus. Further oxygenation, however, produced an acid, providing another example of the dependence of acidic properties upon degree of oxygenation. The example suggested that at a sufficiently low level of oxygenation, acidic properties would not appear. Indeed, all metallic calces likely were examples of this degree of oxygenation, a point that Lavoisier asserted explicitly in the *Traité* with an explanation based on affinity considerations:

> Les substances métalliques pendant leur calcination augmentent de poids à proportion de l'oxygène qu'elles absorbent; en même-tems elles perdent leut éclat métallique & se réduisent en un poudre terreuse. Les métaux dans cet état ne doivent point être considérés comme entièrement saturés d'oxygène, par la raison que leur action sur ce principe est balancée par la force d'attraction qu'exerce sur lui le principe calorique ... Aussi les substances métalliques, en s'oxygénant dans l'air et dans le gas oxygène, ne se convertissent-elles point en acides, comme le soufre, le phosphore et le charbon: il se forme des substances intermédiares qui commencent à se rapprocher de l'état salin, mais qui n'ont pas encore acquis toutes les propriétés salines.[4]

[1] For 1776 as the original date of this memoir, for the appearance of a corrected version in 1779, and for the differences between the versions, see Partington, *History*, III, pp. 410—416. For Lavoisier's explanation of the difference between the analyzed and synthesized acids in the 1779 version (the version appearing in *Oeuvres*, II), see Lavoisier, "Acide nitreux," p. 136. For this same point in the 1778 paper on the nature of acids, see Lavoisier, "Nature des acides," pp. 251—252. Cf. also Lavoisier, *Oeuvres*, II, pp. 550, 552; and *idem*, *Traité*, I, pp. 80—81.

[2] For sulfuric and sulfurous acids, see Lavoisier, "Mémoire sur la dissolution du mercure dans l'acide vitriolique et sur la résolution de cet acide en acide sulfureux aériforme et en air éminemment respirable" (1777), in his *Oeuvres*, II, pp. 194—198, esp. p. 198; and *idem*. *Traité*, I, pp. 76—77.

[3] For arsenic in the memoir on acids, see the passage quoted above, p. 76, accompanied by n. 8.

[4] Lavoisier, *Traité*, I, pp. 82—83.

Like Bergman, Lavoisier made use, in this passage, of the traditional notion of salts to suggest that the calces, though not acidic, represented a transitional saline stage, indicative of the passage from metal to acid. With a sufficiently high degree of oxidation, Lavoisier expected, the acidic characters would have appeared. Oxygen was not merely necessary for the production of acidity, but, in a suitable dose, sufficient. He went on to replace the term "calx" with that of "oxide," to indicate the level of oxygenation preceding that at which fully acidic properties appeared. Since in his view the generic properties of oxygenated substances depended only on the oxygen and not on the other component, all substances, not merely metals, were capable of forming non-acidic compounds, oxides, with small quantities of oxygen. There was now a whole vocabulary expressive of the degrees of oxygenation corresponding to variation in the generic properties, and it was applicable to all substances capable of combination with oxygen:

On voit d'après cela comment le langage que nous avons adopté est si fécond & expressif; un premier degré d'oxygénation constitue les oxides; un second degré constitue les acides terminé en *eux;* comme l'acide nitr*eux,* l'acide sulfur*eux;* un troisième degré constitute les acides en *ique,* tels que l'acide nitr*ique,* l'acide sulfur*ique,* enfin nous pouvons exprimer un quatrième degré d'oxygénation des substances en ajoutant l'epithète d'*oxygéné,* comme nous l'avons pour l'acide muriatique oxygéné.

Nous ne sommes par contentés de désigner sous le nom d'oxides la combinaison des métaux avec l'oxigène; nous n'avons fait aucun difficulté de nous en servir pour exprimer le premier degré d'oxygénation de toutes les substances, celui qui, sans les constituer acides, les rapproches de l'état salin. Nous appellerons donc *oxide de soufre* le soufre devenu mou par un commencement de combustion; nous appellerons oxide de phosphore la substance jaune que laisse le phosphore quand il a brûlé.

Nous dirons de même que le gaz nitreux, qui est le premier degré d'oxygénation de l'azote, est un oxide d'azote.[5]

Lavoisier was offering a classification of oxides: any acidifiable base was at least potentially capable of attaining each of four degrees of oxygenation, each stage representing a genus—an oxide, an acid in *-ous,* an acid in *-ic,* and an acid in *oxygenated -ic.* The "Tableau," of the *Traité,* "des combinaisons binaires de l'oxygène avec les substances métalliques & non-métalliques oxidables & acidifiables" had four columns, one for each of the four "degrés d'oxygénation," and listed as many entries in each as were known or suspected for all simple substances.[6] The principal and archetypical example of the fourth degree was oxygenated muriatic acid (chlorine), believed to be a higher oxide of the muriatic radical, of which muriatic acid was the acid in *-ic.* Oxygenated muriatic acid was itself something of an anomaly, in that it was less markedly acidic than muriatic acid. Rather than acknowledge this as an anomalous effect, Lavoisier declared that all substances capable of combining with oxygen were also capable of this

[5] *Ibid.,* pp. 84—85.

[6] *Ibid.,* p. 203.

same high level of oxygenation, as if the generality of the effect somehow compensated, in Lavoisier's mind, for the reduction of acidity manifest at this level.[7]

Similarly, Lavoisier asserted the general applicability of this pattern to metals. Just as the case of arsenic suggested that there probably existed non-acidic oxygen compounds of combustibles as well as of metals, so it suggested that there should exist acidic oxygen compounds of all the metals as well as of the combustibles. Since three metals, arsenic, molybdenum, and tungsten, formed acids by further oxygenation, it was reasonable to expect that the other metals could be made to yield acids on further oxygenation. Although in this context Lavoisier did not acknowledge the precedents in the ideas of Bergman, he was familiar with them, as pointed out shortly, and their influence is apparent. It was Bergman who suggested, using the model of the metallic acids, that all metals contained a double dose of phlogiston, metallizing phlogiston and coagulating phlogiston, and therefore were subject to a two-stage process of dephlogistication, by which the metal approached the saline and then the acidic state. By removal of the coagulating phlogiston, the calx was made to reveal its radical acid *à nud.* Similarly, Lavoisier expected that a second stage of oxygenation should produce an acid. In the *Traité,* Lavoisier included in his list of acids seventeen metallic acids, one for each of the seventeen metals then known.[8]

d. The oxidation of iron

In his account of combustion, Lavoisier noted that iron, like arsenic, should produce an acid on further oxidation.[9] When he put this suggestion to the test in a memoir on the oxidation of iron (1782), he acknowledged his debt to Bergman, citing the Swede's *De analysi ferri* as one of his chief sources.[1] In the 1782 paper, Lavoisier presented a series of experiments in which he produced both the lower oxide, martial ethiops, and the ochreous material. Accepting Bergman's

[7] For the exceptional oxidation stage of the muriatic radical, see *ibid.*, pp. 76—77. For the general exception of a fourth degree of oxidation for all radicals, see the table cited in the preceding note and the passage quoted above, p. 78, accompanied by n. 5. The table contains a column for the fourth degree with a series of tentative entries and the passage includes the term "acide ... oxygéné" in the nomenclature for acids.

[8] Lavoisier, *Traité,* I, p. 181.

[9] In the passage quoted above, p. 76, accompanied by n. 8.

[1] Lavoisier, "Mémoire sur l'union du principe oxygine avec le fer" (1782), in his *Oeuvres,* II, pp. 557—574; henceforth cited as Lavoisier, "Fer oxygéné." In this memoir, Lavoisier acknowledged frankly that Bergman (and, for the chemistry of certain iron salts, Fourcroy), were his principal sources (p. 568): "Ceux qui ont lu l'excellente dissertation de M. Bergman sur l'analyse des différents fers s'apercevront que j'ai beaucoup profité de ses expériences pour tout ce que je viens d'avancer; j'ai tiré aussi un très-grande parti de celles que M. de Fourcroy a communiquées à l'Académie en 1776 et 1777, sur les précipités martiaux; si je n'ai pas cité à chaque expérience ces chimistes célèbres, c'est pour éviter les répétitions; du reste, il n'est pas presque aucun des faits qu'ils ont rapportés que je n'aie vérifié." For Fourcroy's work on iron salts, see Partington, *History,* III, p. 542, and W.A. Smeaton, *Fourcroy: chemist and revolutionary, 1775—1809* (Cambridge: 1962), pp. 94—95.

view that metals dissolved in acids only in the form of calces, Lavoisier translated it into the terms of the oxygen theory of calcination. To this end he offered a variety of proofs that iron, in its acidic dissolutions, was present in the form of the ethiops, having acquired its oxygen either at the expense of the acid or at that of the water of the solution, depending upon the concentration of the acid.[2] To oppose the contrary opinion that iron could first de dissolved and then calcined, he remarked,

> qu'il existe un degré de calcination qui convient le mieux à la dissolution du fer dans les acides; plus ou moins chargés de principe oxygine, il devient moins dissoluble, et c'est surtout à l'égard de l'ocre que cet effet est sensible, dans cette combination métallique; la tendence du fer à la combinaison est presque complèment satisfaite, elle n'est plus en conséquence dissoluble dans les acides, et elle l'est plutôt, au contraire, dans les substances alcalines.[3]

According to Lavoisier, only the martial ethiops was soluble in acids and therefore was capable, it may be added, of serving as the base in neutral salts. Addition of more oxygen shifted the qualities of the oxide toward the saline and acidic states, as in the case of the ochre. Lavoisier displayed none of Bergman's tentativeness about the nature of the ochre but seized upon the opportunity expressly to apply to the metals his theory of the variability of generic properties in accordance with degree of oxidation.

> On trouve ici un analogie frappante qui lie parfaitement toute cette theorie. J'ai fait voir ailleurs que le principe oxygine était le principe de l'acidité, qu'il n'y avait point d'acide dans la composition duquel ce principe n'entrât; *les substances métalliques,* doivent donc, d'après cette théorie, *se rapprocher de la qualité acide à mesure qu'elles sont plus chargées de principe oxygine,* et c'est par cette raison sans doute qu'elles deviennent indifférents à s'unir aux acides, et qu'elles finissent par acquérir une affinité marquées pour les alcalis.[4]

Thus Lavoisier pressed beyond the bounds of his earlier, cautions induction. With the case of the ochreous oxide, he doubtless believed himself on the threshold of success in his endeavor to reduce the anomaly of the calces, and accordingly here he was most prone to abandon his logical prudence and reveal his true beliefs.

It might appear from this analysis that Lavoisier had in fact succeeded in this effort. By permitting generic properties to vary in accordance with the proportion of oxygen, he could merely label the calces "oxides," defined as compounds containing insufficient oxygen to display acidic properties.[5] By attributing to both combustibles and metals the possibility of forming a sequence of oxidation

[2] For these arguments, see Lavoisier, "Fer oxygéné," pp. 560—568.
[3] *Ibid.*, pp. 564—565.
[4] *Ibid.*, pp. 565. Emphasis added.
[5] Cf. Guyton, in Lavoisier *et al.*, *Méthode,* p. 58: "Après avoir établi la distinction des acides & des oxides métalliques, c'est-à-dire des métaux *oxigênés* & des métaux simplement *oxidés*..."

stages from non-acidic oxide through acid, he preserved both the acidifying role of oxygen and the Stahlian analogy between calcination and combustion, between metals and combustibles. If the acids of all the metals were not yet known, Lavoisier asserted in the *Traité,* it was merely a matter of time until they would be.[6] Moreover, the dissolution of the metals by metallic acids resulting in the production of the metallic salts of the metallic acids seemed to offer a new field of investigation.[7]

e. The oxygen theory of acidity and the qualitative chemistry of metals

The anomaly was not so readily dispatched, however. Lavoisier's principal attempt to produce an acid by further oxidation of a metal was a failure. Like Bergman, he had produced no acidic oxide of iron; in fact, the oxides that he did prepare possess alkaline properties. His discussion of metallic acids in the *Traité* clearly indicates his dilemma. Revealing his hopes, he listed the acids of all seventeen metals then known in his table of acids; reflecting his fears, he conceded grudingly that the acids were "encore peu connus" and that "on ne peut pas encore se flatter sans doute de les avoir tous découverts . . ."[8] The truth would have been more accurately expressed by a flat statement that only three metallic acids were known. He had not abandoned the oxygen theory of acidity, but he had failed to reduce the principal anomaly confronting it. He could only cling to the hope that the acids of the other fourteen metals would be discovered, thus confirming the analogy between combustibles and metals and the generality of the acidifying role of oxygen.

Lavoisier's refusal to abandon the oxygen theory of acidity in the face of these difficulties entailed two major consequences for the chemistry of metals. First, it virtually removed from chemical consideration an entire class of metallic oxides, the higher, non-acidic oxides that Lavoisier believed to be incapable of salt formation. Second, the theory devoted little attention to the positive properties of the acknowledged salifiable oxides (or to those of other salifiable bases). A major achievement of Berzelius was to restore the chemistry of metals to a prominent, indeed, central place in chemistry. He showed that nearly all metallic oxides could serve as bases (and many as acids) and that basicity represented a set of positively defined properties, of status equivalent to those of acids.

Lavoisier's slight of the higher oxides resulted from his failure to assimilate them either to the metallic acids or to the salifiable bases. Evidently non-acidic, the higher oxides must have contained insufficient oxygen to exhibit acidic characters. However, their greater dose of oxygen than that of the salifiable oxides must have removed the oxides from the catagory of bases and placed them on the

[6] Lavoisier, *Traité,* I, p. 182.
[7] *Ibid.*, pp. 182—183.
[8] For Lavoisier on the metallic acids, see *ibid.*, pp. 182—183. The table of acids is on p. 181.

path toward the acidic state.[9] The higher oxides were salts—in the vague, traditional sense of that term. Midway between acids and bases, and playing the role of neither, they had no role left to play; there was no room for traditional salts in Lavoisier's chemistry. Accordingly, Lavoisier relegated the higher oxides to the category of "oxides," naming them according to their radical and their colors (but not their chemical properties), and stopped talking about them.[1] Looking only at the generic role of oxygen, for signs of acidification as the sole result of oxidation, Lavoisier prevented himself from seeing the higher oxides as salifiable bases (as illustrated by his failure to notice the alkaline properties of the iron oxides). Berzelius, who denied the generic role of oxygen, considered acidity and basicity to be specific properties of the radicals. In his view, all oxides of radicals possessed either acidic or basic properties and differed from one another chiefly in the intensity of these properties. Arguing that acidity or basicity of oxides depended first on the identity of their radicals and second on their quantity of oxygen, Berzelius correlated the modifications in the inherent acidity or basicity of radicals with their degree of oxidation.

As for the salifiable oxides, they fared scarcely better in Lavoisier's hands than the higher oxides. Lavoisier merely lumped them together with alkalies and earths in the category of bases, which he regarded as "basic" in the same sense as Rouelle had. His posture is clear from the definition of neutral salts he offered in the *Traité:*

> Je ne rangerai pas non plus . . . les alkalis ni les substances terreuses, telles que la chaux, la magnésie, &c., dans la classe des sels, & je ne désignerai par ce nom que des composés formés de la réunion d'une substance simple oxygénée [that is, an acid] avec une base quelconque.[2]

The role of acid as salifying principle was of vastly greater interest to Lavoisier. In opening the chapter on neutral salts in the *Traité,* he remarked:

> Les substances acidifiables en se combinant avec l'oxigène, & en se convertissant en acides, acquièrent une grande tendance à la combinaison; elles deviennent susceptibles de s'unir avec des substances terreuses & métalliques, & c'est de cette union que résultent les sels neutres. Les acides peuvent donc être regardés comme de véritables principes salifians & les substances auxquelles ils s'unissent pour former des sels neutres, comme des bases salifiables: c'est précisément de la combination des principes salifians avec les bases salifiables dont nous allons nous occuper dans cet article.[3]

Insofar as salt formation was concerned, acids were salifying principles and bases were substrates. Acidifiable substances became highly reactive when they were combined with oxygen and then united with bases. Of the reactivity of the bases,

[9] See above, p. 80, nn. 3—5 and accompanying text.

[1] For the "oxides," see the table of oxides in *ibid.*, p. 203, and for the nomenclature by radical and color, see p. 85.

[2] *Ibid.*, pp. 163—164.

[3] *Ibid.*, pp. 162—163.

however, nothing was said. The acids were active, in this view, and bases were passive and receptive. Though the remainder of this chapter of the *Traité* and all of the next were devoted to an account of salifiable bases, extremely little was offered about the properties of these substances.[4] Even the characteristic corrosive properties of the three alkalies were not described. The bulk of the discussion was devoted instead to methods of preparation, which were offered, not their own sake, but to serve as definitions.[5] The objective of the chapter was solely to identify the different salifiable bases and to note that they were different, but there was no effort to explore the differences for their own sake.

Although the chemical revolution was the outcome of a long tradition of eighteenth-century efforts to restore the possibility of explaining qualitative specificity in chemistry, paradoxically it provided few means to accomplish this. Two-component theories had led their exponents to ignore the properties of many bodies that served as specific ingredients. The two-component approach, in any case, had been narrowly circumscribed in Lavoisier's chemistry. Apart from the caloric theory, which accounted for the gaseous state, the new chemistry had only a pair of two-component theories, the oxygen theory of acidity and the theory of neutral salts. If it is true, as is often remarked, that the chemical revolution eliminated principles from the science, the oxygen theory being only a short-lived vestige, then the revolution had left the science totally without means to account for the qualitative specificity of chemicals. It remained for Berzelius to supply this lack.

4. Guyton's two-fluid radical theory

a. Introduction

As a disciple of Stahl in late-eighteenth-century France, Guyton felt his theoretical commitments endangered by Lavoisier's studies of calcination and acidification. Lavoisier claimed that these processes took place not by loss of phlogiston but by addition of air, which Lavoisier regarded as a chemical having its own reactivity. In response to Lavoisier's work, Guyton devised a theoretical compromise invoking both phlogiston and air as a pair of generic principles that conferred roughly opposing properties on the substances with which they combined. Among the principal features of the compromise was the claim that metals, combustibles, and acids all contained specific components, which Guyton, following Bergman, called "radical" components and later just "radicals." The radicals were the seats of specific properties and were not isolable by analytical procedures. Rather, their presence was inferred by reflection upon the progress

[4] *Ibid.*, Chs. 16 and 17, on pp. 162—187.

[5] E.g., *ibid.*, p. 167, where Lavoisier remarked of his comments on the preparation of potash: "Mon objet n'est point de décrire ici l'art de préparer la potasse, encore moins les moyens de l'obtenir pure: je n'entre même ici dans ces détails que pour obéir à la loi que je me suis faite de n'admettre aucun mot que n'ait été défini."

of certain reactions, notably those that Bergman had studied, leading from metals or combustibles to acids. Guyton's theory amounted to a pair of parallel two-component theories, one for the metals and combustibles and one for the acids. His compromise invoked the old Stahlian technique of discerning the existence of chemicals by reflection upon experience, but the chemicals it discerned were specific components, not generic principles.

Guyton's views were undermined by a serious flaw. His two-component theories failed to meet one of the conditions specified in the last chapter for the viability of two-component theories, in that Guyton had no direct evidence for the existence of radicals. In this respect, Lavoisier's theory was superior to Guyton's. Adopting the term "radical" to label the specific components, Lavoisier offered as radicals a collection of particular chemicals isolated by analysis. Indeed, his radicals were the last term of analysis, and as such they provided the entries in Lavoisier's list of "substances simples." (In Lavoisier's two-component theory of organic chemistry, however, the existence of radicals was not directly demonstrated but was only assumed, as shown in Chapter III.)

This section explores the development of Guyton's radical theory in the period from the mid-1770's through 1786, when Guyton expressed his mature views in the *Encyclopédie méthodique*. It examines Guyton's work as an effort to salvage his Stahlian commitments when threatened by Lavoisier's studies of the chemical role of air. The section also points out Guyton's dependence on Bergman for several leading notions that he tailored to fit his needs. It concludes by examining the fate of the concept of radicals in Lavoisier's science, which set the stage for Berzelius's innovations.

b. Guyton's theoretical compromise

1) *Introduction*

Originally hesitant to accept the fixation of air in calcination, Guyton allowed himself to be persuaded by Lavoisier, whom he visited in Paris in 1775 to consult about the phenomenon.[6] He believed initially that the air involved in calcination was fixed air (carbon dioxide), recognizing only in 1780 that the air fixed in the process was vital air.[7] Unlike Lavoisier, however, he was unprepared to abandon phlogiston, and, accordingly, he responded by devising a compromise.

[6] For Guyton's visit to Lavoisier, see W.A. Smeaton, "Guyton de Morveau's course of chemistry in the Dijon academy," *Ambix*, 9 (1961), 53—69, esp. 61; henceforth cited as Smeaton, "Guyton's course."

[7] Guyton's belief that the air involved in combustion and calcination was fixed air is noted by Partington (*History*, III, p. 519) and by Smeaton ("Guyton's course," 61). Smeaton rightly points out that Guyton obtained his information about the chemical role of air from Lavoisier, who understood only in 1777—1778 that the air involved was oxygen. However, Guyton's visit took place in March or April, 1775, when Lavoisier was just extricating himself from his expectation that the air obtained from heating mercury precipitate *per se* was fixed air (for this, see Partington, *History*, III, p. 405). Guyton may have composed the "Conciliation," the paper in which he first articulated his com-

The process of mutual accommodation began in 1775 and continued until Guyton's conversion to Lavoisier's views in 1787. In 1776 Guyton announced his role as mediator in a brief paper in Rozier's *Observations sur la physique*, entitled, "Conciliation des principes de Sthaal [*sic*] avec les expériences modernes sur l'air fixe."[8] The paper set the theme for Guyton's labors over the next decade. The basic viewpoint of the "Conciliation," and also of the *Chymie de Dijon*,[9] was that metals were compounds of earths and phlogiston, or fixed fire,

promise, as seen shortly, with Lavoisier's expectation in mind, perhaps having left Paris before Lavoisier had persuaded himself that fixed air was not the air involved in calcination and combustion.

Had Guyton left Paris with Lavoisier's new opinion in mind, he would still have been in "error." Lavoisier's revised view, which he committed to print in the famous Easter memoir of 1775, was that the air involved in calcination and combustion was common (atmospheric) air. Guyton's error of 1775 was likely no more than to have accepted at face value the latest news from the research front dominated by Lavoisier.

Nevertheless, even after 1778, when the discovery of oxygen and of its role in calcination became widely known, Guyton did not assimilate these developments. Having accepted Lavoisier's claim that the air in question was the principle of acidity, Guyton was still referring to the acidifying principle as "fixed air," or simply as "air," as late as 1780 (for this, see W.A. Smeaton, "Guyton de Morveau and the phlogiston theory," in *Mélanges Alexandre Koyré*, 2 vols. (Paris: 1964), I, pp. 522—540, esp. p. 536; henceforth cited as Smeaton, "Guyton and phlogiston"). It is possible that Guyton implied by the term "fixed air" merely that the air in question was capable of becoming fixed in bodies, just as he had introduced the term "fixed fire" as a synonym for phlogiston to signify that it was a chemical, *like air*, capable of becoming fixed in bodies by the force of affinity (for this, see above, p. 48, n. 1 and accompanying text). However, there is evidence that Guyton identified the acidifying principle with the specific chemical (carbon dioxide) called "fixed air" since the 1750's, when Joseph Black characterized it. In particular, Guyton supported his claim that fized air was the acidifying principle with a citation of Fontana's work, and Fontana unquestionably was studying the specific chemical, fixed air. In "A note on fixed air: the universal acid," *Ambix*, 20 (1973), 88—94, E.H. Le Grand shows that Fontana was one among several researchers who helped establish a widespread belief in the late 1770's that fixed air was the universal acid. Very probably, Guyton shared this opinion, changing his mind, as is suggested below (p. 91, n. 8 and accompanying text), only in response to Lavoisier's study, published in 1780, of the acidification of nitrous gas. Consistent with this interpretation is the appearance of Guyton's discussion in the *Encyclopédie méthodique* (1786; cited below, p. 90, n. 7) of the question "L'acide méphitique [Guyton's later name for fixed air] est-il le principe acide universel?," where Guyton devoted several paragraphs (pp. 85—86; cf. also pp. 29—30) to the work of the same figures cited by Le Grand.

In the late-1770's, Guyton probably believed that the work of Fontana confirmed what Lavoisier had told him in 1775. Perhaps confusing fixed air with fixation of air, Guyton probably felt no need to adjust his opinions until he began writing the *Encyclopédie méthodique*, which offered responses to Lavoisier's threats to the phlogiston theory of the late-eighteenth-century Newtonian Stahlians. If he was to address Lavoisier's ideas directly, he first had to master them.

[8] Guyton, "Conciliation des principes de Sthaal [*sic*] avec les expériences modernes sur l'air fixe," *Obs. sur la physique*, 7 (1776), 380—395; henceforth cited as Guyton, "Conciliation."

[9] Guyton, H. Maret, and J.F. Durande, *Elémens de chymie, théorique et pratique, rédigées dans un nouvel ordre, d'après les découvertes modernes, pour servir aux cours publics de l'académie de Dijon*, 3 vols. (1777—1778); for full citation and information about authorship, see above, p. 35, n. 3; henceforth cited as Guyton, *Chymie de Dijon*.

and that calces were compounds of the same earths and fixed air.[1] Calcination was described as the simultaneous expulsion of phlogiston from combination with the metallic earth and the union of the earth with air. The metal was a solution of the metallic earth in phlogiston; the calx, a solution of the earth in air.[2] Calcination was a displacement of phlogiston by air. The two fluids, air and phlogiston, were generic solvents, each having its own order of affinities and each conferring its own set of generic properties. The earth was never pure but, like a Stahlian principle, always existed in combination with another body, that is, with one of the two fluids or with an acid. Ultimately, the third alternative disappeared, the air described as "fixed air" in this argument became vital air or oxygen, and the never-isolated earth became a radical.

Originally a response to the pneumatic explanation of calcination, Guyton's theory underwent further evolution in response to Lavoisier's studies of acidification, particularly the study of the acidification of nitrous gas to form nitric acid. Guyton's mature views, expounded in the first part (1786) of the first volume of the *Encyclopédie méthodique,* incorporated his recognition that Lavoisier had made oxygen, not fixed air, the acidifying principle. The revised theory also made use of Bergman's parallel treatment of metals and acids so that Guyton could apply the theory of calcination developed in the "Conciliation" to the acidification of combustibles.

2) *The theoretical roles of phlogiston and of air*

In the "Conciliation," Guyton described the novelty of pneumatic chemistry against the background of the phlogiston theory of the French Stahlians. He first described phlogiston in its dual capacity as a property-bearing principle and as a reagent endowed with affinity:

> Les métaux, suivant Sthaal [*sic*], reçoivent leur forme, leur solidité, leur ductilité, leur éclat, de l'union de leur terre propre avec le phlogistique, qui est un principe fluide, très-subtil, incoërcible, & néanmoins capable de se fixer dans certains corps, qui se trouve abondamment dans le soufre, l'huile, les charbons, que la chymie ôte & rend presqu'à volonté à différentes substances, qu'elle fait passer de l'une dans l'autre, qui dans ces passages, manifeste toujours les mêmes propriétés, qui est par-tout & constamment identique.[3]

The pneumatic account of calcination seemed subversive to Guyton in two respects. First, the explanation of calcination as the fixation of air seemed incompatible with the prevailing account of the process as the loss or expulsion of a property-bearing principle: second, the fixation of an air appeared discordant

[1] For this view in the *Chymie de Dijon,* see Partington, *History,* III, p. 522. Guyton's views in the "Conciliation" are analyzed in detail in the present section.

[2] For Guyton's solution theory, see his *Chymie de Dijon,* I, pp. 43—97; cf. W.A. Smeaton, "Guyton de Morveau and chemical affinity," *Ambix,* 11 (1963), 55—64, esp. 57—59; henceforth cited as Smeaton, "Guyton and affinity."

[3] Guyton, "Conciliation," 389.

with Guyton's views of chemical agency: that is, it did not seem to present a case of a solvent effect, of the action of an agent endowed with the force of affinity and of the attainment of equiponderance of solute and solvent. In the "Conciliation" Guyton gave voice to his keen sensitivity to both subversive aspects. Referring to the modified Stahlian theory that he summarized in this paper, he conceded that

> les expériences modernes sur l'air fixe, ont considérablement ébranlé cette doctrine; car si, à force de multiplier & de varier les procédés les plus ingénieux, on est parvenu à vérifier que les métaux exposés à la calcination, absorbent & s'approprient une portion d'air d'une nature particulière, les conséquences naturelles sont que l'air n'est pas seulement nécessaire à cette opération comme agent mécanique, que l'état des chaux métalliques n'est pas dû à l'absence du phlogistique; bien plus, qu'est-ce que cette union de l'air fixe & de terre métallique? Par quel moyen conçoit-on qu'elle puisse s'opérer entre deux matières, dont l'une reste constamment sous forme pulvérulente, grossière, incapable de devenir homogène avec un fluide? Y a-t-il donc des combinaisons de deux substances sans équipondérance, sans attraction d'affinité, sans dissolution? Comment enfin concilier les faits si étranges avec l'explication si simple, si palpable de la marche univoque de la nature dans tout le reste des opérations chymiques?[4]

Guyton was not pleased, he admitted, to see these novelties, so destructive of received opinion:

> J'avoue que je n'ai pas vu sans peine, ces phénomènes se réaliser à mes yeux, tant que je n'ai pu les envisager, que sous cet aspect, qui menaçoit d'une ruine prochaine la plus belle partie de l'édifice de nos connoissances; mais tandis que j'étois livré à cette inquiétude, j'ai cru appercevoir des rapports, & une analogie marquée entre ces mêmes phénomènes & une infinité d'autres opérations dont la théorie a paru jusqu'à ce jour la moins embarrassante, la plus conséquente aux principes reçus . . .[5]

Guyton's "Conciliation" was the occasion for articulating this analogy that would reduce the subversiveness of pneumatic chemistry. It was based on the attribution, to fixed air in 1776, and later to vital air, of precisely the same two roles that he had attributed to phlogiston: it was simultaneously a chemical agent endowed with affinity and a property-bearing principle. He constructed a theory of calcination reactions in which each served as principles and reagents. Quite appropriately, therefore, Guyton declared himself the author of a "two-fluid" theory:

> Il est facile présentement de distinguer deux fluides essentiellement différens, qui ont en conséquence, un ordre d'affinité propre, l'un est le *phlogistique,* ou la matière du feu; l'autre est ce fluide élastique qui nous nommons *air fixe* . . .[6]

[4] *Ibid.,* 390. The belief that air was a mechanical agent, undermined by the new experiments in pneumatic chemistry, as Guyton noted, was Boerhaave's view of the role of air in combustion (for Boerhaave's view, see Hélène Metzger, *Newton, Stahl, Boerhaave et la doctrine chimique* [1930; rpt., Paris: 1974], pp. 238—243). Guyton employed this theory early in his chemical career, as Partington notes (*History,* III, pp. 520—521), applying it to calcination, which, as a Stahlian, he assimilated to combustion.

[5] Guyton, "Conciliation," 390—391.

[6] *Ibid.,* 391.

Elaborating this assimilation of the new elastic fluid to the category of menstruum or solvent, Guyton asserted that

> L'affinité de l'air fixe avec les différentes terres métalliques, varie comme celle de tous les autres dissolvans, il y en a qu'il attaque spontanément dans l'état métallique solide; tel est le fer dont la rouille est une vraie chaux, & en a toutes les propriétés; il y en a avec lesquelles il ne peut s'unir, du-moins à un certain point, tels sont l'or, la platine, l'argent; il y en a avec lesquelles il se combine si difficilement, qu'elles s'élèvent en vapeurs, sans perdre leur phlogistique, auxquelles il adhère si peu, qu'il s'en sépare sans intermède, c'est le mercure.[7]

Guyton also expanded upon the role of air as a property-bearing principle. In doing so, he asserted the basic tenet of the oxygen theory of acids apparently independently of Lavoisier, who, as noted above, first stated it directly in his 1776 paper on nitric acid, that is, in the same year as Guyton's "Conciliation."

> Cette hypothèse [the two-fluid theory of calcination] nous conduit à penser que les acides contiennent eux-mêmes de l'air fixe, puisqu'ils en fournissent aux terres des métaux, qui leur donne les propriétés de chaux métalliques, sans qu'elles ayant pu recevoir ce principe, ni de l'air libre avec lequel elles n'ont pas été en contact, ni d'aucune autre substance; cela est sur-tout sensible dans les dissolutions où nous disons que l'acide calcine plutôt qu'il ne dissout le métal, comme celle de l'étain par l'acide nitreux, & dans toutes les opérations après lesquelles on retrouve d'un côté, le métal en état de chaux, de l'autre, l'acide vitriolique rendu sulphureux par le phlogistique: dès-lors l'acide le plus simple ne doit plus être considéré comme un principe que la Chymie ne peut décomposer, puisqu'elle lui ôte & rend à volonté, une de ses parties constituantes; cela ne répugne nullement à l'idée que Newton nous a donné de l'acide, en le définissant *ce qui attire fortement, & qui est fortement attiré;* & à en juger par l'acide dont nous pouvons suivre plus aisément la formation, il est très-probable, comme le remarque M. l'Abbé Rozier, que le vin ne tourne à l'acide qu'en absorbant une portion d'air qui augmente son poids.[8]

Thus, an acid was not one of the "élémens chymiques," but a compound of a specific component with the generic aerial fluid that conferred acidity. Though Guyton had hit upon a two-component explanation of acidity, he showed no desire to exploit it until he felt the impact of Lavoisier's studies of acidification. This point is explored in subsection *c*.

3) *The process of calcination*

In Guyton's "Conciliation," he assimilated the events of calcination to his notions of solution and precipitation. As his model, he relied on a well-known reaction, the precipitation of *luna cornea* (horn silver or silver chloride) by addition of muriatic acid to a solution of silver in nitric acid.[9] Because the silver and the muriatic acid were both in solution, the necessary condition for all reactions, they could combine; because the resultant combination was insoluble, it precipitated from solution. Though the suddenness of this effect left the

[7] *Ibid.*, 392.
[8] *Ibid.*, 392—393.
[9] *Ibid.*, 391, for the precipitation of *luna cornea* as Guyton's model.

precipitate in the form of a powder rather than crystals, the reaction had to be understood as taking place in solution:

on n'a pas imaginé pour cela qu'il y eut union sans dissolution, sans fluidité, on dit avec bien plus de fondement, que le nouveau sel se précipitoit, parce qu'il est moins soluble...[1]

The production of an irregular, powdery mass of an insoluble salt, like the production of regular crystals of a soluble salt, could be viewed as a true chemical solution process. The production of an irregular, powdery *calx* could be conceived in exactly the same way:

L'air fixe s'empare des terres métalliques, s'unit avec elles, comme l'acide marin à la lune cornée, à raison d'une affinité supérieure à celle du fluide différent [phlogiston] qui les tenoit en dissolution; il les prend à mesure qu'il les touche, & qu'elles se trouvent dans la condition de cette affinité en s'éprouvant au-moins un commencement de la dissolution; le nouveau composé se précipite de même, parce qu'il est moins soluble; il se précipite sans affecter aucune forme régulière, parce que le passage à l'état solide est trop subite.[2]

The metal was a solution of a metallic earth in the solvent phlogiston, analogous to that of silver in nitric acid. In calcination, the earth left its combination with phlogiston and united instead with the other fluid, the new air, for which it had a greater affinity. The resultant combination, like *luna cornea,* was insoluble, and it therefore precipitated suddenly out of solution in the form of a powdery calx. Guyton's view of the calcination reaction, which has been recognized as a process involving simultaneous expulsion of phlogiston and fixation of air,[3] clearly must be taken as an attempt to see the reaction as dictated by the differing affinity forces of two different, generic substances that were simultaneously principles and reagents. Guyton felt he could succeed in treating calcination and a large number of related reactions as displacements governed by the relative affinities of the two fluids for the metallic earths:

En tenant compte de l'air fixe dans les dissolutions, les réductions, les cémentations, les précipitations alkalines, terreuses & métalliques, il faudra simplement admettre des affinités réciproques & simultanées, c'est-à-dire, décomposition & composition; les acides ne toucheront pas aux chaux métalliques, parce qu'ils seront respectivement comme des corps neutralisés par le même principe; les matières grasses, les flux, les charbons ne lâcheront leur phlogistique aux terres métalliques, qu'autant qu'elles leur céderont leur air fixe; l'alkali & la craie abandonneront ce principe à la terre du métal, en même-temps qu'ils s'empareront de l'acide, & si l'on précipite une dissolution métallique par un métal, ce ne sera plus de l'air fixe qui sera donné en échange, ce sera du phlogistique, parce que c'est du phlogistique que porte le métal; c'est ce qui arrive dans la précipitation du cuivre par le fer, du mercure par le cuivre, &c. Je ne doute pas que lorsqu'on sera parvenu à distinguer deux terres métalliques, dont l'une ait plus d'affinité avec l'air fixe, l'autre

[1] *Ibid.*

[2] *Ibid.*, 392.

[3] By Metzger, in her "La philosophie de la matière chez Lavoisier," *Actualités scientifiques et industrielles,* no. 218 (1935), 1—48 (separate pagination), esp. 20; and by Smeaton, "Guyton and affinity," 59.

plus avec le phlogistique, on ne puisse également les déterminer à cet échange par la voie sèche.[4]

4) *Generic and specific properties of metallic combinations*

In the "Conciliation," Guyton made a thoroughgoing application of the two-component approach to metallic combinations. The specific and generic properties of threee kinds of substances could be explained by treating them as three categories of compounds or solutions of the metallic earths in three different generic components:

> Les terres métalliques sont donc toujours unies ou au phlogistique, ou à l'air fixe, ou à un acide, & il peut arriver que les sels métalliques ne soient corrosifs que parce qu'il y demeure encore une portion d'air fixe qui empêche la parfaire saturation. Quoi qu'il en soit, les propriétés de ces trois composés doivent être fort différentes, & elles le sont en effet dans des degrés qui ne peuvent être plus marqués: l'eau dissout complettement les derniers, & ne touche pas aux deux autres; le feu est le seul dissolvant capable d'agir sur ceux-ci; mais il fait simplement du premier un métal fluide, il forme un verre avec le second: cette propriété constante n'exclut pas les différens degrés de fusibilité, à raison de la nature particulière de la terre; la chaux de plomb, par exemple, est bien plus soluble que la chaux d'étain.[5]

Metals, calces, and salts were different genera of compounds of the specific metallic earths with different property-bearing generic principles (or with a substance, like an acid, containing such a principle).

This idea, although it entailed decomposition of the acids, was almost entirely an elaboration of the theory of metals and only marginally an elaboration of the theory of acids. Though (like Lavoisier) Guyton pointed in 1776 to the generic role of air in acid formation, he did not inquire into the specificity of the acids. Lavoisier had approached the acids directly, from the phenomenon of the combustion of certain non-metals; beginning with the specific component, he had added the generic and produced an acid. Guyton, however, started with the acids and considered the removal of the generic component; the specific component was left behind, but in less evident form.[6]

c. The combustion reaction: radicals and the specificity of the acids

Guyton did not address the issue of the specificity of the acids until he had experienced forcefully the impact of Lavoisier's ideas about combustion and those of Bergman and Scheele on metallic acids. He developed his views on acids in the first volume (1786) of the *Encyclopédie méthodique.*[7] In the article "Acide,"

[4] Guyton, "Conciliation," 393.

[5] *Ibid.*, 393—394.

[6] E.g., in the passage quoted above, p. 88 accompanied by n. 8.

[7] Details about the publication of the chemical volumes of the *Encyclopédie méthodique* are given by W.A. Smeaton in his "L.B. Guyton de Morveau (1737—1816): a bibliographical

which introduced most of the subject matter of the first volume, Guyton signalled that his approach was dictated by the fusion of precedents from Bergman and Lavoisier. He remarked that Lavoisier's studies of acidification, especially the acidification of nitrous gas, had persuaded him that vital air was the principle of acidity:

Les auteurs des élémens de Chymie de l'académie de Dijon [Guyton, Maret, and Durande], assurés par les belles expériences de M. Lavoisier de la présence de l'air vital dans l'acide nitreux, & ayant de plus observé que le soufre & le phosphore absorboient une quantité d'air en passant à l'état acide, commencèrent à soupçonner que ce fluide entroit comme partie constituante *dans la composition de tous les acides,* qu'il étoit essentiel à leur état combinaison aqueuse, en un mot qu'il pourroit bien être le vrai *acide universel,* l'élément acide, au lieu de l'acide du soufre qui n'existe pas lui-même tout formé [that is, as a preexisting ingredient] dans le soufre.[8]

A few pages later, Guyton called attention to the ideas of Bergman and Scheele, "ces deux célèbres Chymistes," and remarked of their theory of metallic acids

study," *Ambix,* 6 (1957—1958), 18—34, esp. 25. Smeaton points out that Guyton was commissioned in 1780 by the publisher Pancoucke to write the chemical volumes. Of the six volumes that appeared, Guyton wrote only the first (or at least the bulk of it), which Smeaton describes thus (25n36): "*Encyclopédie Méthodique, Chymie, Pharmacie et Métallurgie. La Chymie, par M. [Guyton] de Morveau. La Pharmacie, par M. Maret. La Métallurgie, par M. Duhamel.* Vol. I, 4to, pp. viii & 774; Paris, *Pancoucke,* and Liège, *Plomteux.* The first part was published in 1786, the date on the title-page; the colophon at the end of the second part (p. 774) is dated 1789, and shows that it was printed in Dijon by L.N. Frantin. Generally, both parts are bound together in one volume, but one copy (PO [Patent Office, London]) is in two parts, in the original boards; part two, which has no separate title-page, begins with the article 'Acidifiable' on p. 417." Cf. Denis I. Duveen and Herbert S. Klickstein, "A letter from Guyton de Morveau to Macquart relating to Lavoisier's attack against the *phlogiston* theory (1778); with an account of de Morveau's conversion to Lavoisier's doctrines in 1787," *Osiris,* 12 (1956), 342—367, esp. 350n31, 353n47, 364n88, 365n89; (henceforth cited as Duveen and Klickstein, "Letter from Guyton"); and Partington, *History,* III, p. 524. The first volume of the *Encyclopédie méthodique,* written primarily by Guyton, is cited henceforth as Guyton, *Encyclopédie méthodique, chymie,* I.

Of the remaining volumes, Smeaton says (*loc. cit.*): "Volumes II—V, written by Fourcroy, appeared in 1792, 1796, 1805 and 1808, and Volume VI, completed by Vauquelin after Fourcroy's death, in 1815; the volume of plates was published in 1813." Cf. Smeaton, *Fourcroy: chemist and revolutionary, 1755—1809,* p. 229.

[8] Guyton, *Encyclopédie méthodique, chymie,* I, p. 29. Three of Lavoisier's memoirs served as Guyton's chief sources for Lavoisier's oxygen theory of combustion and acidity. Aspects of all three memoirs are treated in section 3. In the first of the memoirs, Lavoisier's "Nature des acides" (1777), cited by Guyton himself in the paragraph following the one quoted here, Lavoisier illustrated the oxygen theory of acidity by producing oxalic acid from the oxidation of sugar. The second of Lavoisier's memoirs of importance to Guyton was the paper on the production of nitric acid by acidification of nitrous gas. Lavoisier first stated the oxygen theory of acidity in the original (1776) version of this paper and he restated it in the corrected and better known version of 1779; (for the two versions, see above, p. 77, n. 1). The third of Lavoisier's papers was the (1778) "Mémoire sur la combustion en général." Appearing in 1780, the same year in which Guyton began working on the *Encyclopédie méthodique,* this memoir perhaps occasioned Guyton's accepting vital air instead of fixed air as the principle of acidity. Moreover, because the memoir emphasized the specificity of the acids by claiming that the combustion of each different combustible produced a different acid, the memoir likely induced Guyton to pursue the question of the specificity of the acids that he ignored in 1776.

(as well as of Lavoisier's 1778 claim that arsenic and iron should be acidifiable by oxygen) that

s'il n'est pas prouvé que tous les métaux sont des acides particuliers unis au phlogistique, cette idée, l'une des plus belles que la Chymie ait enfantée, se trouve déjà appuyée & presque vérifiée par plusieurs faits très-concluans, comme nous le verrons plus particulièrement aux articles: *Acide Arsenical, Acide molybique, métaux,* &c. En rapprochant ce que dit M. Lavoisier, que quelques chaux métalliques, comme l'arsenic, le fer, & peut-être plusieurs autres combinées avec le principe acidifiant, acquièrent les propriétés communes aux acides (Acad. des Sci. ann. 1778. [the "Nature des acides"]), on voit avec satisfaction que les observations qui doivent produire enfin cette grande vérité, n'avoient pas moins fait d'impression en France qu'en Suède.[9]

Guyton's response to Lavoisier and to Bergman and Scheele was to elaborate a parallel treatment of combustibles and metals, just as Bergman had done. Accepting the need to explain acidity, however, Guyton invoked earthy specific components, like those of the 1776 "Conciliation," instead of the acidic ones that Bergman had used. He also applied the two-fluid displacement theory, devised in 1776 to explain calcination, to the acidification of combustibles. Like calcination, combustion involved the removal of phlogiston and the addition of an air, which Guyton now knew to be pure or vital air, the acidifying principle.

The point in the *Encyclopédie méthodique* at which Guyton's theory of the acidification of combustibles emerged most clearly was the article "Acide nitreux." There, he made clear that he deemed the radical theory a signal achievement of the contemporary revolution in chemistry:

[Q]uand la Chymie aura achevé la révolution qu'elle a si heureusement commencée, que sa lumière aura pénétré avec le temps jusque dans ces laboratoires, ou l'on ne connoit encore que les formules des anciens, la marche de celui qui voudra expliquer la nature & développer les propriétés de l'acide nitreux sera moins timide & moins incertaine.

Il fixera d'abord l'atttention sur le principe le plus simple, quoique d'un ordre de composition très-avancé; c'est la base acidifiable ou pour mieux dire le principe *radical* de l'acide nitreux, & il le suivra dans toutes ses surcompositions.

Il considérera successivement ses combinaisons plus essentiels, & formera le tableau suivant:

Le *radical nitreux* uni au		phlogistique	forme	*le gas nitreux*
	à	l'air vital l'eau		*l'acide nitreux*
	à	l'air vital l'eau le phlogistique		*l'acide nitreux phlogistiqué*
	à	l'air vital l'eau la potasse		*le nitre*

[9] Guyton, *Encyclopédie méthodique, chymie,* I, p. 31.

à	l'air vitale l'eau la potasse le phlogistique	*le nitre phlogistiqué*

Et de cette simple exposition naîtra l'explication de tous les phénomènes que présente ce principe modifié par ses diverses compositions.[1]

The radical was not truly elementary, being only a stable chemical, one incapable of isolation. It was the seat of specific properties, on which generic properties could be imposed. The basic pattern exhibited in the table was that of a sequence of reactions by which one could proceed from the completely phlogisticated radical, by means of displacement of phlogiston with vital air, to the completely oxygenated, dephlogisticated radical; that is, from the genus of combustible to that of acid. The specificity of both acid and combustible was determined by the radical. Any stage intermediate between combustible and acid was a compound of the radical with both phlogiston and vital air, and the generic properties of the intermediate stage were midway between those of combustible and acid. In the present instance, the radical was the nitrous radical; the combustible formed by its union with phlogiston was nitrous gas, the substance acidified by Lavoisier in his paper on nitric acid; the intermediate stage was phlogisticated nitrous acid; and the acid was simply nitric acid.

The factual underpinning that Guyton adduced for his theory pertained largely to the same phenomena reported by Lavoisier in his 1776 paper on nitric acid. Guyton was unprepared to accept either that the acid was simply the combination of nitrous gas with vital air, as Lavoisier had maintained, or that phlogiston was not a participant in the calcination or acidification reactions, as Lavoisier had gingerly suggested. In reference to the recomposition of the acid from nitrous gas and vital air, Guyton asked, "est-ce le gas nitreux tout entier qui entre dans la composition de cet acide?"[2] Here, he took his stand:

> Ceux qui admettent le phlogistique n'hésitent pas de répondre que le gas nitreux est composé ... de deux ... substances, savoir, le phlogistique du métal dissous [the mercury that Lavoisier dissolved in nitric acid, producing mercury calx and nitrous gas; see the passage quoted above, p. 73, accompanied by n. 7], & la base acidifiable ou principe *radical* nitreux; que le phlogistique & l'air vital ne se rencontrent que pour mettre en jeu une affinité qui décompose le dernier [in accordance with Crawford's idea, mentioned by Guyton on the same page, that phlogiston removed caloric from vital air, thus "decomposing" it], il est impossible qu'ils co-existent essentiellement dans l'acide nitreux pur; que dès lors il n'y a que le radical nitreux qui entre en combinaison avec l'air vital [ou] principe acidifiant pour régénérer l'acide; que l'effervescence, la chaleur, la couleur rouge qu'occasionne le mélange des deux fluides attestent la décomposition réciproque.[3]

Thus, Guyton perceived the acidification of nitrous gas as proceeding generally in

[1] *Ibid.*, pp. 141—142.
[2] *Ibid.*, p. 158.
[3] *Ibid.*

accordance with his displacement scheme. Nitrous gas did not enter "tout entier" into nitric acid; only the nitric radical did. Thus Guyton found he was not in accord with "les Chymistes (s'il en est) qui, tirant déjà d'une hypothèse que M. Lavoisier n'a lui donner jusqu'à présent que comme probable, oseroient affirmer que la théorie du phlogistique est renversée."[4]

This response to Lavoisier's ideas about the acidification of combustibles was paralleled by a similar treatment of the acidification of metals in response to Bergman. Bergman's role as a source for Guyton's ideas on metals appeared most prominently in the articles of the *Encyclopédie méthodique* on the three metallic acids, arsenic, molybdic, and tungstic, and above all in the general article "Acides métalliques."[5] In the last article, Guyton made his debt entirely explicit:

> *Acide métallique.* C'est le nom que l'illustre Bergman donne à quelques acides, qui relativement à leur origine, meritent en effet d'être considérés sous des rapports communs, & de former par conséquent une classe nouvelle.
>
> Ces acides sont tirés des métaux, & sont susceptibles de reprendre l'état métallique complet; ils passent enfin par l'état intermédiare de chaux métallique. Voilà des caractères bien marqués, & qu'il n'est pas possible de révoquer en doute, puisqu'on les voit successivement neutraliser les alkalis, s'allier aux métaux, & teindre les flux vitreux.[6]
>
> Suivant M. Bergman [here Guyton referred to the *De acidis metallicis*], c'est la présence ou l'absence du phlogistique qui produit tout ces changemens; dans l'arsenic, par exemple, il y a une terre radicale essentiellement acide, qui quand elle est pure, en manifeste tous les caractères; est-elle pourvue d'une certaine quantité de phlogistique? les caractères acides s'affaiblissent, c'est une chaux métallique; est-elle saturée de phlogistique? c'est complettement un métal.[7]

Having accepted the oxygen theory of acidity, however, Guyton could not simply adopt Bergman's views intact. He continued:

> j'ai fait voir à l'article *Acide* que cette condition unique de l'addition ou de la sustraction du phlogistique ne suffisoit pas pour expliquer tous les phénomènes, & que l'air vital étoit une partie constituante essentielle de tous les acides.[8]

The attribution of generic properties to the specific component was avoided by invoking the oxygen theory that showed the acids to be compound. The earth in arsenic was not then "essentiellement acide," but was a primitive matter capable of receiving forms:

> Ainsi nous considérons dans l'arsenic une substance primitive, simple, une élément chymique, qui n'est pas par lui-même ni acide, ni métal, mais qui est à la fois *acidifiable,* & si j'ose le dire, *métallisable.*[9]

[4] *Ibid.*

[5] See *ibid.*, pp. 39—42 for "Acide arsénical"; pp. 117—120 for "Acide molybdique"; pp. 330—339 for "Acide tungstique"; and pp. 115—117 for "Acide métallique."

[6] *Ibid.*, p. 115.

[7] *Ibid.*

[8] *Ibid.*

[9] *Ibid.*, pp. 115—116.

The two-fluid theory emerged as the basis for both acidification and metallisation, that is, for the informing of the specific, radical earth:

> La grande affinité de cette matière [the radical earth] avec ces deux fluides si universellement répandus, est cause que nous ne pouvons la saisir dans son état de pureté; elle est donc toujours unie ou à l'air vital ou au phlogistique, ou à l'un ou à l'autre en même temps, dans certaines proportions. Avec l'air vital ou principe acidifiant commun, c'est un acide dans toute la rigueur de l'acceptation; avec le principe métallisant commun, c'est un métal complet; retenant partie de l'un & prenant partie de l'autre, c'est un chaux métallique . . .[1]

The basic structure of Guyton's radical theory already deployed in the article on nitric acid in response to Lavoisier is unmistakably clear in these passages on the metallic acids. The justification for applying the radical theory to both the metallic and the non-metallic acids lay in the essential similarity between metals and combustibles that Bergman, in the Stahlian tradition, had continued to uphold in elaborating his own views. Guyton, too, made this justification explicit as he continued his discussion of the metallic acids:

> Cette théorie est à présent trop bien prouvée pour qu'on puisse la regarder comme une simple hypothèse. Non-seulement elle embrasse toutes les vues que le savant Suédois développe à ce sujet sur l'acide radicale des métaux; mais elle les généralise encore, en établissant une analogie entre tous les métaux & tous les acides. Tous ces êtres ne sont dans l'origine que des bases disposées à subir les mêmes changemens par des affinités semblables; les métaux sont des soufres moins inflammables et plus ductiles, les soufres sont des métaux moins ductiles et plus inflammables.[2]

Thus, like Bergman and Lavoisier, Guyton had invoked the Stahlian analogy between metals and combustibles to generalize his theory. In the chemistry of both substances, the same fundamental process was at work: the informing of a primitive matter (a radical) by generic principles (phlogiston and oxygen). By depriving the radical components of their own properties (the acidity that Bergman saw in them), Guyton further emphasized the passivity of radicals as form-receiving substrates. In adopting this posture, Guyton carried the generic bias of two-component theories to its extreme.

d. Radicals in the work of Guyton and Lavoisier

Unlike Lavoisier's theory, Guyton's failed to meet one of the two criteria for the viability of two-component theories: it could not demonstrate directly the existence of the specific, radical components, and it could not show concretely that the radical indeed entered into the composition of acids, metals, and combustibles. In the article "Acidifiable, Acidifiant, Acidifié" of the *Encyclopédie méthodique*, Guyton himself pointed out the difference between his system and Lavoisier's:

[1] *Ibid.*, p. 116. [2] *Ibid.*

Dans le système de M. Lavoisier, c'est le soufre lui-même qui est la base acidifiable de l'acide vitriolique; le gas nitreux, celle de l'acide nitreux; le sucre, celle de l'acide saccharin, &c. En rapprochant tous les faits, il paroît plus sûr de dire que les substances contiennent seulement cette base acidifiable, & avec elle un principe essentiellement combustible, une matière phlogistique, qui les met actuellement en état de soufre.

La question [whether Guyton's theory was correct] seroit bientôt décidée, si l'on pouvoit facilement obtenir ces bases isolées; mais jusqu'à présent on n'est pas parvenu à faire cesser leurs combinaisons que par voie d'échange, c'est-à-dire, en leur rendant un principe à la place de celui qu'on leur enlève.[3]

Guyton knew that he could make his theory prevail over Lavoisier's only if he could show the existence of the thus far unisolated radicals.

In the article on nitric acid, Guyton revealed that he was troubled by this weakness when he launched into a lengthy discourse on the problem of finding direct evidence for the nitrous radical.[4] Though he opened his discussion by admitting frankly that little could be said of the radicals themselves,[5] he recounted numerous experiments by a variety of figures to provide suitable material for speculation. Pausing in the midst of this exercise, he acknowledged the difficulty involved, and he conceded that the question of establishing foundations for his theory, in the form of direct analytical evidence, was at stake:

Les Chymistes qui sont assez avancés pour sentir toutes les difficultés, tout l'intérêt de ces recherches, ne seront point étonnés de me voir insister sur ces probabilités; ils savent que l'on n'aura une analyse complette de l'acide nitreux que lorsqu'on connoîtra la substance qui le caractérise essentiellement acide nitreux, qui soit par conséquent distinct de l'air vital, du phlogistique, du calororifique, puisque celles-ci entrent aussi dans la composition des autres acides, & qu'il est bien difficile, pour ne pas dire impossible, de se persuader qu'une simple différence dans les proportions puisse produire des propriétés aussi invariables, qui subsistent jusqu'à l'entière désunion des principes, & ne laissent jamais appercevoir aucunes traces de passage à une autre espèce du même genre.[6]

With his conversion in 1787 to Lavoisier's views, Guyton abandoned the two-fluid theory, eliminating his problem of finding proof that the radicals existed.[7] However, the term "radical" was retained in the antiphlogistic system, where it served its original function as the label for the specific components. The combustibles were called radicals insofar as they were the specific components

[3] *Ibid.*, p. 417.

[4] The discussion appeared in *ibid.*, pp. 159—165 and opened with the question, "Quelle est maintenant cette substance qui dans le gas nitreux se trouve unie au phlogistique?"

[5] *Ibid.*, p. 159: "On a déjà remarqué qu'il y avoit des corps que nous ne connoissons que par les compositions dans lesquelles ils entroient & les propriétés qu'ils leur donnoient, que nous ne pouvions recueillir séparées de tout autre corps, & qui, quand nous parviendrions à les isoler, seroient encore pour nous des élémens chymiques. Cette vérité s'applique sur-tout aux principes radicaux des acides & des métaux; il semble donc que nous n'ayons qu'à répéter ici ce que nous avons dit des autres principes absolument analogues, que c'est une substance propre, de son genre, formant tel acide avec l'air vital, tel soufre avec le phlogistique: voilà en effet tout ce que nous savons des radicaux des acides, & il est raisonnable de s'arrêter où finit la lumière."

[6] *Ibid.*, p. 160.

[7] For Guyton's conversion, see McKie and Klickstein, "Letter from Guyton."

of acids (whereas in Guyton's theory they were specific components of combustibles, metals, and acids). In the conclusion of the chapter of the *Traité* on the formation of acids, Lavoisier made clear this application of the term retained from Guyton's former views. Having discussed the production of phosphoric, sulfuric, and carbonic acids by the combustion in oxygen of phosphorus, sulfur, and carbon, respectively, Lavoisier remarked that

> Les trois exemples d'ailleurs que je viens de citer, suffisent pour donner une idée claire & précise de la manière dont se forment les acides. On voit que l'oxygène est un principe commun à tous, & que c'est lui qui constitue leur acidité; qu'ils sont ensuite différenciés les uns des autres par la nature de la substance acidifiée. Il faut donc distinguer dans tout acide, la base acidifiable à laquelle M. [Guyton] de Morveau a donné le nom de radical, & le principe acidifiant, c'est-à-dire, l'oxygène.[8]

The term "radical" or "radical principle" was a synonym for the expression "acidifiable base," which Guyton had introduced in the *Encyclopédie méthodique,* as noted in subsection 4.*c.* The synonomy was an equation of the term "base," which Guyton had used in the *Chymie de Dijon* as the label for any specific component held in solution by a generic menstruum, with Bergman's term "radical," now also used as a noun. Guyton insured the assimilation of this usage into the antiphlogistic chemistry by giving it prominence in his contribution to the *Méthode* of 1787. Section II of his "Mémoire," in that work, "sur la développement des principles de la nomenclature méthodique,"[9] was entitled, "Des bases acidifiables ou principes radicaux des acides."[1]

In the same language Lavoisier declared any substance that could form an acid by oxidation to be an acidifiable base or radical.[2] In accordance with the dictates of the oxygen theory, however, Lavoisier deployed this identification quite generally. As his views on the metals indicated, he believed that any substance capable of combining with oxygen at all, without undergoing decomposition, was capable of combining with a sufficient quantity of oxygen to become acidic. In Lavoisier's view any combustible substance was an acidifiable base, a radical, and that included the metals. By taking this position, Lavoisier was disagreeing with two of his collaborators in the *Méthode.* Neither Guyton nor Fourcroy went so fat as to label the metals acidifiable bases, even though Guyton had borrowed the term from Bergman's discussion of metals and had applied it to both metallic and non-metallic acids. Fourcroy went so far as to suggest, *horribile dictu,* that the metals may represent a class of substances not capable of acidification.[3] Despite their hesitations, however, and despite the later vicissitudes of the

[8] Lavoisier, *Traité,* I, p. 69.

[9] Lavoisier *et al.*, *Méthode,* pp. 26—74.

[1] *Ibid.*, p. 36.

[2] For example, in *Traité,* I, pp. 194, 195, 196, where, tabulating the "substances simples" and "radicaux composés," Lavoisier made clear his equation of radical with oxidizable substance. However, Fourcroy (Lavoisier *et al.*, *Méthode,* p. 79) labelled only twenty-two non-metals out of the fifty-five "substances simples" as acidifiable radicals.

[3] Lavoisier *et al.*, *Méthode,* p. 88.

oxygen theory, Lavoisier's equation of the word "radical" with the notion of a "combustible," understood to be anything capable of union with oxygen, persisted in nineteenth-century chemistry.

The radicals of a number of the acids appearing in Lavoisier's table in the *Traité* were known: sulfur, phosphorus, nitrogen, carbon, and the seventeen metals. They were isolable, manipulable, characterizable substances. In contrast to the hidden radicals of Guyton's chemistry, at least some of Lavoisier's were directly manifest and thereby acquired a measure of specific distinction denied them by Guyton. Yet, Lavoisier's insistence on the generic role of oxygen diminished his interest in the specific components to the vanishing point. In the *Traité,* for example, the only chapter in which three of the archetypical radicals were the focus of the discussion was entitled, "De la décomposition du gaz oxygène par le soufre, le phosphore & le carbon, & de la formation des acides en général."[4] Neither the properties of the radicals themselves, nor those of the distinct acids obtained from them, but the common pattern of acidification was the topic of this chapter. What mattered to Lavoisier was very much like what mattered to Guyton in his speculations on the radical of nitric acid: only that the radicals *existed* and that they differed from one another *at all.* Differences among the radicals were what gave rise to the existence of differences among the acids. What these differences were, what their mode of production might have been, were questions of scant importance.

The unconcern shown by the founders of the antiphlogistic system for specific properties is even more apparent in the treatment given to those acids of which the radicals were considered unknown. The utility of the term "radical" to Lavoisier and his collaborators in the *Méthode* becomes particularly clear in its application to these acids. When introducing the nomenclature of the acids in the *Méthode,* Guyton noted the necessity of distinguishing the known from the unknown acidifiable bases.[5] He cited the known radicals of four of the archetypical acids, nitrogen, carbon, sulfur, and phosphorus, and observed that these substances and their compounds provided the basis for the nomenclature of the acids.[6] He continued, saying that in regard to the acids with unknown radicals,

> telles que les bases de l'acide marin, de l'acide du borax, de l'acide du vinaigre, &c. &c. nous nous sommes contentés de designer l'être simple qui y modifie l'oxygène, par l'expression de *base acidifiable,* ou, pour abrégér, de *radical* de tel acide . . .[7]

A few pages later, Guyton offered the example of muriatic acid: —

> L'analogie nous porte à croire que l'acide muriatique a une base acidifiable, de même que les acides carbonique, sulfurique & phosphorique, qui sert également à donner un caractère propre et particulier au produit de la combinaison oxigène: nous n'avons dû

[4] Lavoisier, *Traité,* I, Ch. 5.

[5] Lavoisier *et al.*, *Méthode*, p. 37.

[6] *Ibid.*

[7] *Ibid.*, pp. 37—38.

designer cette substance que par l'expression de *radical* muriatique ou principe radical muriatique . . .[8]

Acids of unknown composition were assumed by analogy to have the same two-component composition exhibited by the archetypical acids. That is, each was expected to contain a specific, radical component, acidified by oxygen. The term was useful, Guyton affirmed,

afin de ne pas donner un nom à un être inconnu, & pour nous renfermer dans l'expression de la seule propriété que nous lui connoissons, qui est en effet de produire cet acide. Nous avons eu la même circonspection par rapport à tous les acides sur lesquels nos connoissances ne sont pas plus avancées, & dont il est très-possible que dans la suite on découvre les bases parmi des substances déjà nommées. Nous sommes forcés de comprendre dans cette classe jusqu'aux bases des acides végétaux et animaux, dont nous n'avons encore d'analyse exact, malgré la facilité avec laquelle on résout ces composés en leurs élémens.[9]

The term "radical" was therefore a convenient device for labeling the supposed specific component of an acid having unknown or poorly understood composition. The technique permitted signifying the role of the component as an acidifiable base while avoiding premature attribution of specific properties to it. After all, just as had been the case in Guyton's theory, the specific component was known only in its role as a base.

And only in that role was it of interest. Earlier in the *Méthode,* Guyton remarked of the acids with unknown radicals that

Il est probable que plusieurs de ces acides ont des bases composées, ou même qui ne diffèrent entr'elles que par des proportions diverses des mêmes principes; quand l'analyse aura démontré leur premier élément & leur ordre de filiation, il sera juste sans doute de les ramenés à ce type originel; mais il ne cessera pas pour cela d'être utile d'étudier leurs propriétés, leurs attractions dans leur état actuel de composition, & nous ne pouvions dès-lors nous dispenser de les comprendre dans le système de nomenclature.[1]

Many such acids could be and were studied without a thorough understanding of their composition. It seems clear that their radicals were considered less for their intrinsic interest than for the possibility that they might further confirm the two-component theory of acidity, thereby permitting inclusion of acids with unknown radicals in the general system of chemistry. The possibility of studying compounds of radicals without attending significantly to the properties of the radicals themselves underlay the older radical theory of organic chemistry. Particularly in the light of the conceptual problems posed by organic matter, as discussed in the next chapter and in Part II, this possibility was the only basis for the early development of a science of organic matter.

[8] *Ibid.*, p. 46.
[9] *Ibid.*, pp. 46—47.

[1] *Ibid.*, p. 38.

CHAPTER III

Mineralogy and Organic Chemistry in the Late Eighteenth Century: Access to the Specific Level

1. Introduction

In his chemistry, Berzelius emphasized that acids owed their acidity to the character of their radicals, not to their possession of oxygen. Acidity was a specific property and basicity an equivalent, opposing one. For Berzelius, the specificity of the acidic and basic oxides was established by two facts: the identity of the radical of the oxide and the quantity of oxygen with which it was combined. The specificity of salts, the compounds of oxides with one another, was established by a similar pair of facts: the identification of certain oxide-components as acids and of others as bases and the conformity of the proportions of the acidic and basic components to certain stoichiometric laws. Berzelius assimilated many complex compounds to this saline model by applying this combination of qualitative identification of saline components with quantitative constraints on their proportions. This approach served to distinguish compounds from one another and from mixtures, that is, to specify them.

Having developed these techniques of specification in his inorganic chemistry, Berzelius attempted to deploy them in mineralogy and organic chemistry. Though lying outside the boundaries of inorganic chemistry, these fields shared with it the fundamental need for criteria that permitted discrimination of one substance from another. Berzelius hoped that his innovations in inorganic chemistry could be exploited to answer this need in mineralogy and organic chemistry. These deployments are analyzed in the last chapter of this study. The background to the deployments in the mineralogy and organic chemistry of the eighteenth century are explored in this chapter.

Berzelius's innovations in mineralogy must be assessed in the light of prevailing belief about the notion of "species" in its application to mineral matter. At the end of the eighteenth century, the major concepts of mineral species issued from the Wernerian and crystallographic schools, which vied for domination of mineralogy. Under the leadership of Abraham Gottlob Werner (1749—1817), famed professor at the mining academy at Freiberg, the Wernerian school developed a morphological concept of species, akin to the substantial forms of Aristotelian tradition. Like a species of plant, a mineral species was distinguished by a collection of readily accessible characters possessing certain values. René Just Haüy (1743—1822), leader of crystallography around the turn of the century, combined stoichiometric with morphological criteria in defining species, claiming that a species was represented by a unique geometric form and a unique definite composition (but in practice he relied primarily on crystal morphology). Both Werner and Haüy argued that the form (substantial or geometric, respectively) directly reflected the essence of minerals, that is, their chemical composition. Accordingly, morphology permitted discrimination of species because macroscopic (or external) criteria reflected internal essence (composition).[1]

Berzelius allied himself with Haüy in requiring stoichiometric foundations for discrimination of species, but unlike Haüy he emphasized the stoichiometry rather than the morphology. In taking this stoichiometric stance, Berzelius placed himself in direct opposition to the Wernerian school, which, through its criticism, compelled Berzelius to elaborate his doctrine. In sections 2 and 3 of this chapter, the Wernerian and crystallographic approaches to species are explored to provide background for the analysis in Chapter VI of Berzelius's mineralogy.

In the organic chemistry of the eighteenth century, natural-historical methods were also deployed to discriminate kinds of organic matter, but they never became as elaborate and formal as they had in mineralogy. Reflecting the rudimentary state of organic analysis, the number of known organic substances was far smaller than the number of known minerals, and the pressure to classify them was correspondingly less intense. Progress in extractive analysis of vegetables did prompt developments in organic classification, but classifications of organic matter differed in an important respect from mineral taxonomies. In mineralogy and in natural history generally, the specific level was considered fundamental and was made the focal point of taxonomic systems, but in organic chemistry at the end of the eighteenth century no taxonomic level was considered specific in a fundamental sense. The definition of the specific level in organic chemistry and the formulation of the problem of gaining access to it were first set forth by Berzelius. The concluding section of this chapter describes the state of organic analysis inherited by Berzelius and explores several antecedents for his achievements in the work of Lavoisier.

[1] An informative contemporary account of the differences between the Wernerian and crystallographic schools is Thomas Thomson, "Some observations in answer to Mr. Chevenix's attack upon Werner's mineralogical method," *Ann. Phil.*, 1 (1813), 241—258. For the history of mineralogical systematics and for divergent schools in mineralogy, see Franz von Kobell, *Geschichte der Mineralogie von 1650—1860*, Geschichte der Wissenschaften in Deutschland. Neuere Zeit, Vol. II (Munich: 1864); E.H.M. Beekman, *Geschiedenis der systematische mineralogie* ([the Hague: 1906]); Paul Groth, *Entwicklungsgeschichte der mineralogischen Wissenschaften* (1926; rpt., Wiesbaden: 1970); and Charles S. St. Clair, "The classification of minerals: some representative mineral systems from Agricola to Werner" (Doctoral dissertation, University of Oklahoma: 1966). Groth's book is henceforth cited as Groth, *Mineralogische Wissenschaften*.

2. Natural-historical method in mineralogy: Linnaean example and Wernerian practice

a. Introduction: mineralogy as natural history

In the prevailing eighteenth-century view, the natural order had three realms or kingdoms, animal, vegetable, and mineral.[2] Corresponding to these realms were the three branches of natural history, zoology, botany, and mineralogy.[3] As a branch of natural history, mineralogy shared the goals and characteristics of the entire natural-historical enterprise.[4] The principal goal of natural history was taxonomic, that is, the development of a classification system or method by which the objects of the natural order, animals, plants, and minerals, could be arranged in hierarchically ordered groups.[5] The classicification was to exhibit

[2] So powerful was the prevailing tripartite conception that Linnaeus, primarily a botanist, felt compelled to treat all three realms of the natural order. His *Systema naturae* (1735; followed by numerous revised editions), laid out the plan of each realm, as Henri Daudin points out in his *De Linné à Jussieu: méthodes de classification et idée de série en botanique et en zoologie (1740—1790)* (Paris: [1926]), p. 177; henceforth cited as Daudin, *Linné*. The tripartite conception persisted in the nineteenth century, as is indicated by Playfair's 1812 definition of natural history as "the branch of knowledge which collects and classifies facts... Its objects are confined to what are called the three kingdoms, the Mineral, the Vegetable, and the Animal." (Quoted by Charles C. Gillispie, *Genesis and geology: a study in the relations of scientific thought, natural theology, and social opinion in Great Britain, 1790—1850* [1951; rpt., New York: 1959], p. 17; henceforth cited as Gillispie, *Genesis and geology*).

[3] As is indicated, for example, by Playfair's definition, quoted in the preceding note, and by d'Holbach's article "Minéral règne," in *Encyclopédie*, X (1765), p. 534, which opens with these statements: *"Mineral règne (Hist. nat.)* c'est ainsi qu'on nomme l'assemblage total des corps qui appartiennent à la terre, & qui se forment dans son sien. Ces corps s'appellent *Minéraux*, ou *substances du règne minéral;* ils font une des trois branches dans lesquelles il a plu aux Physiciens de partager l'histoire naturelle. Le règne minéral est l'objet d'une étude particulière, qu'on nomme *Minéralogie*." For attribution of this article to d'Holbach, see Richard N. Schwab, Walter E. Rex, and John Lough, *Inventory of Diderot's* Encyclopédie, 6 vols., Studies on Voltaire and the eighteenth century, ed. Theodore Besterman, nos. 80, 83, 85, 91—93 (I, no. 80, Geneva: Institut et Musée Voltaire, 1971; II—VI, nos. 83, 85, 91—93, Banbury, England: Voltaire Foundation, 1971—1972), esp. III, p. 655, and, for symbols identi fying the contributors to the *Encyclopédie*, II, pp. 11—12; henceforth cited as Schwab *et al.*, *Inventory*. For d'Holbach's contributions to the *Encyclopédie*, especially regarding subjects such as mineralogy, mining, and metallurgy, cf. John Lough, *Contributors to the* Encyclopédie (London: 1973), pp. 5, 10, 14, 15, 30—31, 35, 36, 84; henceforth cited as Lough, *Contributors*.

[4] For mineralogy as part of the overall natural-historical enterprise, see R. Hooykaas, "De invloed der biologie op de anorganische natuurwetenschappen," *Geloof en wetenschap (Christlijke vereeniging van natuur- en geneeskundigen orgaan)*, 101 (1936), 101—104, esp. 111—114; *idem*, "De natuurlijke klassificatie der chemische substanties," *Chemisch Weekblad*, 33 (1936), 599—610, esp. 599, 601, 605—606; *idem*, "The species concept in eighteenth-century mineralogy," *Arch. int. d'hist. sci.*, 5 (1952), 45—55; and Groth, *Mineralogische Wissenschaften*, pp. 151—155. Hooykaas's papers henceforth are cited, respectively, as Hooykaas, "Invloed"; Hooykaas, "Natuurlijke klassificatie"; and Hooykaas, "Species concept."

[5] Natural history was typically equated with classification, as Playfair's definition (quoted on this page, in n. 2) indicates. Cf. also Julius von Sachs, *History of botany (1530—1860)*, rev. Isaac Bayley Balfour, tr. Henry E.F. Garnsey (Oxford: 1890), pp. 84—85; and James L. Larson, *Reason and experience: the representation*

both the perceived similarities or affinities, as they were called, and differences between individuals or groups. Similarities and differences were estimated by comparing certain characters or marks. Each taxon was characterized by the appearance, in each of its members, of certain values of the characters. The characters were usually external and readily accessible, a choice dictated by both the professional need for rapid and easy classification, that is, the need for practical convenience, and an unwillingness to disturb the integrity of natural objects by dissecting them.[6] The model for this procedure was Linnaean botany,

of the natural order in the work of Carl von Linné (Berkeley: 1971), pp. 35, 40. (Sachs's book is cited henceforth as Sachs, *History of botany;* Larson's, as Larson, *Reason and experience.*) The enterprise of constructing and deploying classifications or methods was called "systematics." Daudin (*Linné,* p. ii) defines succinctly the "idée d'une classification 'systématique' ou 'méthodique'" as one "qui, procédant d'après des caractères détérminées, distribue un ensemble d'êtres donnés en fractions de plus en plus petites, toujours subordonnées, définies et circonscrites d'après des règles fixes..." The present section of this chapter investigates several aspects of systematics in natural history and in mineralogy.

6 Daudin (*Linné,* pp. 24—25) points out that the professional need for practical convenience was a major motive for the choice of external characters. In this connection, see also Larson, *Reason and experience,* pp. 47, 72. Daudin (*Linné,* pp. 52—53) also points out that internal characters seemed insufficiently discriminating to botanical practitioners who were preoccupied with phanerogams, all of which shared similar internal structural characteristics. He also notes (pp. 53—58) that internal characters were of importance to zoology from the first, but he later points out (pp. 150, 159) that zoologists, concentrating on man and the higher vertebrates, felt they could follow the example of botany and rely all but exclusively on external characters.

The reluctance of the naturalist to dissect is suggested by this remark of Daubenton, in the article "Histoire naturelle," *Encyclopédie,* VIII (1765), pp. 225—230, esp. p. 228, where he sought to distinguish chemistry, an analytical science, from natural history, a descriptive one: "Cette science [chemistry] commence au point où l'Histoire naturelle termine. Le naturaliste recherche toutes les productions de la nature dans son propre sein; il leve avec précaution le voile qui les couvre; il les observe d'un oeil attentif sans oser y porter un main téméraire; s'il y est obligé de les toucher, il est toujours dans la crainte de les déformer; s'il est forcé de pénétrer dans l'intérieur d'un corps, il ne le divise qu'à regret, il n'en rompt l'union que pour en mieux connaître les liens, & pour avoir une idée complette de la structure intérieure aussi bien que de la forme extérieure. Le chimiste au contraire ne voit les opérations de la nature que dans les procédés de l'art; il décompose toutes les productions naturelles; il les dissout, il les brise; il les soumet à l'action du feu pour déplacer jusqu'aux plus petites molécules dont elles sont composées, pour découvrir leurs élémens & leurs premiers principes." Cf. his observation in the same article (p. 226) on the difference between the naturalist and the anatomist. Though in Schwab *et al., Inventory,* III, p. 512, this article is listed without attribution, two facts may be cited in support of attribution to Daubenton. First it is well known that Daubenton wrote the bulk of the articles on natural history in the *Encyclopédie;* (for this, see Lough, *Contributors,* pp. 35, 76—77); it seems reasonable to expect that he should have written the major article on the subject of natural history. Second, the same ideas as those just cited appeared in print under Daubenton's name, in his "Introduction à l'histoire naturelle," in *Encyclopédie méthodique, histoire naturelle des animaux,* I (Paris: 1782), pp. i—ix; (for this reference, the author is indebted to his colleague, Professor Richard W. Burkhardt, Jr.). The distinction between the naturalist and the anatomist was a theme expressed by others in the eighteenth century, such as Lamarck and Vicq d'Azyr (for them, see Daudin, *Linné,* pp. 190n2, 224n6).

Hooykaas, "Invloed," 107, traces the naturalist's nearly exclusive concern with the undisturbed object to the persistent force of Aristotelian hylomorphism and its preoccupation with the unitary character of the forms imposed upon matter. Cf. also Hooykaas, "The discrimination between 'natural' and 'artificial' substances and the development of the corpus-

which practitioners of zoology and mineralogy hoped to emulate. As a taxonomic science, mineralogy was dedicated to the classification of minerals just as botany was dedicated to the classification of plants.[7]

b. Classification schemes and the idea of series in botany and mineralogy

1) *Natural "methods" and artificial "systems" of classification*

The objective of a classification scheme or taxonomy was to sort out the objects of natural history into hierarchically ordered sets of categories or taxa in a way that expressed the affinities or relations subsisting among the members of a group.[8] The notion of affinities had its origins in the earliest categorizations of natural-historical objects. Aristotle, for example, recognized that some natural-historical entities were perceived immediately and without effort as displaying great similitude and belonging to readily labelled groups, such as birds or fish. The example of botany in the Renaissance shows that other, less immediately striking groups could also be perceived by observers accustomed to close scrutiny of natural objects. Whether spontaneously evident or the outcome of sustained observation, groups identified by observation were perceptual constructs largely devoid of explicit intellectual characterization. By exhibiting many more or less inclusive groups of plants whose members were similar to one another but different from those of other groups, the German and Dutch "fathers" of botany established the notion of natural affinity.[9]

A taxonomy of natural groups depended on more than their perception by observers, however. The taxonomist's first task was to undergird the perceptual groups by determining the characters and the values of the characters that were shared by the members of the group and that occasioned the perception of affinity among them. The same kind of intellectual effort that was expended to characterize explicitly the perceptual groups was then devoted to construction of additional groups that had not emerged perceptually among observers. This

cular theory," *Arch. int. d'hist. sci.*, 1 (1947—1948), 640—651; henceforth cited as Hooykaas, "Discrimination."

[7] The analogy between the goals and methods in mineralogy and those prevailing in the other two branches of natural history was sustained conceptually by the explicit assimilation of homogeneous objects of the mineral kingdom to articulated natural-historical objects of the plant and animal kingdoms, that is, by assimilation of minerals, aggregates of like parts (the particles of the mixt), to plants and animals, aggregates of unlike parts (organs). For this, see Hooykaas, "Naturlijke classificatie," 599—600. Cf. also Hooykaas, "Invloed," 103, and Groth, *Mineralogishe Wissenschaften*, p. 153.

[8] See Daudin's definition of classification, above p. 103, n. 5. Cf. also Daudin, *Linné*, p. i. where he defines "le problème de la 'methode naturelle'" as "la représentation scientifique des affinités des animaux et plantes."

[9] For the notion of natural affinities, see Sachs, *History of botany*, pp. 4—9, 17, 31—32; Daudin, *Linné*, pp. 13—19, 23; Larson, *Reason and experience*, pp. 6—19.

task entailed not only isolating and defining the characters, but determining the order in which they were to be invoked. This task was the preoccupation of botanists from Cesalpino to Linnaeus.[1]

The goal of taxonomies was to represent the natural affinities. Two approaches to this goal characterized botany to Linnaeus: either a multiplicity of characters, drawn from several parts of the plant, were taken together, in an effort to represent the totality or, as it was called, the "habitus" of a plant from which the perception of affinity was taken to derive; or else a limited number of characters was employed, drawn from only one part of the plant, in hope of isolating particular features, rather than the totality, that would succeed in representing natural affinities. The "natural methods" that exponents of the former view constructed were undermined by the arbitrary criteria they invoked to subordinate the characters to one another, while the avowedly "artificial systems" relying on few parts of the plant for characters were vitiated by the failure to recognize clearly that a limited choice of characters could never do justice to natural affinities.[2] By the end of the eighteenth century, particularly in the light of Linnaeus's criticism of earlier ostensibly "natural" taxonomies, noted in the next subsection, few practitioners believed that a natural method could be established in the short run. Most naturalists still considered a natural method to be their ultimate goal, but were ready to adopt an artificial system to serve until a natural method could be established.

2) *Foundations of artificial methods*

According to Sachs, it was Linnaeus who set the example for this procedure.[3] He recognized that his predecessors in systematics could not have achieved an exhibition of natural affinities by relying on characters selected *a priori*. Unable to devise a natural method himself, Linnaeus chose an artificial system that he believed was a close approximation to the ideal of a natural method.[4] His was a "sexual system," in that it drew its characters from the morphology of the organs of generation of plants, the flower and fruit, which Linnaeus labelled the "fructification."[5] That the sexual system went far in approximating the ideal of a natural method, in Linnaeus's view, is suggested by his explicit comparison of the sexually based taxa with the set of intuitively perceived natural ones that he published as his *Fragmenta Methodi Naturalis*, a part of the 1738 *Classes*

[1] Daudin, *Linné*, pp. 19—33; Sachs, *History of botany*, pp. 7—9, 38.

[2] Daudin, *Linné*, 26—28; Sachs, *History of botany*, pp. 41, 53.

[3] Sachs, *History of botany*, pp. 8, 81, 92.

[4] For Linnaeus's tabling the attempt to achieve a natural method and for his development of an artificial system to serve until a natural method could be constructed, see Sachs, *History of botany*, p. 92; Daudin, *Linné*, pp. 114—115; Larson, *Reason and experience*, pp. 61—72.

[5] For the sexual system, see Sachs, *History of botany*, pp. 82—83, 96—97; Daudin, *Linné*, pp. 36—37; Larson, *Reason and experience*, pp. 57—58, 60, 76—83, 115—118. Cf. also H.K. Svenson, "On the descriptive method of Linnaeus," *Rhodora*, 47 (1945), 273—302, 363—388, esp. 282—292.

plantarum.[6] Though the natural genera rested on intuitively perceived affinities rather than on explicit characterizations, Linnaeus endeavored to find underpinnings for the intuitive categories in marks drawn from the fructification. This effort was most successful on the generic level, where Linnaeus showed that the intuitive genera largely coincided with those based on sexual characters. This coincidence of categories led him to claim the title "natural" for his genera based on the sexual system.[7]

Linnaeus's claim to the naturalness of the genera rested on more than this coincidence. Also supporting it was his attitude toward the significance of the characters drawn from the fructification. The latter belonged to the "essence" of the plant, and a taxon characterized by its features *ipso facto* reflected that essence. In the Aristotelian language of eighteenth-century natural history, an "essence" of a thing was whatever constituted its own being as distinct from the being of other things, for example, the being of a particular genus as such. Linnaeus did not go so far as to claim that characters drawn from the fructification completely distinguished the essence of a taxon. He did believe that because the fructification was essential, characters drawn from it reflected the essence. To the degree that this reflection was faithful and complete, a taxon so characterized was natural for Linnaeus. The intuitive confirmation of the sexually determined genera supported the viability of this idea.[8] In mineralogy, practitioners agreed that essence lay in material composition, but they disagreed about whether material essence was reflected in various characters and therefore whether taxa based on those characters were natural. Moreover, with the appearance of stoichiometric principles, some practitioners, notably Haüy and later Berzelius, argued that material essence required a much narrower definition. They distinguished essential from accidental ingredients, arguing that the essence of a mineral lay solely in the constituents present in definite proportions. Other constituents were mere accidental impurities.

3) *The conflict of method and series*

Whether natural or artificial, a classification scheme was intended to distinguish taxa sharply from one another. Conflicting with this objective was the prevailing notion of series.[9] The idea of series was the claim that members of the natural order—individuals or groups—stood in a linear sequence in which each

[6] For the *Fragmenta* and other sources of Linneaus's views of the natural method, see Larson, *Reason and experience*, pp. 62—66.

[7] For the comparison of natural with the artificial groups, see *ibid.*, pp. 73, 74, 83—84, 90—93.

[8] For the "essential" nature of the fructification as a pillar of the naturalness of the sexually determined genera, see Daudin, *Linné*, p. 40; and Larson, *Reason and experience*, pp. 65, 71—72, 77.

[9] This conflict is the theme of Daudin, *Linné* (in which, see p. i).

member differed from the ones next to it by only the smallest degree.[1] There were barely perceptible gradations, or, as they were called, transitions, by which individuals or groups of one sort could be observed to pass progressively into those of others. The transitions were expressed in the gradual changes in the values of the characters and indicated nothing about genetic relationships among members of the series. Though the distance between proximate members of the series was minimal, the difference between distant members was very perceptible. The idea of series implied the absence of sharp boundaries between kinds, because intermediates could always be found. In emphasizing affinities subsisting between kinds, rather than among members of given kinds, the idea of series opposed itself to those of system and method, which sought to establish distinct boundaries between kinds.[2] As noted below, the doctrine of seriation became a major preoccupation of Wernerian mineralogists and was a target of Berzelius, whose laws of stoichiometry led him to deny transitions in minerals.

The pervasiveness of the doctrine of series among naturalists intent upon devising classification schemes suggests that the opposition of series and systematics was not total.[3] Daudin observes that both Donati and Linnaeus found in the ideas of series and transitions notions compatible with their view of the goal of taxonomy. For these men, natural-historical individuals or groups were connected by numerous, close affinities, expressed in a multiplicity of characters.[4] Artificial systems could not do justice to this image; only a natural method could express it. That is, for Linnaeus and Donati, the natural method was defined as the one that expressed, rather than violated, natural affinities as they were evoked by the idea of series. In Linnaeus's mind, the path to a natural method was obstructed by more than the two obstacles already noted—the difficulty of choosing proper characters and the problem of subordinating them rationally. Naturalists were still ignorant of numerous genera, the absence of which inevitably disfigured any attempt to exhibit affinities in series. Nevertheless, the goal of a natural method remained ever in view, and all artificial systems, including that of Linnaeus, were constructed in the effort to approach this goal as closely as possible. In mineralogy, too, the goal of naturalists was the representation of natural affinities by a natural method.

[1] For an exploration of the notion of series, see Daudin, *Linné,* Ch. 2. Cf. also Arthur O. Lovejoy, *The great chain of being: a study in the history of an idea* (1936; rpt., New York: 1960), esp. Chs. 6—8. As Daudin points out (*Linné,* pp. 101—105) the idea of series contained two notions, that of the continuity of the series and that of the ascendancy of the series through hierarchical stages of increasing perfection. He also points out (*Linné,* pp. 111, 115—116) that only the notion of continuity, not that of hierarchy, influenced the naturalists of the eighteenth century.

[2] See Daudin, *Linné,* pp. 110—116, for a characterization of the relations of systematics and series.

[3] For this paragraph, see *ibid.* Cf. also Larson, *Reason and experience,* pp. 41—49.

[4] This claim corresponded to a shift in the metaphor describing seriation from that of a unilinear chain to that of a net or map. For the net analogy, see Daudin, *Linné,* pp. 110—112; for the map, see Larson, *Reason and experience,* pp. 69—70.

4) *Seriation in natural-historical mineralogy*

Wernerians were concerned with classification in accordance with similarities and differences among external characters, which they analyzed along lines drawn by Werner in his 1774 *Von den äusserlichen Kennzeichen der Fossilien.*[5] The Wernerian approach to characters reflected a preoccupation with transitions in seriated bodies. Werner himself had argued that seriation was the distinguishing characteristic of the natural order, and he claimed that whatever exhibited seriation reflected essence. He used this view in the introduction to the *Aeusserliche Kennzeichen* to argue that essence lay in composition, for it was in composition that seriation in minerals took place—a statement easily made in the absence of stoichiometric laws.[6] In the first chapter of the book, he affirmed that external characters reflected essence, and he established the foundations for external characterization of mineral species. His students, equally convinced of the significance of seriation, hoped to overcome the conflict of method and series by using transitions as indices for the construction of boundaries between neighboring taxa, especially species. The Wernerian doctrine of series and the use of seriation as a means for construction of species are illustrated here with the

[5] Werner, *Von den äusserlichen Kennzeichen der Fossilien* (1774; rpt., Amsterdam: 1965); henceforth cited as Werner, *Aeusserliche Kennzeichen.* On Werner, see Richard Beck, *Abraham Gottlob Werner: eine kritische Würdigung des Begründers der modernen Geologie; zu seinem hundertjährigen Todestage* (Berlin: 1918); T.L. Hasse, ed., *Denkschrift zur Erinnerung an die Verdienste des in Dresden am 30. Juni 1817 verstorbenen K.S. Bergrath's Werner und an die Fortschritte bei der Bergakademie zu Freiberg* (Dresden and Leipzig: 1848); and Georges Cuvier, "Eloge historique d'Abraham Gottlob Werner," in *idem, Recueil des éloges historiques lus dans les séances publiques de l'Institut de France,* 2 vols. (Paris: 1819), II, pp. 305—336, esp. p. 313, where Cuvier compared Werner to Linnaeus. Cf. also Cuvier, "Eloge historique de René-Just Haüy," in Cuvier, *Eloges historiques, précédés de l'éloge de l'auteur par M. Flourens* (Paris: n.d.), pp. 247—282, esp. p. 274.

[6] Werner, *Aeusserliche Kennzeichen,* p. 25, in the note beginning on p. 20. Having argued in the introduction that essence lay in composition and having denied that external characters were reliably correlated with composition (pp. 26—27, in the same note), Werner paid lip service to a chemical species definition, saying (p. 20): "Meine Meynung ist: *Die Fossilien müssen bis auf ihre Gattungen herunter nach ihrer Mishung eingetheilet werden.* Denn ein Mineralsystem hat keinen andern Zweck, als die natürliche Folge oder Reihe der verschiedenen Fossilien zu bestimmen, und je genauer diese darinnen geschiehet, je vollkommner wird das Mineralsystem seyn: nun liegt aber die wesentliche Verschiedenheit der Fossilien in ihrer Mischung, so wie sie bey den Thieren und Pflanzen in ihrer Zusammensetzung [that is, the mode of association of the organs, or articulation] liegt, und erstreckt sich bis auf ihre Gattungen herunter: es müssen also auch die Fossilien bis auf ihre Gattungen herunter, nach dem Grunde ihrer wesentlichen Verschiedenheit, d.i., nach ihrer Mischung geordnet werden."

However, these thoughts did not accurately reflect Werner's deepest convictions. He affirmed the reliability of the correlation of material essence and external characters in the first chapter of the book (pp. 36—38) and claimed that classification must proceed by external characters (pp. 44—46). In the latter passage he even argued that classification by external characters must *precede* chemical analysis, for only by external classification could the chemist be certain of the identity of the species to which his specimens belonged.

example of C. A. S. Hoffmann, author of a multi-volume Wernerian handbook explicitly designed to exhibit the views of the master.[7]

The expectation that minerals displayed seriation was a consequence of Werner's well-known ideas about mineral formation. In his version of an early geology, or geognosy as he called it, minerals were formed by precipitation or sedimentation from a primordial sea.[8] The seriation of minerals as revealed in the external characters was taken to derive from gradual changes in their composition that occurred as the process took place. Smooth transitions and indistinct species boundaries were the inevitable consequences of this view of mineral genesis:

> Uebrigens können bey den Fossilien die Gattungen nie so scharf abgeschnitten und so ausgezeichnet sein, wie bey den organisirten Körpern, da ihre Entstehung auf eine ganz andere Art erfolgt, und die Bildung der Individuen bey ihnen weit mehreren Zufälligkeiten unterworfen ist. Die aufeinander folgenden ungleichartigen Niederschläge, wodurch die verschiedenen Fossilien gebildet worden sind, haben ihre Natur nicht immer auf einmal und plötzlich geändert; die Veränderung ist zuweilen ganz allmählig und gradweise vor sich gegangen. Nur erst in den Extremen einer Reihe solcher Bildungen treten die Verschiedenheiten stärker hervor: in diesen sind die Gattungen vollständig ausgezeichnet; die mittleren Glieder aber geben eine Folge von Individuen, die sich von der einen Gattung immer weiter entfernen, und sich der andern allmählich nähern, die indes zu einer von beyden gerechnet werden müssen, wo es denn oft sehr schwer fällt, diejenige Gattung anzugeben, zu welcher sie vorzugsweise gehören. Bey den Gattungen der organisirten Körper hingegen sind die Verschiedenheiten der Bildung bestimmter und ausgezeichneter, und die Individuen einer jeden karakterisiren sich dadurch ganz genau, dass sie immer wieder Individuen von derselben Art und Beschaffenheit hervorbringen.[9]

This geologically based belief in the proximity of species was unaffected by the stoichiometric ideas that had been articulated by this time (1811). Stated in the immediate aftermath of the famous Proust-Berthollet controversy (1802—1808) about definite proportions, Hoffmann's conception of mineral genesis was consistent with Berthollet's view of precipitation, not with Proust's. For both Hoffmann and Berthollet, the composition of each new portion of a precipitate differed slightly from that of the preceding portion. Composition changed

[7] Hoffmann's work is Carl August Siegfried Hoffman, *Handbuch der Mineralogie,* 4 vols. (Freyberg: I, 1811; I,a, 1812; II,b—IV,b, fortgesetzt von August Breithaupt, 1815—1818); henceforth cited as Hoffman, *Handbuch der Mineralogie.* Hoffmann remarked in his preface (I, pp. vii—ix) that "[die] *wernerischen* Bestimmungen nun in ihrer ganzen Reinheit zu liefern, und dadurch einer Menge zu Tadel und Verunglimpfungen Veranlassung gebenden Missverständnissen abzuhelfen, war einer meiner ersten Zwecke bey Ausarbeitung des gegewärtigen Handbuches, und ich habe mich mit der angestrengsten Sorgfalt bemühet, alles fremdartige, was von andern hinzugesetzt worden, und von dessen Richtikeit ich nicht vollkommen überzeugt war, wieder davon abzuscheiden."

[8] For a brief account of Werner's theory that rocks and minerals were precipitates from a primordial brine, see Archibald Geike, *The founders of geology* (2nd ed., 1905; rpt., New York: 1962), pp. 215—221; and Gillispie, *Genesis and geology,* pp. 43—46.

[9] Hoffmann, *Handbuch der Mineralogie,* I, p. 316.

gradually, through transition, not through saltation, as Proust, champion of definite composition, would have had it; and transition in composition, the essence of minerals, was reflected in transition in external characters. Haüy and later Berzelius, as advocates of definite proportions, argued that in the mineral realm composition changed only through saltation. Wernerians countered this argument by citing cases of apparent transitions that later were recognized as examples of isomorphic substitution. This criticism elicited the mature expression of Berzelius's dualistic doctrine.

As dictated by the mechanism of mineral genesis, the seriation of bodies should have made it difficult to discriminate species. Wernerians, whose abilities in discrimination of species were legendary, nevertheless believed that they had overcome this problem. For example, Hoffmann introduced his discussion of classification with the assurance that species could be distinguished unambiguously by use of suitable procedures based on the doctrine of seriation:

> Das erste also, was der Klassifikator der Fossilien in der Oryktognosie bey Entwerfung des Systems zu thun hat, ist die Aufsuchung der in ihren Kennzeichen, und zwar in der Gesammtheit ihrer äussern Kennzeichen, sich findenden wesentlichen Verschiedenheiten oder ihre Gattung und so viel sich dergleichen wesentliche Verschiedenheiten unter ihnen finden, so viel giebt es Gattungen; alle Fossilien hingegen, die in ihren äussern Kennzeichen keine wesentlichen Verschiedenheiten, sondern die höchste Uebereinstimmung, und also den höchsten Grad der oryktognostischen Verwandtschaft zeigen, bilden zusammen eine Gattung. Die eine Gattung zugehörenden Fossilien lassen zwar wohl in manchen ihrer Kennzeichen Abänderungen bemerken:
>
> diese verlaufen sich aber stets in einander, bilden unmittelbar zusammenhängende, durch keine jähen Sprunge unterbrochene geschlossene Suiten, die von den Suiten derselben Kennzeichen bey andern Fossilien-Gattungen immer ausgezeichnet verschieden sind, und stören also die Gattungs-Identität keines weges.[1]

Precisely what could be taken as an "essential" difference as exhibited by external characters alone, without reference to chemical composition, Hoffmann did not state directly. In the latter portions of the passage, he implied that a species was defined by observable transitions among its characters, indeed, that the characters in which transition suites could be observed were *ipso facto* the specific characters. He also implied that the collection of minerals in which a smooth series of varieties in the characters could be observed was the species. Moreover, Hoffmann stated here that the species so defined was well defined. Though they elsewhere insisted on the existence of *inter*specific transitions as well, Wernerians believed that a species consisted of a closed transition suite.[2] Because Berzelius was led by his stoichiometric doctrines to draw species boundaries inside Wernerian species,

[1] *Ibid.*, pp. 312—313.

[2] Hoffmann denied the adequacy of the chain metaphor and, to illustrate his claim that species participated in multiple seriations, he offered a plate displaying schematically the multiple affiliations of the species and varieties of the silver genus both among themselves and with other genera (*ibid.*, p. 327 and accompanying pl.).

he elicited their criticism. His response was to deny transitions in mineral matter but concede them in organic matter.

3. The crystallographic tradition in mineralogy: specificity of form and definite proportions

Under Haüy's leadership, crystallography was deeply concerned with the proper means for discrimination of species. Haüy was a pioneer in both stoichiometric and morphological determinations of species, arguing that both approaches should converge. The law of definite proportions, which Haüy (and before him Romé de l'Isle) postulated prior to Dalton, guaranteed specificity of form and permitted determination of species by morphological methods.[3] In tying morphology to stoichiometry, Haüy committed himself to a unique correlation of composition and form (apart from certain highly regular forms, the "limiting forms," such as the cube and regular octahedron, shared by many widely differing minerals). To uphold specificity of form, Haüy had to uphold its foundation, the mineralogical application of the law of definite proportions. His posture invited the criticism of C. L. Berthollet (1748—1822), who opposed definite proportions. Haüy's theories and their criticism at the hands of Berthollet provided precedents for Berzelius's stoichiometric mineralogy and for his conflict with Wernerian mineralogists.

a. Haüy and the specificity of crystal form

Haüy's morphology focussed on the microscopic level, rather than on the macroscopic cleavage forms.[4] A specific mineral consisted of a regular aggregation of specific "integrant molecules." An integrant molecule was the smallest bit of the substance possessing the characters of the substance. In turn, the integrant molecule was composed of "constituent molecules," that is, the minimal particles of each of its elementary constituents. An integrant molecule was distinguished by a specific geometric form, subject to description in terms of precise geometric parameters that Haüy developed. The macroscopic form of a crystal was determined jointly by the shape of the integrant molecule and the mode of its aggregation into a crystalline mass. Moreover, the shape and mode of aggregation dictated the cleavage characteristics of the crystal, so that the

[3]For precedents provided Haüy by Romé, see Hooykaas, "Species concept," 46—48. For Romé's accomplishments, see *idem, La naissance de la cristallographie en France au XVIIIe siècle* (Paris: 1953), pp. 8—12 (henceforth cited as Hooykaas, *Naissance*); and *idem.* "De kristallografie van J.B. de Romé de l'Isle (1783)," *Chemisch Weekblad,* 47 (1951), 848—855.

[4]For Haüy's crystal theories, see these publications by Hooykaas: "Species concept," 48—55; *Naissance,* pp. 12—26; "Les débuts de la théorie cristallographique de R.J. Haüy, d'après les documents originaux," *Rev. d'hist. sci.,* 8 (1955), 319—337; and "Haüy, René-Just," *DSB,* VI, pp. 178—183.

form of the integrant molecule could be deduced from observation of cleavage (or, in the absence of cleavage, of striation), which indicated the "natural joints" of the specimen. Haüy concisely expressed this microscopic focus of his thought in his 1801 *Traité de minéralogie:*

> La théorie qui a servi à développer ces lois [of crystallography], repose toute entière sur un fait dont l'existence avait été jusqu'alors plutôt entrevue que démontrée. Il consiste en ce que ces petites solides, qui sont les élémens des cristaux, et que j'appelle leurs *molécules intégrantes* ont, dans tout ceux qui appartiennent à une même espèce de minéral, une forme invariable, dont les faces sont dans le sens des joints naturels indiqués par la division mécanique de ces cristaux, et dont les angles et les dimensions respectives sont donnés par le calcul combiné avec l'observation. De plus, les molécules intégrantes relatives à différentes espèces ont aussi entre elles des diversités plus ou moins marquées, excepté dans un petit nombre de cas [the limiting forms], où leurs formes ont des caractères de régularité, d'où résultent comme des points de contract entre certains espèces.[5]

Underlying the specificity of the integrant molecule was the law of definite proportions. Haüy made the point in the *Minéralogie,* where he claimed that the specificity of the integrant molecule

> nous conduit a une considération intéressante relativement à la composition chimique des minéraux. C'est que les principes qui concourent à former leurs molécules intégrantes doivent, ce me semble, être constans, quant à leurs qualités et à leurs quantités . . .[6]

Accordingly, he defined species by uniting specificity of composition and form:

> D'après ces considérations [of definite proportions as underlying the specificity of the integrant molecule], il me semble qu l'on peut définir l'espèce, en minéralogie, une collection de corps dont les molécules intégrantes sont semblables, et composées des mêmes élémens unis en même proportions. Cette dernière condition généralise la définition, et l'étend aux substances qui ayant leurs molécules configurées de la même manière, diffèrent essentiellement par les principes qui composent ces molécules.[7]

A mineral species was defined by two things: the form of the integrant molecule, expressed in terms of exact geometric parameters, and the composition, expressed as the identity and definite proportions of the constituents. All samples having the same integrant molecule and the same composition belonged to the same species.

In the "preliminary discourse" to the *Minéralogie,* Haüy foresaw a fruitful conjunction of chemical and morphological methods in the determination of species. Asserting that chemistry "nous dévoile la composition intime de et le fond de chaque substance,"[8] Haüy argued that chemical analysis must be invoked to fix the position of a mineral species in a system of classification. He pointed out, however, that chemical methods were too inconvenient for identi-

[5] Haüy, *Traité de minéralogie,* 4 vols., atlas (1801; rpt., Brussels: 1968), I, p. xiv; cf. pp. 18—19, 29—30; henceforth cited as Haüy, *Minéralogie.*

[6] *Ibid.,* p. 161.

[7] *Ibid.,* pp. 162—163.

[8] *Ibid.,* pp. xxi—xxii; see also pp. xxii—xxiii.

fying (as opposed to calssifying) mineral species.[9] More suitable was determination of the (specific) form of the integrant molecule.[1] An assemblage of minerals, each having the same integrant molecule (and therefore belonging to the same species), could be classified by the chemical analysis of but a single sample,[2] a claim leading Haüy to offer this optimistic conclusion:

> On conçoit par là dans quel sens il faut entendre ce que j'ai annoncé plus haut, que c'est à la chimie qu'appartient la détermination des espèces. Il seroit peut-être plus vrai de dire qu'elle complète cette détermination, en faisant connoître les molécules principes, dont les molécules intégrantes sont les assemblages. Déjà il est aisé de sentir (et la suite en offrira plusieurs exemples), combien il est intéressant que les recherches relatives à ces deux sortes de molécules conspirent vers un but commun; que le chimiste et le minéralogiste s'éclairent mutuellement dans leurs travaux, et que le gonyomètre, qui fournit des données pour soumettre les formes cristallines au calcul soit associé à la balance qui pèse les produits de l'analyse.[3]

Because definite proportions guaranteed the unique correlation of form and composition, chemical and morphological methods should have led to identical results in the determination of species.

Aware that the convergence of chemical and morphological methods was scarcely universal, Haüy accommodated his theory to the observed anomalies. He knew that many minerals shown to be crystallographically identical exhibited differences in composition. The quantities and often the identities of the constituents varied from one sample to another; but the different samples were isomorphic. Haüy addressed this problem in the *Minéralogie*. If the results of chemical analysis differed from one sample to another of a single crystallographic species, Haüy claimed that the differences resulted from the presence of impurities that were not chemically combined but only mechanically mixed in the mass of the sample. Viewed apart from the impurities, the mineral itself was nevertheless a distinct compound in definite proportions. Upholding the law of definite proportions and, with it, the specificity of crystal form, thus occasioned the distinction between two kinds of ingredients in minerals, accidental ones, present as immixed impurities, and essential ones, combined in a stoichiometric compound. Haüy made this point while elaborating the claim quoted above (p. 112 accompanied by n. 6) that specificity of the integrant molecules implied the presence of the same ingredients in the same proportions. Accordingly,

> les substances qui font varier les produits de l'analyse sont etrangères aux molécules, et seulement interposées entre elles dans la masse du mineral (1). On peut comparer une substance mélangée de ces principes additionels, à certains sels auquels d'autres sels sont unis accidentalement, par exemple au nitre de la première cuite. Lorsque l'on fait subir à ce sel des dissolutions et des cristallisations successives, pour l'épurer, le liquide n'altère aucunement la figure de ses molécules; il ne fait que les séparer les unes des autres, et les débarrasser de celles des autres sels, qui leur étoient associées, et qui n'étoient entrées

[9] *Ibid.*, p. xxxvii; but see pp. 158—159.

[1] *Ibid.*, xiv—xvi.

[2] *Ibid.*

[3] *Ibid.*, pp. xvi—xvii.

pour rien dans leur composition. De même les principes d'où dépendent les différences entre les analyses des diverses morceaux, d'un même minéral, ne forment avec la substance propre de celui-ci qu'un simple mixte [mécanique], dont les molécules intégrantes sortiroient intactes, s'il nous étoit donné d'en faire, pour ainsi dire, le départ. (1)[4]

Haüy could preserve definite proportions, the basis for his belief in the specificity of the integrant molecule, and account for variability of analytical results by invoking this distinction between kinds of ingredients. Differences in composition among samples of a single crystallographic species resulted from different impurities, in different proportions, present because of the accidents of formation. The effect was similar to the production of impure precipitates in artificial processes. Faced with the same problem, varying analyses within a species, Berzelius used a similar argument, bolstering it with his vast experience as an analyst and controlling it by invoking his elaborate stoichiometric laws.[5]

b. Difficulties confronting a chemical species definition

Many of the minerals in which Haüy and later Berzelius saw immixture of accidental impurities were actually cases of isomorphic substitution. Substitution is the replacement in a sample of all or part of an ingredient by one or several substances possessing the same form as the replaced substance, that is, by isomorphs of the replaced material. Because the new or vicarious constituents shared the form of the replaced ingredient, the crystal form of the sample was undisturbed (within narrow limits). Accordingly, the integrant molecule of a substituted sample as determined by Haüy was the same as that of a pure substance present in the sample, permitting Haüy to assign the sample to the species of that pure substance and to regard the vicarious constituents as accidental. However, as shown by Mitscherlich in his 1819 paper announcing the discovery of isomorphism, the supposed impurities were actually essential, that is, chemically combined with the other ingredients in proportions varying within stoichiometrically determined limits.[6]

[4] *Ibid.*, pp. 161—162.

[5] The quotation accompanied by the last note indicates another precedent that Haüy provided Berzelius. In his second note to the passage (p. 162n), he argued that several external characters were decisively determined by the presence or absence of accidental ingredients, thereby pointing out that Wernerian characterizations reflected nothing essential.

Elsewhere in the *Minéralogie* (III, 241—242), Haüy pursued the point that among minerals change occurred by saltation. Change by transition implied gradual change in the angles and relative dimensions of the integrant molecules of one species into those of another. Since no intermediates were observed in nature but only "un saut brusque" from one to another, change was by saltation. Because Haüy's doctrine of specificity of form rested on definite proportions, saltation in form rested on saltation in proportions from one definite compound to another.

[6] Mitscherlich's paper is "Ueber die Krystallization der Salze, in denen das Metall der Basis mit zwei Proportionen Sauerstoff verbunden ist," *Abhl. Akad. Berlin*, 1818—1819 (1819), 427—437. For an analysis of this paper, see the present author's "Mitscherlich's discovery of isomorphism," *Hist. Stud. Phys. Sci.*, 11,i (1980), 87—123, henceforth cited as Melhado, "Mitscherlich's discovery."

As Hooykaas argues, Haüy ruled out isomorphism as a theoretical possibility because of his commitment to definite proportions.[7] By insisting upon a unique correlation of form and composition and by explaining differing analytical results in terms of accidental immixture, he denied the very possibility of isomorphism (apart from minerals possessing the limiting forms). Consequently, though he insisted in his theory on the coincidence of stoichiometric and morphological determinations of species, he relied on morphology in the practice of his science.[8] As shown in Part II, Berzelius was also blinded to isomorphism at least partly because of his insistence that all true compounds were rigorously stoichiometric. Nevertheless, the phenomena of substitution were sufficiently striking to arouse critics of both Haüy and Berzelius. Haüy was criticized by C. L. Berthollet, champion of indefinite composition, in the light of cases later recognized as exhibiting substitution. Berzelius was the target of Wernerian mineralogists, who relied on similar evidence, as shown in Chapter VI. Berthollet's critique, a dress rehearsal for Berzelius's conflict with the Wernerians, reveals the severe difficulties confronting a stoichiometric approach to mineral species.

Berthollet offered his critique in a note to his 1803 *Essai de statique chimique.*[9] There, he established his own stance by contrasting it with the implications of Haüy's commitment to definite proportions. Noting Haüy's conclusion that definite proportions implied the absence of natural-historical transitions, Berthollet pointed out that as a consequence, "... Haüy regarde les parties colorantes de quelques minéraux, par exemple, celle de l'oxide de chrôme qui colorent l'émeraude verte, comme simplement disséminées, de manière qu'elles ne nuisent pas à la transparence."[1] That is, Haüy denied that impurities, responsible for transitions in such characters as color, were chemically combined. However, Berthollet felt no difficulty in claiming that this kind of union was no mere juxtaposition of impurities and essential ingredients, but a true chemical combination:

> L'uniformité dans la composition malgré la différence de pésanteur spécifique des parties élémentaires, la transparence qui prouve qu'elles n'exercent plus une action séparée sur les rayons de la lumière, des propriétés communes, mais différentes de celles des parties élémentaires séparées, sont cependant une indication irrécusable de la combinaison.[2]

Moreover, Berthollet considered fine mixture a special sort of combination that permitted seriation in the composition of samples with preservation of form. Thus, he remarked,

[7] Hooykaas, "Species concept," 47—48, 54—55. See also *idem, Naissance,* p. 25.

[8] For Haüy's reliance on morphology, see Hooykaas, "Species concept," 52, 54.

[9] C.L. Berthollet, *Essai de statique chimique,* 2 vols. (Paris: 1803), I, pp. 443—449; henceforth cited as Berthollet, *Statique chimique.*

[1] *Ibid.,* p. 439, citing Haüy, *Minéralogie,* IV, p. 415.

[2] Berthollet, *Statique chimique,* I, p. 439.

Les sels même les plus constants dans leurs proportions peuvent se surcomposer ou se combiner ensemble, sans que leur forme cristalline et leur transparence [the sign of true composition] soient altérées, ils peuvent varier dans leurs proportions, sans que leur forme subisse de changement . . .[3]

Berthollet offered a series of examples in which transitions in composition produced no change of form, that is, cases exhibiting isomorphic substitution. Among the most striking cases that he presented was the following, offered in italics:

Une dissolution à parties égales de sulfate de fer et de sulfate de cuivre, donne des prismes tétraëdres rhomboidaux d'un bleu verdâtre; la forme de ces cristaux est parfaitement bien déterminée, et il aisé de reconnaître à l'oeil simple l'homogénéité de leurs substances. On peut les faire dissoudre et cristalliser à plusieurs reprises, sans que cette substance, ni la configuration de ses cristaux, soit changée en aucune manière . . . Un mélange de trois parties de sulfate de fer et d'une partie de sulfate de cuivre, donne des cristaux d'un vert d'emeraude et de même forme que les précédents; seulement quelque différence dans la couleur distingue ces deux espèces surcomposés. [4]

Another one of Berthollet's examples was connected with alums, which were cited later by Berzelius's critics and by a supporter, J. N. Fuchs, who clearly anticipated the principle of isomorphism:

Vauquelin [who did mineral analyses for Haüy] a fait voir que le sulfate d'alumine contenait indifféremment sept parties sur cent de potasse ou d'ammoniaque, sans qu'on appercoive aucune différence dans la cristallisation; la proportion de l'acide lui-même peut changer, ainsi que je m'en suis assuré, et Leblanc a prouvé que le sulfate d'alumine pouvait se surcomposer d'une quantité considérable de sulfate de fer, et cependent fournir, par la cristallisation, des octaëdres réguliers.[5]

From these and similar cases Berthollet concluded that

Quoique les chimistes aient jusqu'à présent négligé de porter une attention particulière sur les formes des sels surcomposés, il serait facile d'accumuler les observations qui prouvent que les sels peuvent se surcomposer sans éprouver dans leurs formes un changement qui réponde à la surcomposition; et cependant ces surcompositions qui conservent leur transparence, sont l'effet de l'action réciproque des éléments que les composent.

Si cette vérité est incontestable pour les substances saline qui ont une solubilité considérable, et qui par conséquent éprouvent de la part de l'eau une action énergique, relativement à leur force de cohésion, l'affinité réciproque des substances qui ont peu de solubilité, doît être beaucoup plus efficace pour les réunir dans un état de combinaison.

Il n'est donc pas étonnant que l'on trouve, dans les minéraux des variétés considérables dans les proportions des éléments qui les composent, quoiqu'ils présentent les indices d'une combinaison complète, telle que la transparence; et ce serait se fonder sur un système arbitraire, que de méconnaître dans ces combinaisons l'action réciproque des parties qui les constituent.[6]

A practitioner did not have to be as thoroughgoing an advocate of indefinite

[3] *Ibid.*, p. 440.

[4] *Ibid.*, pp. 440—441.

[5] *Ibid.*, p. 441.

[6] *Ibid.*, pp. 441—442.

composition as Berthollet was to accede to the force of these arguments against definite proportions in mineralogy. Prior to the discovery of isomorphism by Mitscherlich in 1819, cases such as those Berthollet cited, exhibiting persistence of form with change of composition, constituted a grave obstacle to the acceptance of a stoichiometric definition of mineral species. Had Haüy admitted isomorphism, the problem of identifying essential ingredients would have been minimized and the difficulties confronting the stoichiometric species definition would have been largely overcome. To the end of his career, however, Haüy denied isomorphism, still rejecting it following the publication of Mitscherlich's 1819 paper.[7] In the face of evidence such as Berthollet had adduced, Haüy offered a modest defence of immixture but left the work of upholding this doctrine to his partisans, notably Beudant.[8] Berzelius, confronting Wernerian criticism similar to Berthollet's, responded by elaborating his dualistic view of combination into its classic form.

4. Organic chemistry in the late eighteenth century: a unitary theory of organic composition and a two-component theory of organic specificity

a. Introduction: Differing criteria for discrimination of substances in organic and inorganic chemistry

1) *Stoichiometry versus natural history*

The prevalance of two-component theories in the late eighteenth century, exhibited in Chapter II, depended on the gradual resolution of the traditional elements and principles into collections of particular chemicals. Stahl's chemistry, having relied on a small number of principles, had to assume that many compounds contained the same ingredients and differed in the proportions of the ingredients. The Stahlian distinction among the acids according to the phlogiston they contained is a prime example. With the increase in the number of particular chemicals, differences among compounds could be treated instead by two-component theories, which relied more on the qualitative distinctions among particular ingredients of different compounds than on proportions of ingredients believed common to many compounds.

However, the Stahlian explanation in terms of proportions was still needed in the late eighteenth century. Many cases appeared in which the same set of particular chemicals gave rise to several distinct compounds. For example, Lavoisier's work on oxidation offered dramatic evidence that a given radical could have several distinct oxides. Another example is provided by Proust's famous con-

[7] For this, see R. Hooykaas, "Haüy, René-Just," *DSB*, VI, pp. 178—183, esp. p. 182.

[8] For this, see Melhado, "Mitscherlich's discovery."

troversy with Berthollet about definite composition.[9] Proust argued that cases of apparently indefinite composition were really mixtures of several compounds, formed from different proportions of the same ingredients. Thus, late-eighteenth-century chemists accounted for the specificity of distinct compounds having the same ingredients just as Stahl had explained the specificity of acids, by invoking differences in the proportions and referring changes in properties to changes in proportions. Lavoisier's correlation of degree of oxidation with degree of acidity is one of the most prominent examples of this approach.

In the late eighteenth century, the Stahlian explanation of specificity was often coupled with the law of definite proportions. That chemists readily sided with Proust against Berthollet, despite the latter's prestige and the cogency of his arguments, suggests that Proust was merely making explicit a common (though often tacit) assumption that combination took place in definite proportions. Lavoisier, for example, appears to have assumed definite proportions in devising his correlation of degree of oxidation with degree of acidity. Another example is provided by Haüy, whose claim that inorganic matter changed only by saltation depended explicitly on his coupling the correlation technique with the assumption of definite proportions. Berzelius, relying on the same coupling of the correlation technique with definite proportions, recapitulated this argument.

The ideas described in the preceding paragraphs concerned chiefly inorganic matter. Opinions about organic matter in the late eighteenth century were very different. As Holmes shows, the entire antecedent tradition of fire analysis suggested to many practitioners the apparent uniformity in the composition of even widely differing vegetables, offering little basis for referral of qualitative differences to differences in proportions.[1] Though Lavoisier made a fundamental advance, exhibited in the next subsection, in the understanding of fire analysis, he did not thereby provide the means to connect composition with properties. He showed that vegetables were composed chiefly of three elementary substances, carbon, hydrogen, and oxygen, and he argued that these elements, rather than enduring, compound proximate constituents, were the immediate ingredients of vegetables. Each of the three elements combined directly with the other two, forming a "unitary entity," as it was called in the nineteenth century. Substances obtained from fire analysis were not proximate principles, but artifacts of the technique. Though essential to the understanding of fire analysis, Lavoisier's

[9] On the Proust-Berthollet controversy, see Ida Freund, *The study of chemical composition: an account of its method and historical development* (1904; rpt., New York: 1968), Ch. 5; and S.C. Kapoor, "Berthollet, Proust, and proportions," *Chymia,* 10 (1965), 53—110.

[1] Frederick L. Holmes, "Analysis by fire and solvent extractions: the metamorphosis of a tradition," *Isis,* 62 (1971), 129—148, esp. 136—137. There was actually considerable diversity in the results, Holmes remarks (136), but no way to reduce these differences to generalizations. Nevertheless, the overall qualitative uniformity of the results was their striking feature, and it was to this task that Homberg, a leading figure in the second generation of chemists at the Paris academy, addressed himself, as Holmes observes (136).

views did not diminish the appearance of uniformity in the composition of vegetable products. All vegetable matter contained the same three elements, and all animal matter, as Berthollet and others suggested, contained four elements, nitrogen plus the three found in vegetables. The proportions of the elements varied from one substance to another, but in no discernable patterns. Moreover, Lavoisier's analytical techniques did not permit very accurate or reliable determination of the proportions of the elements, and the substances he analyzed were often mixtures. Accordingly, no arguments connecting composition and properties were possible in organic chemistry as they were in inorganic chemistry.

Animal and vegetable materials were characterized not by elemental analysis, but by the methods of natural history. In successive steps, fire analysis and, more often in the late eighteenth century, extractive analysis were used to separate animal or vegetable matter into more and more animal or vegetable substances, such as oils, gums, resins, or starches. Discrimination of these substances was achieved by invoking several sorts of characters, including the analytical techniques by which the substances were obtained and the reactivities and other notable properties of the products of analysis. Any taxonomy of extracts or distillates was unstable, for the substances classified were themselves the subjects of further analysis and increasingly were resolved into further products. This continual advance of the frontier of analysis resulted in a proliferation of taxa. By the opening of the nineteenth century, the five principles of the iatrochemists had become Fourcroy's twenty categories of vegetable substances.[2]

At first, these categories paralleled those of botanical and zoological taxonomies by referring substances to the species of the plant or animal of origin. As the techniques of analysis increasingly were deployed to characterize the substances that they had separated from organic beings, a distinct chemical classification replaced the naturalists'. Fourcroy's *Connaissances chimiques* of 1800 is taken to be a watershed in this connection. There he replaced the older botanical taxa with chemically characterized categories of the "principes immédiats" of vegetables, that is, the proximate principles obtained by analysis (but he retained the zoological taxonomy of animal substances).[3]

2 Presented in his *Système des connaissances chimiques, et de leurs applications aux phénomènes de la nature et de l'art,* 11 vols. (Paris: An 9—10 [1801—1802]), VII, pp. 125—126; henceforth cited as Fourcroy, *Connaissances chimiques.* Fourcroy also classified animal substances, *Connaissances chimiques,* IX, pp. 121—124. However, as noted shortly, unlike his classification of vegetable matter, Fourcroy's classification of animal matter offered no significant innovation.

3 For Fourcroy's views on the "principes immédiates" of vegetables, see his *Connaissances chimiques,* VII, pp. 51—56, 111—113, 120—127. Examples of immediate principles (pp. 125—126) were sugar, vegetable albumin, extractive matter, tannin, resin, camphor, and vegetable oil. These substances were obtained by moderately strong analytical techniques (e.g., expression, extraction, gentle heating) and were destroyed by more vigorous techniques (e.g., strong heating, corrosive reagents).

Fourcroy's definition of immediate principles rested upon Lavoisier's prior conceptual in-

2) *Material foundations of distinctions among organic substances*

Substances with sharply distinguished qualitative characters were readily taken to owe their differences to significant differences in composition. But what of

novations in the theory of organic analysis, explored in subsection 4.*b*. Here may be noted that Fourcroy considered the immediate principles as substances that preexisted in the vegetable as enduring ingredients—as Lavoisier and Fourcroy put it, they were "tout formé" in the vegetable prior to analysis—and were not artifacts of the analytical techniques used to obtain them (for this, see especially VII, p. 112).

Initially making use of the term"principes immédiates" (VII, p. 51), Fourcroy later (p. 113) declared that the term "matériaux immédiats" was more appropriate. The word "principle," he remarked, implied participation in mixtion, and he did not wish to claim that the vegetable principles were combined with one another, forming mixts. The principles merely coexisted with one another in various parts of the plant and were products of the "travail de la végétation."

Classification of the immediate principles was to be divorced from the botanical taxa (VII, pp. 121—122): "En supposant qu'on connaisse et qu'on puisse recueillir en particulier tous les matériaux immédiats dont la réunion forme l'ensemble des végétaux; on se figure, pour représenter cet ensemble et pour procéder avec ordre à l'analyse végétale, toutes les plantes mêlées et broyées pour ainsi dire, ne formant ensemble qu'une seule masse produite par la végétation considérée dans son intégralité, soumise ensuite aux différens moyens d'extraction que j'ai fait connaître, et séparable en une série de matières diverses; on se figure cette séparation ou cette extraction poussée jusqu'au point où l'on ne pourrait plus aller au-delà sans détruire la composition même et sans dénaturer ces diverses matières. Telle est l'idée qu'il faut se former des produits de l'analyse immédiate ou des matériaux immédiates des végétaux. Chacun d'eux n'est plus alors considéré comme appartenant en particulier à telle ou telle plante, mais à toute l'ensemble de la composition végétale, mais à toute la masse des plantes existantes. Chacun d'eux, d'une nature générale identique, représente en quelque sorte la même matière existante dans tous les êtres végétaux. Par exemple, le suc gommeux et le suc sucré sont les mêmes dans tous les végétaux; à quelque plante ou partie de plante qu'on les ait empruntés, ils sont très-exactement les mêmes: en sorte qu'avec quelques végétaux seulement, pourvu qu'on les choisisse de manière à ce qu'ils puisse fournir tous les composés qu'on a reconnus jusqu'ici parmi les matériaux immédiats des plantes, on est censé procéder à l'analyse végétale dans toute son étendue."

Since the priciples could be explored without reference to plant species of origin, characterization could be chemical, not botanical. Evaluating several methods of distinguishing the principles, Fourcroy had this to say of chemical classification (VII, p. 124): "Une troisième méthode de distinction entre ces corps ou composés végétaux distincts [the immediate materials] pourrait être puisée dans des caractères vraiment chimiques, ou dans les propriétés intimes qui les séparent et les font reconnaître. Par exemple, il y a des matériaux immédiats muqueux, d'autres sont sucrés, ceux-là sont acides, ceux-ci huilex et inflammables; tels colorés, tels solides et indissolubles, tels autres rapprochés des matières animales. On pourrait même combiner avec cette première division celle qui considérerait ces matériaux par leur état ou leur consistance comme liquides, moux, fibreux, glutineux, lamelleux, pulvérulens; celles encore qui les partageraient par la saveur en doux et fades, doux et sucrés, acides, acerbes, amers, âcres; par l'odeur, en odorans, inodores, fétides, aromatiques. Cette mainère de disposer les matériaux des végétaux est une des plus utiles et des plus philosophiques; elle tient aux propriétés chimiques d'une part, tandis, que de l'autre elle conduit à la connaîssance des propriétés avantageuses ou usuelles: elle mérite donc, à tous égards, la préférence ... et l'on verra bientôt que c'est celle dont je me rapproche le plus."

E. Hjelt, *Geschichte der organischen Chemie von ältester Zeit bis zur Gegenwart* (Brunswick: 1916), pp. 20—22, and Partington, *History*, III, p. 543, agree that Fourcroy's chemical classification was a model for the organization of textbooks in organic chemistry for decades. (Hjelt's book is cited henceforth as Hjelt, *Organische Chemie*). Cf. also Marcel Florkin, *A history of biochemistry*. Comprehensive biochemistry, Vol. XXX (Amsterdam: 1972), pp. 112—117, esp. p. 114. For Fourcroy's more traditional classification of animal substances, see the preceding note.

small qualitative differences between very similar substances? What about the differences among the sugars or starches, for example, obtained from various plant and animal sources? What about the various organic acids? Whether these differences were considered fundamental varied with the sensibilities of the observer—as in natural history. An example is Lavoisier's treatment of vegetable oxides in the *Traité*. Calling attention to the vegetable oxides then known, he cited one genus of similar substances along with two others that he did not label as generic:

> Les oxides végétaux à deux bases [carbon and hydrogen; the idea of organic radicals is considered in subsection 4.*c*.] sont le sucre, les différents espèces de gommes que nous avons réunies sous le nom générique de *muqueux*, & l'amidon.[4]

He went on to describe the composition of "ces trois substances" as if the members of the generic category of gums might just as well be considered the same substance.[5] The distinctions among the members of this genus were not so sharp that Lavoisier felt he needed to treat them as distinct substances.

Another example is the division of opinion around the turn of the century about the individuality of the organic acids. Originally the acid of vinegar was the sole organic acid clearly recognized as a distinct compound, but Scheele had discovered several new ones in the 1780's. These discoveries were part of the unanticipated late-century proliferation of known acids. Though Scheele marvelled about the number of new organic acids, he considered them distinct.[6] Many contemporaries, however, received his work with skepticism. The similarity of new organic acids with acetic acid led many to claim that the new substances were merely modifications of acetic acid.[7] Other chemists acknowledged the distinctiveness of some of the organic acids, but invoked the modification theory for others. Fourcroy and Vauquelin, for example, confirmed the distinct identity of malic acid but supposed that some organic acids, especially those that did not crystallize, such as lactic acid, were acetic acid contaminated by tenacious impurities.[8] Lavoisier, whose oxygen theory of acidity led him to emphasize the potentially limitless number of acids, was perhaps more ready than many to acknowledge the diversity of organic acids. In the *Traité*, he listed all the organic acids, unlike the species of gums, as distinct substances.[9] The characters of the

[4] Lavoisier, *Traité*, I, p. 125.
[5] *Ibid.*, pp. 125—126.
[6] For Scheele's discoveries of organic acids and their novelty, see Olsson, *Kemiens historia*, pp. 351—360; Maurice P. Crosland, "Lavoisier's theory of acidity," *Isis*, 64 (1973), 306—317, esp. 311; and Hjelt, *Organische Chemie*, pp. 10—11. For more about organic acids in the late eighteenth century, see Partington, *History*, III, pp. 108, 544—545, 555—556, 579.
[7] For reactions to Scheele's discoveries, see Hjelt, *Organische Chemie*, p. 11.
[8] For Fourcroy and Vauquelin on organic acids, see Partington, *History*, III, p. 555. In 1808, Berzelius proved the separate identity of lactic acid, as he reported to Gahn in a letter of 29 March 1808 (*Bref*, V,ix, pp. 32—36, esp. p. 34), but he did not publish his study of the acid. For an English translation of the relevant passage from the letter, see J. Erik Jorpes, *Jac. Berzelius: his life and work* (Berkeley: 1970), p. 36.
[9] Lavoisier, *Traité*, I, pp. 129—130.

acids were sufficiently distinct to permit this judgement, but, as the opinions of other chemists suggest, insufficiently distinct to compel it.

If to many observers some similar organic substances were insufficiently sharply characterized to permit their discrimination as fundamentally distinct, to what cause did they refer the modest differences that were present? In some cases, practitioners cited impurities, as just seen with the example of the organic acids. In others, the answer is not yet clear and awaits further scholarship. One very real possibility is that the doctrine of definite proportions was not assumed for organic matter. The composition of two gums, for example, expressed as the percentage composition of carbon, hydrogen, and oxygen, might indeed differ slightly, but the gums were still all gums.[1] In the light of the overall uniformity in the composition of all vegetable substances, what could justify deferring to a miniscule difference in composition, a difference only assumed and incapable of detection by available techniques, to require the discrimination of two gums as distinct substances? Perhaps with larger differences in composition, gums would have turned into something else, sugar for example.[2] However, that transformation would have represented a large jump in composition between two extremes connected by a series of intermediates. In organic chemistry, characters and therefore also composition changed by transition, not saltation. The occurrence of transitions was possible because composition was indefinite. In inorganic chemistry, transitions were impossible because composition was definite.

3) *Novelty and tradition in Berzelius's organic chemistry*

The best evidence supporting the preceding observations about organic chemistry at the turn of the eighteenth and nineteenth centuries is provided by the work of Berzelius. As shown in Part II, he explicitly raised the issue of specificity in organic chemistry, bringing it into sharp focus. His concern for the specific was first manifested in his early extractive analyses and was later given theoretical formulation in terms of the theory of proportions. In the course of his extractive analyses, which represent the highest level of competence with this technique attained up to his time, Berzelius criticized his predecessors for failing to attend sufficiently to evident differences between generally similar analytical products. For Berzelius, even slight differences constituted marks of fundamental underlying material distinctions. In his later quantitative organic analyses, which were

[1] See Fourcroy's statement, quoted above, p. 120, in n. 3, that the immediate principles were the same regardless of plant of origin. If the older botanical taxonomy of the products of vegetable analysis had made chemists at all aware of specific distinctions between closely allied substances from different plant species, abandonment of the botanical taxonomy entailed loss of that awareness. It was Berzelius who combined a chemical taxonomy of the products of vegetable analysis with a sensitivity to specific distinctions.

[2] Fourcroy, for example, was happy to explain differences between *different* immediate principles in terms of proportions. For this, see his *Connaissances chimiques*, VII, p. 54.

conducted on the assumption of the observance of definite proportions in organic chemistry, Berzelius showed that a sufficiently large number of atoms entered into organic compounds to permit the manifestation of apparent transitions occasioned by relatively small differences in composition. The transitions nevertheless were observed between substances of definite composition.

Berzelius, then, differed from his predecessors in organic chemistry in two respects: taking the specific level seriously and (later) defining it in terms of definite proportions. He shared his predecessors' perception of the similarity of organic chemistry with natural history, however, and, articulating this perception clearly, offered the means to explain it. Moreover, he used the perception to locate the boundary between the organic and the inorganic. In inorganic chemistry and mineralogy, transitions did not occur and species boundaries, dictated by definite proportions, were sharp. In organic chemistry, however, transitions did occur, and species boundaries were vague. Accordingly, Berzelius asserted the need to remain with the methods of the naturalist in organic chemistry, and he defined the obstacles to further progress in the field. Chief among these obstacles was the problem of achieving sure access to the specific level. A fire analysis of a generic mixture, Berzelius warned repeatedly, indicated nothing fundamental. A fire analysis of a specific compound, however, did indicate something fundamental, the definite proportions of the elements that defined the compound and distinguished it from all others.

Isolation of specific compounds and determination of their composition by elemental analysis were preconditions for the emergence in the 1830's of the "older radical theory" of organic chemistry.[3] Once specific compounds had been isolated and defined stoichiometrically, it became possible to observe reactions connecting distinct compounds of comparable complexity. To distinguish these reactions from both decompositions by art and syntheses by nature, contemporaries designated them as "metamorphoses." In the older radical theory, the radicals, though not isolated, were identified by reflection on the progress of metamorphoses, just as in Stahl's chemistry the principles had been identified by reflection on the progress of reactions involving mixts. Like Stahl's principles, the radicals were observed to undergo displacement from one compound to another.[4]

The interpretation of metamorphoses as migrations of compound radicals rested on conceptual foundations provided by Berzelius long before the older radical theory was articulated. Once chemists had isolated specific compounds,

[3] For the older radical theory, see Ernst von Meyer, *Geschichte der Chemie von den ältesten Zeit bis zur Gegenwart* (Leipzig: 1889), pp. 205—212; Hjelt, *Organische Chemie*, Ch. 5; and *idem*. "Berzelius—Liebig—Dumas. Ihre Stellung zur Radikaltheorie 1832—1840," *Samml. chem. u. chem.-tech. Vorträge*, 12 (1908), 447—482.

[4] Cf. Hélène Metzger, *Newton, Stahl, Boerhaave et la doctrine chimique* (1930; rpt., Paris: 1974), p. 120.

analyzed them, and observed their reactions, they could turn to Berzelius's chemistry for the theoretical postulates that undergirded the older radical theory. Berzelius's ideas in turn rested on precedents in the chemistry of Lavoisier, which are explored in the subsections that follow. Most significant among these precedents were Lavoisier's novel theory of fire analysis and his proposal of a radical theory of organic chemistry parallel to that of his inorganic chemistry.

b. Fire analysis and organic composition: Lavoisier's novel views

1) *Introduction*

Lavoisier's ideas about organic composition emerged as the theoretical underpinning for his investigation of organic fire analysis. With these studies he solved the problem of the extent to which fire analysis created its products instead of revealing them. As soon as the composition of water became known, Lavoisier used this knowledge to deploy his oxygen theory of combustion in a study of organic distillation. Water was the source of two elements, either or both of which could be added to the elements of the material analyzed, producing new combinations. Thus, distillation was a creative rather than analytical process. The quantities and proportions of the elements in the new products were determined by the reaction conditions. To sustain his observations about the character of the decompositions, Lavoisier also articulated new ideas about chemical combination peculiar to organic matter.

As expounded in the *Traité* and in memoirs that preceded it, Lavoisier's leading ideas about organic matter and organic decomposition provided major precedents for Berzelius's work in organic chemistry and laid the conceptual foundations for the organic chemistry of much of the nineteenth century. To analyze these ideas and to investigate their origins are the objects of this subsection. The ideas may be summarized as a set of five principal contentions.

First, organic compounds, whether of vegetable or animal origin, were not considered to be composed of proximate constituents, themselves compound, that endured in the compound and were revealed by analysis. Rather, they were considered to be immediate compounds of each of the elementary constituents with all the others. Each constituent exerted its affinity directly upon all the others, resulting in a state of static equilibrium among the ingredients. In the nineteenth century, this arrangement was called a "unitary" compound to distinguish it from substances composed of proximate constituents, and this term is employed in the present chapter. Second, the equilibrium of the constituents was thought to be stable only under mild conditions—that is, affinity effects were weak—and could be readily upset by even modest heating. Decomposition then ensued, producing first simpler organic products and ultimately simple inorganic substances. Third, decompositions of organic substances, especially when caused by heat, were the striking phenomena, accessible both intellectually and practically. They took place by addition of new material, either oxygen or the elements of water, and rearrangement of the material undergoing decomposition. Accordingly, fire analysis did not merely separate proximate constituents but added new ingredients and created new substances. Fourth, synthesis was impossible: the chemist did not have it in his power to create a substance characterized by unstable equilibrium among several elementary constituents.

Though the four ideas just listed distinguished organic matter from the inorganic, the fifth idea asserted that organic compounds shared one characteristic of the inorganic. Because organic compounds typically contained oxygen, they could be conceived as oxides of compound radicals and classified in the same way as inorganic oxides. With this approach, the first radical theory of organic chemistry, Lavoisier hoped to treat organic compounds in terms of a two-component theory in which the radicals were the specific components. The concept of organic radicals conflicted with the unitary conception, because, at least conceptually, the radical theory regarded the oxygen as combined with the radicals as wholes, as proximate constituents, not directly with the elements, carbon and hydrogen. Nevertheless, by the 1830's it became possible to interpret many organic reactions as migrations of radicals, whether or not the radicals could be isolated.

2) *Early foundation of Lavoisier's thought*

a) The chemical, carbon

The genesis of these ideas depended upon Lavoisier's contribution to the recognition that carbon was a chemical and that it played the role of an element in organic composition. The notion of the widespread existence of carbon as an always self-identical, ontologically secure chemical, one that was a proximate ingredient of many substances and that could be extracted from them predictably and used in synthesis, emerged only in the late eighteenth century. Recorded in the 1787 *Méthode* of Lavoisier and his collaborators and in Lavoisier's *Traité* by the appearance of carbon in the list of simple substances, this insight was further developed shortly thereafter by recognition of the chemical identity with carbon of the two forms now called allotropes, diamond and graphite. Previously, substances in which the modern mind recognizes carbon as an ingredient, such as plant and animal substances, or as the principal constituent, such as coal, charcoal, and burned sugar, had been lumped together in the category of combustibles along with non-carbon-containing substances such as sulfur. During the eighteenth century, their common susceptibility to combustion was seen as the result of their common possession of a principle of combustibility. Even what may now be recognized as the purest samples of carbon were believed to consist of phlogiston in combination with some other substance. Though the various threads leading from this conception to that of elementary carbon are too numerous, too complex, and too removed from the concerns of this study to be considered here,[5] Lavoisier's contribution was a major one and must be analyzed.

In brief, Lavoisier recognized the chemical carbon by his understanding that the gas then called "fixed air" (carbon dioxide) was produced by the combination of oxygen with the carbonaceous matter obtained from charring vegetables. That is, he recognized that carbon was the specific component of the acid gas, "fixed air." Previously, the specific component was uncertain, because in several other methods of obtaining fixed air, for example by effervescence of mild alkalies with acids, the presence of carbonaceous matter was not immediately evident. With Lavoisier's insight, it became clear that whatever else produced that gas must also contain carbonaceous matter. In the 1775 "Mémoire sur la nature du principe qui se combine avec les métaux pendant leur calcination & qui en augment le poids,"[6] Lavoisier showed that revivification of metallic calces in the presence of charcoal consumed the charcoal and produced fixed air. The air was indentical to that obtained by revivification of *minium,* from combination of calcareous earths or effervescent alkalies with acids, and from fermentation of vegetable matter.[7] In addition, he pointed out that the gas was a compound of charcoal and the principle fixed in the calx, pure air, later called oxygen. In the 1777 "Mémoire sur la combustion des chandelles,"[8] he announced his acceptance of the name that Bucquet had given this gas, the "chalky acid,"[9] and remarked that it could be produced by union of pure air with the inflammable air (carbon monoxide) given off by a candle flame.[1] In his paper of the same year on Homberg's pyrophorus,[2] Lavoisier again maintained that the chalky acid was a combination of the carbonaceous matter found in the pyrophorus (or in the charred sugar that he had used in its stead) and pure air. He cited as proof that the revivification of mercury calx by addition of carbonaceous matter produced the acid.[3] He added that the inflammable air obtained in the combustion of the pyrophorus (again, carbon monoxide) was distinct from that obtained in metallic dissolutions.[4]

Reiterated in the 1777 "Mémoire sur la combustion en général,"[5] this view of carbonaceous matter as the base of the chalky acid reached a temporary cul-

[5] For a general discussion of the emergence of carbon as a chemical and as an element, see T.A. Wertime, "The discovery of the element carbon," *Osiris,* 11 (1954), 211—220; cf. also Mary Elvira Weeks and Henry M. Leicester, *Discovery of the elements* (7th ed., Easton, Pa.: 1968), pp. 70—74.

[6] For citation of Lavoisier's memoirs in this study, see p. 41, n. 2. The memoir here cited appears in *Oeuvres,* II, pp. 122—128.

[7] Cf. Lavoisier's sealed note of 24 March 1774, for which, see Partington, *History,* III, p. 403.

[8] Lavoisier, "Mémoire sur la combustion des chandelles dans l'air atmosphérique et dans l'air éminemment respirable" (1777), in Lavoisier, *Oeuvres,* II, pp. 184—193.

[9] *Ibid.,* p. 185.

[1] *Ibid.,* p. 193.

[2] Lavoisier, "Expériences sur la combustion de l'alun avec les matières charbonneuses et sur les altérations qui arrivent à l'air dans lequel on fait brûler du pyrophore" (1777), in his *Oeuvres,* II, pp. 199—208.

[3] *Ibid.,* pp. 206—207.

[4] *Ibid.,* p. 207.

[5] In Lavoisier, *Oeuvres,* II, pp. 225—233.

mination in the "Considérations générales sur la nature des acides,"[6] portions of which were analyzed above (Chapter II, subsection 3.*b*.). Having described in that paper the formation of the acid of sugar by what he believed to be simple addition to sugar of the acidifying principle, Lavoisier noted that by carrying the reaction further, the acid of sugar could be analyzed.[7] Using nitric acid as the source of oxygen, Lavoisier remarked of the latter,

> qu'il commençait par se combiner avec le surce pour former avec lui un acide particulier, et l'expérience dont il est ici question nous apprend que, si, après que l'acide du sucre a été formé, on le pousse à un degré de feu modéré, il se résout presque en entier en air fixe ou acide crayeux aériforme et en air inflammable [the latter similar to that obtained from candles and the pyrophorus]; or qu'est-ce que l'acide crayeux? J'ai fait voir ailleurs que c'était un résultat de la combinaison de la matière charbonneuse avec le principe acidifiant ou oxygine, d'où il suit que le sucre est composé d'un peu d'air inflammable et de beaucoup de matière charbonneuse; cette dernière s'unit avec le principe acidifiant ou oxygine fourni par l'acide nitreux [nitric acid], et forme avec lui la grande quantité d'acide crayeux aériforme qu'on a obtenue sur la fin de l'opération.[8]

Carbonaceous matter, the base of the chalky acid, was a constituent of a vegetable substance, sugar. This perception led Lavoisier to suggest in his summary that carbonaceous matter was a stable enduring constituent of vegetables, revealed, but not created by fire. From the experiment just described it followed

> Qu'il ne serait pas impossible ... que la matière charbonneuse fût toute formée dans les végétaux, et qu'elle ne fût point l'ouvrage du feu, comme le chimistes l'ont pensé jusqu'ici.[9]

The expression "tout formé" was Lavoisier's term to indicate that the substance pre-existed as an enduring ingredient of the material analyzed. It was already formed before the analysis and was not a creation of the decomposition. Lavoisier was realizing that carbon was one of the three apparently elementary constituents of organic matter.

Lavoisier solidified his view that the chemical carbon was the base of the chalky acid in some experiments deriving from his studies with Meusnier on the decomposition and synthesis of water.[1] Synthesis of water by combustion of

[6] Lavoisier, "Considérations générales sur la nature des acides et sur les principes dont ils sont composés" (1778), in his *Oeuvres*, II, pp. 248—260.

[7] *Ibid.*, pp. 256—258.

[8] *Ibid.*, p. 258.

[9] *Ibid.*, pp. 259—260.

[1] Lavoisier and Jean Baptiste Marie Charles Meusnier de la Place (1754—1793), "Mémoire où on prouve, par la décomposition de l'eau, que ce fluide n'est point une substance simple, et qu'il y a plusieurs moyens d'obtenir en grand l'air inflammable qui y entre comme principe constituant" (1781), in Lavoisier, *Oeuvres*, II, pp. 360—373; henceforth cited as Lavoisier and Meusnier, "Décomposition de l'eau." This memoir followed Lavoisier's own "Mémoire dans lequel on a pour objet de prouver que l'eau n'est point une substance simple, un élément proprement dit, mais qu'elle est susceptible de décomposition et de recomposition" (1781), in his *Oeuvres*, II, pp. 334—359; henceforth cited as Lavoisier, "L'eau décomposable." This paper also reported in part some of the experiments done in collaboration with Meusnier in "Décomposition de l'eau." The experiments with Meusnier were undertaken to remove doubts about the compound nature of water and to oppose

hydrogen having revealed its ingredients, its decomposition could be achieved by placing it in contact with substances having more affinity for oxygen than did hydrogen.[2] Chiefly metals exhibited this greater affinity, and Lavoisier and Meusnier showed that water could be decomposed by using it to calcine metals.[3] Since combustibles ought to have worked as well, they tried charcoal, first heating it to drive out any inflammable air it might have contained, and they succeeded in producing the chalky acid.[4] They claimed that the phenomenon provided the first example of a combustion achieved without the presence of air.[5] With this success behind him, Lavoisier wrote the 1781 "Mémoire sur la formation de l'acide nommé air fixe ou acide crayeux, & que je désignerai desormais sur le nom d'acide du charbon."[6] There he distinguished ordinary "charbon," from "substance charbonneuse," the purified chemical.[7] He studied the combustion of this chemical, of substances containing it, such as wax, and the amounts and composition of the resultant acid.[8] Moreover, in the paper he elaborated the view of aqueous decomposition by carbon and by metals as species of combustion reactions.[9] Indeed, it was this notion that had attracted Lavoisier to organic matter.

b) Organic combustion by aqueous decomposition

Partington points out the principal early papers in which Lavoisier's views of organic matter were taking shape.[1] In these papers, the leading idea was that organic phenomena may be interpreted as combustion processes, taking place by decomposition of water which supplied the oxygen. In particular, Lavoisier realized that vegetative metabolism, fermentations, and similar processes were natural methods for the decomposition of water. In his ideas about the course of the decomposition of water and subsequent oxidation of organic matter lay the germ of his belief that organic substances were not composed of compound proximate constituents, but were immediate unions of all their elements with one another, that is, were unitary substances.

The first of the papers leading to this view was the 1781 "Mémoire dans lequel on a pour objet de prouver que l'eau n'est point une substance simple, un élément proprement dit, mais qu'elle est susceptible de décomposition et de recomposition."[2] Citing, among other precedents, Bergman's studies of the calcination of iron by water,[3] Lavoisier recounted in this memoir several ex-

arguments that had been raised against it (for this, see Lavoisier and Meusnier, "Décomposition de l'eau," pp. 360—361).

[2] Lavoisier and Meusnier, "Décomposition de l'eau," p. 361.

[3] *Ibid.*, pp. 362—365; cf. Lavoisier, "L'eau décomposable," pp. 341—343.

[4] Lavoisier and Meusnier, "Décomposition de l'eau," pp. 370—371.

[5] *Ibid.*, p. 371.

[6] In Lavoisier, *Oeuvres*, II, pp. 403—422.

[7] *Ibid.*, p. 404.

[8] *Ibid.*, pp. 404—412.

[9] *Ibid.*, p. 418.

[1] Partington, *History*, III, pp. 468—469.

[2] See above, p. 127, n. 1.

[3] Lavoisier, "L'eau décomposable," p. 341.

periments in which the oxygen of the water was removed to perform calcinations, the hydrogen becoming free.[4] Still other means for the decomposition of water could be found: "... la nature nous en offre un grand nombre d'autres et nous n'avons à cet égard, qu'à suivre ses operations."[5] These methods included the processes of vegetation, in which not the hydrogen, but the oxygen became free. The hydrogen remained in combination to form the combustible matter of plants. The combustible matter of vegetables was not phlogiston-containing, but consisted of unions of hydrogen and carbonaceous matter capable of undergoing oxidation.[6]

Vinous fermentation, too, was a process involving decomposition of water, but in this case, both constituents were consumed:

> La fermentation spiritueuse est encore un moyen de décomposer l'eau par la voie humide: le sucre, comme je l'ai fait voir, contient une quantité très-considérable de matière charbonneuse toute formée; puis donc que la matière charbonneuse a plus d'affinité avec le principe oxygine que ce dernier n'en a avec le principe inflammable aqueux [hydrogen], puisque, en vertu de cet excès d'affinité, le charbon décompose l'eau par la voie sèche, pourquoi ne la décomposerait-il pas par la voie humide?[7]

As the oxygen left the hydrogen and combined with the carbonaceous matter, the hydrogen was not liberated as a gas, but like the oxygen, combined with the same carbonaceous matter:

> Il paraît donc que, dans la fermentation spiritueuse, la matière charbonneuse du sucre ou du corps sucré se combine avec le principe oxygine de l'eau, et que le principe inflammable aqueux, devenu libre [from its combination with oxygen], se fixe dans la combinaison en s'unissant avec une portion assez considérable du principe charbonneux, et que c'est ce principe inflammable qui forme la partie spiritueuse, l'esprit-de-vin. La décomposition de l'eau dans la fermentation spiriteuse se fait donc en vertu d'une double action: d'une part, la matière charbonneuse tend à se combiner avec le principe oxygine; de l'autre, cette même matière charbonneuse tend à se combiner avec le principe inflammable aqueux.[8]

Lavoisier went on to cite evidence in support of the existence of affinities between the carbon of the carbonaceous matter and both oxygen and hydrogen.

> Cette double combinaison me paraît déjà établie par des expériences décisives; celle du principe oxygine avec le charbon est prouvée par la quantité énorme d'air fixe qui se dégage pendant la fermentation; or on ne peut plus douter aujourd'hui que l'air fixe ne soit un composé de principe charbonneux et de principe oxygine: la combinaison du principe inflammable aqueux avec la matière charbonneuse est prouvée, parce que l'esprit-de-vin, en brûlant, donne de l'air fixe; donc il contient le principe charbonneux, qui seul forme de l'air fixe en brûlant.
>
> L'existence du principe inflammable aqueux dans l'esprit-de-vin n'est pas moins certaine, parce qu'il se reforme de l'eau dans sa combustion; or il n'y a que le principe inflammable aqueux qui, combiné avec le principe oxygine, ait cette propriété.[9]

[4] *Ibid.*, pp. 341—351.
[5] *Ibid.*, p. 355.
[6] *Ibid.*, pp. 355—356.
[7] *Ibid.*, p. 356.
[8] *Ibid.*
[9] *Ibid.*, pp. 356—357.

Lavoisier's belief that both the oxygen and the hydrogen were in combination with the carbonaceous matter, simultaneously, implies that he did not view the decomposition of sugar in vinous fermentation as proceeding by simple addition of water to the carbonaceous matter of sugar and its persistence, as water—as a proximate constituent—in the combination. Instead he believed that the water was decomposed, and that its elements, as such, combined with the carbonaceous matter, and therefore that spirit-of-wine was a unitary compound.

c) Decomposition by addition

Lavoisier's interest in organic matter had been sparked by his perception of certain organic processes as means for the decomposition of water and for the combustion of organic matter with the oxygen thereby made available. Though this idea led him in the direction of adopting a unitary view of organic compounds, he did not pursue this implication readily. His views matured slowly through a series of papers. Before this evolution may be analyzed, one other ingredient in Lavoisier's thinking must be signalled: in all these cases of organic decompositions, organic substances underwent change by *addition* of something: either the oxygen of the water, the hydrogen of the water, or both. In this respect, they were fully equivalent to direct combustions, for the weight of the products exceeded that of the organic substances employed by the weight of the material added. Having adduced the products of the combustion of spirit-of-wine as proof of the affinity of carbonaceous matter with hydrogen and oxygen in the 1781 memoir on the decomposition of water, Lavoisier presented a brief account of the process of its combustion in which he emphasized the aspect of addition:

On peut brûler ... autant d'esprit-de-vin qu'on le juge à propos, et chaque livre de seize onces donne, quand on opère avec toutes les précautions convenables, dix-huit onces quatre à cinq gros d'eau très-pure, ce qui fait deux onces et demie d'augmentation par livre; c'est à très-peu un septième.
... nous ne verrons autre chose, dans cette expérience, que la preuve qu'il s'ajoute quelque chose à l'esprit-de-vin dans sa combustion, et que ce quelque chose est de l'air. Nous en concluons que l'augmentation de poids, la fixation de l'air, est un phénomène général de toute combustion; que tout concourt à prouver que la partie inflammable de l'esprit-de-vin est toute formée dans l'eau, qu'il ne s'agit que de la dégager d'avec le principe oxygine avec lequel elle est combinée; enfin, que l'eau est un composé du principe oxygine uni à un principe inflammable.[1]

This emphasis on the quantitative aspect of organic combustion was immediately juxtaposed, in the conclusion of the paper, with a similar comment on fermentation:

Une autre circonstance très-remarquable de la fermentation spiritueuse, c'est que, si on en rassemble soigneusement les produits, on voit clairement qu'en réunissant le poids d'air fixe qui s'est dégagé, celui de la portion du sucre qui reste sans être décomposée,

[1] *Ibid.*, p. 358.

enfin, la partie spiritueuse, on a un produit en poids beaucoup plus considérable que celui du sucre qu'on a employé, tandis qu'au contraire on trouve un manquant égal sur le poids de l'eau.[2]

In its application to organic analysis, this understanding of the role of addition was a major novelty, for the antecedent debate about the possibly creative role of fire considered only the reorganization of matter already present, not the addition of new matter. According to Lavoisier, organic decomposition, like combustion and calcination, proceeded by addition. He later showed that, in general, both addition and rearrangement occured, and that they did so on the chemical, rather than sub-chemical level of material organization.

3) *Unitary view of organic matter*

By pursuing this quantitative line of thought about organic change, Lavoisier was moved further in the direction of his unitary views of organic composition, but he had to struggle to detach himself from other modes of thought. In a 1784 paper, for example, while presenting a quantitative study of the combustion of alcohol and other organic products, Lavoisier claimed that spirit-of-wine did indeed contain water as such.[3] The object of the paper was to demonstrate simultaneously the theory of combustion and the compound nature of water by showing that the weight of the material burned plus that of the oxygen used equaled the weight of the products.[4] For the case of spirit-of-wine, Lavoisier employed known amounts of alcohol and oxygen, measured the resultant carbonic acid, and deduced the weight of the water formed by calculation. The weight of the water equaled the weight of the starting materials less that of the carbonic acid evolved.[5] Lavoisier also showed, by subtracting the oxygen of the carbonic acid from the total amount of the oxygen used in the experiment, that the oxygen remaining, that derived from the spirit-of-wine, was insufficient to produce the quantity of water obtained. Instead of merely suggesting, to explain this, that the elements of the remaining water were available in the spirit-of-wine, he assumed that the excess water was contained in it "toute formée."[6] In this view, alcohol appeared to be composed of two parts, a unitary organic compound and water. This idea persisted in Lavoisier's next (1786) paper on oxidation of organic substabces by aqueous decomposition,[7] even though Lavoisier there gave voice to a unitary view of other organic substances, but it was abandoned by the time of the *Traité*.[8]

[2] *Ibid.*

[3] Lavoisier, "Mémoire sur la combinaison du principe oxygine avec l'esprit-de-vin, l'huile et différents corps combustibles" (1784), in his *Oeuvres*, II, pp. 586—600; henceforth cited as Lavoisier, "L'esprit-de-vin."

[4] *Ibid.*, pp. 587—588.

[5] *Ibid.*, pp. 589—594. Cf. Partington, *History*, III, p. 470.

[6] Lavoisier, "L'esprit-de-vin," p. 592.

[7] Examined here shortly.

[8] As Partington (*History*, III, p. 480) points out.

Lavoisier's persistence in seeing water as a proximate constituent of spirit-of-wine detoured, but did not derail the rain of his thought toward a unitary view. His route was determined by his belief that organic decompositions took place by addition, either in direct combustion or aqueous decomposition, as may be demonstrated by examination of his 1786 "Réflexions sur la décomposition de l'eau par les substances végétales & animales."[9] In this paper, Lavoisier developed the conclusion that if organic change took place by addition, the compounds, especially carbon dioxide, produced from analysis by thermal decomposition could not be seen as the ingredients of organic matter. He began by investigating the calcination of carbon itself, that is, the phenomena occasioned by exposure of carbon to violent heat.[1] Showing that the carbon consumed in the process combined with water that it had acquired from the atmosphere, Lavoisier applied the results of this study to organic decompositions, of which he claimed the calcination to be the core. Fixed air and hydrogen gas, Lavoisier indicated, were the products of the calcination of carbon, which left the residual carbon modestly reduced in weight. When exposed to the air, the carbon regained the weight. Calcination again produced more carbonic acid and hydrogen. The process could be repeated until all the carbon was consumed. Noting that previous observers of this process had not recognized that the products far outweighed the carbon employed, Lavoisier indicated that an addition must be assumed to take place during the exposure of the carbon to the air. Since the atmosphere contained only air and water, he recalled, one of these had to be responsible for the augmentation in weight. He showed that if it was conducted in dry air, the calcination produced no hydrogen and that calcination of moist carbon out of contact with air produced both hydrogen and carbonic acid in abundance. Quantitative investigation of these phenomena confirmed the decomposition of water. "Cette expérience, dans laquelle il n'entre que deux substances," Lavoisier observed, "m'a éclairé sur des distillations beaucoup plus compliquées, dans lesquelles on obtient également des quantités très-considérables de gaz acide carbonique et de gaz hydrogène."[2]

Such "more complicated" processes were the dry distillations of vegetable matter conducted by Hales, who had claimed that carbon dioxide was an ingredient of vegetable matter (as noted above, in Chapter I, subsection 5.*c*.). Repeating them, Lavoisier stated that he always obtained a mixture of carbonic acid and hydrogen. The relative quantities of these products were nearly the same for all vegetable matter and were approximately equal to those obtained by the distillation of water and carbon. Only vegetables containing "huile toute

[9] In Lavoisier, *Oeuvres*, II, pp. 656—670; henceforth cited as Lavoisier, "Décomposition organique."

[1] For the discussion of the calcination of carbon, Lavoisier, "Décomposition organique," pp. 656—659

[2] *Ibid.*, pp. 659—660.

formée'' gave more hydrogen.[3] Accordingly, dry distillation of vegetable matter was essentially the same process as the calcination of moist carbon:

> Une conformité aussi grande dans les résultats annoncait une identité dans la cause qui les produisait, et je n'ai plus douté, dès lors, qu'une grande partie du gaz hydrogène et du gaz acide carbonique qui se dégagaient lorsqu'on distille les végétaux à feu nu ne fussent un effet de la décomposition de l'eau, et que la matière charbonneuse ne fût toute formée dans les végétaux, comme je l'avais annoncé dès 1778, et je n'ai plus vu, dans la décomposition des végétaux par le feu, qu'un jeu de l'affinité de l'oxigène qui entre dans la composition de l'eau, et qui quitte le gaz hydrogène pour s'unir au charbon, et former l'acide carbonique.[4]

In support of this opinion, Lavoisier showed that slow heating of vegetable matter, a process that drove out the available water before the affinity of carbon for oxygen was sufficiently enhanced to decompose the water, produced almost no carbonic acid and little hydrogen; but rapid heating produced the gasaeous products in abundance.[5] With this demonstration Lavoisier could deny the opinion that the products of organic analysis were the proximate constituents of organic compounds:

> Ces expériences, ou plutôt les conséquences naturelles qui en résultent, renversent entièrement le système que le docteur Hales et après lui un grand nombre de physiciens s'étaient formé sur la constitution des végétaux. On s'était persuadé, d'après la quantité énorme des fluides élastiques qui s'en dégagent lorsqu'ils se résolvent dans leurs principes, que l'air était le ciment des corps, que c'était lui qui liait entre elles les molécules des autres éléments. Nous voilà aujourd'hui forcés de reconnaître que cet air fixe, l'acide carbonique, auquel on faisait jouer un si grand rôle, n'existe même pas dans les végétaux, et qu'il est un produit, un résultat de la distillation; en sorte que les anciens supposaient dans les végétaux ce qui n'y est pas, tandis qu'ils n'y reconnaissaient pas le charbon tout formé, qui cependant y existe.[6]

4) *Complex composition*

If Lavoisier had concluded from his studies of organic decomposition that organic substances did not contain (and therefore were not composed of) the compounds produced from their decomposition, then how was their decomposition to be described? To answer this question, Lavoisier cited the example of sugar:

> L'analyse la plus rigoreuse n'y découvre, en dernier résultat, que de l'eau et de charbon, autrement dit, que de l'hydrogène, de l'oxygène et du carbone; ... le point important serait de connaître l'ordre d'après lequel ces principes sont combinés entre eux, et voici l'idée que je m'en forme. Il paraît d'abord qu'il y a dans le sucre une portion d'oxygène et d'hydrogène combinés dans l'état d'eau, qui n'est pas essentielle à la constitution du sucre, et qui forme, en quelque façon, son eau de cristallisation; mais le sucre contient en outre une grande quantité d'oxigène et d'hydrogène unis au carbone, et qui paraissent former une combination triple. Cette combinaison, qui s'opère par la végétation, et que l'art ne paraît pas encore avoir imitée, est très-commune dans le règne végétal; elle est en

[3] *Ibid.*, p. 660.
[4] *Ibid.*
[5] *Ibid.*, pp. 661—662.
[6] *Ibid.*, p. 662.

général très-commune sous le nom *corps sucré, corps muqueux,* etc. Le charbon est dans un excès considérable dans ce genre de combinaison; à l'égard de l'hydrogène et de l'oxigène, ces deux principes y sont à peu près dans la proportion nécessaire pour constituer l'eau; il y a seulement un légér excès d'oxygène. Ainsi, quoique le sucre, et en général les matières végétales, contiennent les matériaux de l'acide carbonique, ceux de l'huile et ceux de l'eau, elle ne contiennent réellement aucune de ces substances toute formée, parce que ces principes n'y sont point combinés deux à deux, mais, comme je l'ai déjà dit, qu'ils y forment une combinaison triple.[7]

Vegetable substances were not agglomerations of binary unions but were unitary compounds of all their elements. The view that those elements could be present in proportions suitable to form some of the typical decomposition products permitted Lavoisier to account, as he pointed out,[8] not only for decompositions by addition, but also for simple destruction of the compound by removal of ingredients.

A process of this kind was the thermal decomposition of sugar, which Lavoisier went on to describe assuming first slow, and then rapid heating.[9] In the case of slow heating, Lavoisier indicated, the oxygen and hydrogen left their triple compound with carbon and united with one another to form water. As the water distilled, it carried off the excess oxygen of the sugar and left an oily acid formed chiefly from the residual carbon and hydrogen, called *acide syripeux,* or, in the new nomenclature, *pyromucique.* The acid carried with it some oil, some carbonic acid arising from decomposition of some of the water by carbon, and some hydrogen gas, this last carrying what Lavoisier took to be carbon in solution (it was carbon monoxide gas, which Lavoisier had ceased to regard as a distinct chemical). The residue was carbon. In the case of rapid heating, on the other hand, more oxygen combined with carbon, producing much carbonic acid and evolving a greater quantity of carbon-laden hydrogen gas. The yield of carbonic acid and hydrogen could be increased at will, Lavoisier noted, by recohobation of the water and any pyro-mucic acid formed with the residual carbon. Last, Lavoisier asserted that the problem of the conversion of sugar into oil could be resolved by perceiving the process as an abstraction of the oxygen from the sugar.[1] Lavoisier, then, was able to account for the variation in the decomposition products with the conditions of the reaction by rejecting the assumption that the products were present as ingredients and by holding that only the constituents of the products were available. These constituents underwent separation and recombination into new substances, the identity of which was determined by the conditions of the decomposition.

5) *Weak affinity*

The views just summarized were a powerful means for explaining organic decomposition by addition and subtraction of ingredients. In both cases, fire

[7] *Ibid.*, pp. 662—663.

[8] *Ibid.*, p. 663.

[9] *Ibid.*, pp. 663—664.

[1] *Ibid.*, p. 664.

created the compounds produced by analysis, doing so by separation and recombination of elementary ingredients. Its action, that is, was on the chemical level of organization. To sustain this view, however, Lavoisier needed to articulate suitable notions of the character of organic combinations. He pointed to the need for recognizing that different modes of combinations prevailed in organic matter:

> On objectera peut-être que, si cette explication est vraie, on devrait obtenir de l'huile en recohobant de l'eau sur du charbon; et, en effect, dans cette opération, le carbone décompose l'eau, il lui enlève son oxygène pour former de l'acide carbonique; alors l'hydrogène qui devient libre se trouve en contact avec le carbone, et il semble qu'il devrait se combiner avec lui et former de l'huile. Cette objection m'avait d'abord paru sérieuse; mais la réflexion a bientôt dissipé ce qu'elle m'avait d'abord presenté d'imposant. On ne peut conclure qu'un effet arrivera constamment qu'autant que les circonstances dans lesquelles il a coutume de se produire seront absolument les mêmes, et c'est ici que ce principe trouve son application. Quoique l'eau et le charbon contiennent tous les mêmes principes que le sucre, de l'eau et du charbon ne sont pas du sucre, et j'en ai déjà donné la raison: c'est que le sucre est une combinaison triple de carbone, d'oxygène et d'hydrogène, et que le mixte n'est plus le même dès que les principes se sont combinés deux à deux. Or il ne faut qu'un degré de chaleur très-médiocre pour détruire la combinaison triple, et à ce degré de chaleur il peut se former de l'huile; il n'en est pas de même de la décomposition de l'eau par le charbon: elle ne peut avoir lieu qu'à une chaleur rouge, et non-seulement cette chaleur est supérieure à celle nécessaire pour former les huiles, mais elle suffit même pour les décomposer. Il n'est donc pas étonnant qu'il se forme de l'huile dans la décomposition du sucre, qui se fait à un feu doux, et qu'il ne s'en forme pas dans la décomposition de l'eau par le charbon, qui exige un feu vif.[2]

Likely because of his lack of sympathy with affinity theory, Lavoisier did not offer these remarks in terms of affinities. As noted in Part II, however, Berzelius articulated these same points in the language of affinities and used them as criteria to distinguish organic matter from inorganic. Organic compounds were unitary entities, in which the affinity of each elementary constituent was exerted upon all the others. Those containing three elements were ternary compounds; those containing more were higher compounds. Present in a relatively unstable equilibrium, the affinity forces were subject to disruption by even mild heating, resulting in the patterns of decomposition that Lavoisier described in this passage. Here was the source of a fundamental distinction between organic and inorganic matter that remained a dominant theme of organic chemistry through much of the nineteenth century. Intimately connected with it was another notion: that the generation of these multiple compounds sustained by weak affinities was the work of nature, not of art.[3]

[2] *Ibid.*, pp. 666—667.

[3] Fourcroy also articulated the view of organic matter as composed of three elements forming unitary compounds dependent upon an equilibrium that was easily upset. He also argued that nature, not man, could make these complex compounds. For these ideas, see the passages on "immediate principles" cited above, p. 119, in n. 3. For Fourcroy's ideas on the vegetative processes that produced organic compounds, see his *Connaissances chimiques*, VIII, pp. 287—317.

6) *Organic products the work of life processes*

Adumbrated in the last quoted passage, the contention that only nature, not art, could accomplish organic synthesis was made explicit at the conclusion of the paper on the decomposition of water by organic matter. To this end, Lavoisier recalled a vocabulary that he claimed belonged to the old doctrine of mixtion and to its corresponding theory of affinities. Each constituent of a mixt offered a "side" for attack by the reagents with which it experienced a mutual affinity:

> Je terminerai ce mémoire par quelques réflections relative à la végétation. Toutes les fois qu'on se propose de décomposer une substance formée de la réunion de deux principes [as in inorganic chemistry], l'on peut attaquer séparément l'un ou l'autre de ces principes: c'est ce qu'experimaient les anciens chimistes, lorsqu'ils disaient que les mixtes ont différentes côtés, différents *latus*, et que les combinaisons se forment par les *latus* analogues. L'eau étant composée de deux substances, l'oxigène et l'hydrogène, elle est susceptible d'être décomposée par l'un ou l'autre de ces *latus*. Dans toutes les combustions, soit ardentes, soit obscures [the vocabulary is one that Lavoisier coined in this paper to distinguish an ordinary combustion from one taking place by decomposition of water or acids], c'est principalement par ce *latus* de l'oxigène que s'opère la décomposition; mais il est une opération de la nature dans laquelle cette même décomposition s'opère par le *latus* opposé, ou plutôt par une double affinité, c'est la végétation.[4]

As far as vegetable matter was concerned, "Nature makes, man breaks." Elaborating this point, Lavoisier went on to suggest the character of the natural process:

> Pour se faire une idée de ce qui se passe dans cette grande opération, que la nature semblait avoir, jusqu'ici, environnée d'un voile épais, il faut savoir qu'il ne peut y avoir de végétation sans eau et sans acide carbonique; ces deux substances se décomposent mutuellement dans l'acte de la végétation, par leur *latus* analogue; l'hydrogène quitte l'oxigène pour s'unir au charbon, pour former les huiles, les résines, et pour constituer le végétal; en même temps, l'oxigène de l'eau et de l'acide carbonique se dégage en abondance, comme l'ont observé MM. Priestley, Ingenhousz et Sennebier, et il se combine avec la lumière pour former du gaz oxygène.[5]

By the agency of the life processes, in this case, the process of vegetation, nature could produce the triple and higher compounds of organic chemistry from simple inorganic substances. It was not within the power of chemists to do so. The same considerations, Lavoisier believed, applied equally well to animal matter. Though Berthollet had argued for the presence of nitrogen as a character of animal as opposed to vegetable matter,[6] Lavoisier did not believe that the presence of a fourth element rendered animal matter fundamentally different from vegetable matter. Recognizing Berthollet's ideas, he was undeterred by them from re-

[4] *Ibid.*, p. 669.
[5] *Ibid.*, p. 670.
[6] For Berthollet's and Fourcroy's (later) views on the distinctions between vegetable and animal matter, see D.C. Goodman, "The application of chemical criteria to biological classification in the eighteenth century," *Med. Hist.*, 15 (1971), 23—44, esp. 27—34.

garding decomposition of animal matter as more complex, but not different in kind, from that of vegetable matter:

> Toutes ces réflexions s'appliquent également aux matières animales comme aux végétales; ces substances sont également le résultat d'une combinaison triple d'oxigène, d'hydrogène et de carbone; elles ne contiennent ni eau, ni acide carbonique, ni huile toute formée; mais elles en contiennent tous les éléments. Le moindre degré de chaleur, pourvu qu'il soit un peu supérieure à celui de l'eau bouillante, suffit pour réunir l'oxygène et l'hydrogène, l'hydrogène et le carbone, et pour former de l'huile et de l'eau; mais aussi les phénomènes se compliquent un peu davantage, parce qu'il existe, comme M. Berthollet le démontre, un quatrième principe dans les matières animales, l'azote, qui, avec l'hydrogène, forme de l'acali volatil ou ammoniac. L'observation qu'à fait M. Fourcroy, de la conversion de la partie musculaire des animaux en graisse, après un laps de temps très-considérable, dans les cimetières, vient à l'appui de cette théorie: l'oxigène a été soustrait par une circonstance quelconque, et il n'est resté que de l'hydrogène et du carbone, qui sont les matériaux qui composent la graisse.[7]

Though exhibiting differences in composition and therefore in decomposition products, animal and vegetable matter were essentially similar, and shared common characteristics that were distinct from those of inorganic matter. Organic products were multiple compounds of weakly united elements, incapable of artificial synthesis, and subject to ready decomposition. The ultimate source of this distinction lay in the life processes. Organic matter, the product of life function, was distinguished by the characters imposed by its provenance. Such powers lay outside the reach of art.

7) *Limits on artificial control of organic decomposition*

Unable to synthesize organic compounds from simple substances, the chemist attended to what he could accomplish. As Lavoisier's work showed, the decomposition most accessible to quantitative study was the production of simple inorganic substances either by decomposition without addition or by oxidation. Less accessible and less subject to fine control were intermediate decompositions, giving rise to simpler organic products. These decompositions remained the focus of much of organic chemistry through the early decades of the nineteenth century. The doctrine that only decompositions to simpler products were accessible to investigation received its most elegant expression shortly before its downfall after mid-century in Charles Gerhardt's "échelle de combustion."[8]

The effects of Lavoisier's insights about organic decomposition on his views of synthesis were manifest in his revision, in the paper on the decomposition of water by organic matter, of his earlier views on the production of the acid of sugar. Though in the 1778 "Considérations générales sur la nature des

[7] Lavoisier, "Décomposition organique," pp. 665—666.

[8] For which, see Edouard Grimaux and Charles Gerhardt [fils], *Charles Gerhardt [père], sa vie, son oeuvre, sa correspondance, 1816—1856: document d'histoire de la chimie* (Paris: 1900), pp. 336—338.

acides'' he had represented the formation of the acid of sugar as a synthesis by direct oxidation of sugar without decomposition,[9] the work of Scheele and Berthollet had since shown this acid to be a common, but by no means exclusive product of various organic decompositions.[1] Lavoisier accommodated his thinking to this development by describing organic acidification as a decomposition by addition and added these remarks to his discussion of the thermal decomposition of sugar:

> Mais c'est surtout par addition qu'on peut le plus aisément changer la proportion des principes qui composent le sucre: si on l'oxygène soit par l'acide nitrique, soit par l'acide muriatique oxygéné, soit par quelque autre procédé que ce soit, on le convertit en un acide dont la nature varie suivant la proportion de carbone, d'hydrogène et d'oxygène qui s'établit, et qui est ou tartareux, ou oxalique [the acid of sugar], ou malique, ou acéteux.
>
> Loin que cette propriété soit particulière au sucre quand on l'oxigène; elle est, au contraire, commune à presque toutes les substances animales et végétales, comme l'ont fait voir, principalement pour l'acide oxalique, MM. Scheele et Berthollet, non pas que ces acides y soient réellement contenus antérieurement à l'oxygénation, comme ils l'ont supposé; mais le carbone et l'hydrogène entrant dans la composition de toutes les substances animales et végétales, et à peu près dans la proportion nécessaire pour constituer l'acide oxalique, il en résulte qu'on ne peut les oxygéner sans former une quantité plus ou moins grande de cet acide.[2]

The acid of sugar was a common product of oxidation of organic matter, as Berthollet's work had shown, because much organic matter contained its combustible elements in suitable proportions. Oxalic acid was not the only organic product of organic oxidation, however. Using his earlier quantitative study of the combustion of spirit-of-wine, Lavoisier rationalized the generation of several products from oxidation of a single substance.

> L'esprit-de-vin, comme je l'ai fait voir ... est composé d'hydrogène, de carbone et d'eau [still regarded as present ''toute formée''], à la différence des huiles, qui contiennent les mêmes principes (hydrogène et carbone), mais dans une proportion un peu différente, et privés d'eau. Il n'est donc point étonnant qu'en oxygénant l'esprit-de-vin on détruise la partie spiritueuse, qu'il se produise de l'eau, et qu'on forme en même temps de l'acide acéteux et de l'acide oxalique. M. Berthollet me paraît donc s'être trompé, lorsqu'il a conclu que ces acides existaient tout formés dans l'esprit-de-vin; cette liquer spiritueuse n'en contient que les radicaux, et les acides eux-mêmes sont formés par l'acte de l'oxygénation.[3]

Some results of organic oxidation were therefore predictable. Processes having at least partially foreseeable outcomes could be rendered intelligible by recognizing that they proceeded by alteration in the proportions of the chemical constituents—that is, on the chemical level of organization—of which the

[9] For this, see above, Chapter II, subsection 3.b.

[1] For Berthollet on oxalic acid, see Partington, *History*, III, 514; for Scheele, *ibid.*, p. 232, and Olsson, *Kemiens historia*, pp. 353—355.

[2] Lavoisier, ''Décomposition organique,'' pp. 664—665.

[3] *Ibid.*, p. 665.

predominant one was the carbonaceous matter.[4] Hence, Lavoisier indicated that the combustion of spirit-of-wine began with the formation of water by union of the hydrogen with the oxygen employed in the reaction. This combination altered the initial proportion of carbon and hydrogen. To this altered proportion could be referred first, the production of ether, and, as the oxidation proceeded, the production of oxalic, acetic, and other acids.

Though these oxidations could be explained *a posteriori,* none was subject to fine control. Generally more accesible to study were the complete destructions of organic matter. Returning to the original pattern of aqueous oxidation with which Lavoisier had opened this paper, he went on to argue that the predominant proportion of carbon in organic matter was the chief cause of the action of organic compounds on water and explained the production of simpler products, especially inorganic ones. Production of simple inorganic products was also the outcome of fermentation, both vinous, which Lavoisier indicated he was continuing to study, and putrid fermentation of animal matter, which Berthollet was investigating.

Lavoisier had come full circle. The chemist could produce organic compounds not by synthesis from simple products, but by decomposition from more complex ones. Decompositions were intelligible and their outcomes in some measure foreseeable, but their course was capable of only crude control. Only full decomposition of organic matter, whether by art or by nature, was easily achieved and easily studied.

c. A two-component theory of organic matter

In these papers on organic composition all but one of the leading ideas emerged that served as the foundations for the organic chemistry of the early nineteenth century, and, in particular, for Berzelius's thinking: the unitary organization of organic compounds, sustained by weak affinities; the ease and accessibility of decomposition, and the impossibility of laboratory synthesis; the account of decomposition as addition and reorganization of matter on the chemical level; and the common submission of animal and vegetable matter to these principles. All these served to distinguish organic matter from the inorganic. The one notion not yet considered, the analogy of the organic with the inorganic oxides, emphasized the similarity of the two kinds of matter. This idea was nascent in Lavoisier's 1778 discussion of the production of the acid of sugar by oxidation of sugar and in his later duscussions of the production of organic acids as decomposition products of many organic materials. It came to fruition, as suggested earlier, in conjunction with Guyton's radical theory. The concept provided the basis for the integration of Lavoisier's ideas on organic composition with the rest of his chemical thought, and in general it constituted the major foundation for the first tentative stages of the development of nineteenth-century organic

[4] *Ibid.*, pp. 667—668.

chemistry, especially the older radical theory. The integration was developed in the five chapters of the *Traité* devoted to organic chemistry, where Lavoisier expounded his mature views on that subject and on its relations with inorganic chemistry.[5] Only the first of the chapters need be considered here.[6]

The discussion opened with an account of organic compounds as oxides of multiple acidifiable bases and of the comparability, in this respect, of organic with inorganic substances.[7] To signal the similarity, Lavoisier began by pointing to *aqua regia,* as an acidic inorganic combination of two acidifiable bases, exhibiting properties distinct from those of the individual acids.

> Mais ce phénomène, qui n'a été observé que pour l'acide nitromuriatique [dans le règne minéral] se présente continuellement dans le règne végétal; il est infiniment rare d'y trouver un acide simple, c'est-à-dire qu'il ne soit composé que d'une seule base acidifiable. Tous les acides de ce règne ont pour base l'hydrogène & le carbone, quelquefois l'hydrogène, le carbone & le phosphore, le tout combiné avec une proportion plus ou moins considérable d'oxygène. Le règne végétal a également des oxides qui sont formés des mêmes bases doubles & triples, mais moins oxygénés.
>
> Les acides et les oxides du règne animal sont encore plus composés; il entre dans la combinaison de la plupart quatres bases acidifiables; l'hydrogène, le carbone, le phosphore & l'azote.[8]

The nomenclature of these substances, no less than that of inorganic compounds, was to reflect their composition.[9] Since the specific component was compound, its name had to reflect its composition. Beginning with vegetable substances, Lavoisier cited what he considered to be three known oxides "à deux bases," noted above: sugar, mucous, and starch. "Ces trois substances ont pour radical l'hydrogène & le carbone combinés ensemble, de manière à ne former qu'une seule base [acidifiable], & portés à l'état d'oxide par une portion d'oxygène; ils ne diffèrent que par la proportion des principes qui compose la base." Addition of oxygen, Lavoisier remarked, produced acids, which were distinguished by the differing proportions of hydrogen, carbon, and oxygen. Explicitly following Rouelle's precedent, Lavoisier used the names of the two components of the radical to form its specific name, he used their order to exhibit their relative proportions, and he listed the possible names in a table:

Oxide hydro-carboneux.
Oxide hydro-carbonique.

Oxide carbone-hydreux.
Oxide carbone-hydrique.

Acide hydro-carboneux.
Acide hydro-carbonique.
Acide hydro-carbonique oxygéné.

[5] Lavoisier, *Traité*, I, Chs. 11—15.

[6] *Ibid.*, Ch. 11, "Considérations sur les oxides & les acides à plusieurs bases, & sur la composition des matières végétales & animales," pp. 123—131.

[7] *Ibid.*, pp. 123—124.

[8] *Ibid.*, pp. 124—125.

[9] *Ibid.*, pp. 126—127, for this paragraph.

Acide carbone-hydreux.
Acide carbone-hydrique.
Acide carbone-hydrique oxygéné.

The only departure from his practice with inorganic nomenclature was the intentional provision of a systematic name for more than one basic oxide. Having no example of a clearly salifiable organic oxide and no example of a radical with hydrogen in excess, Lavosier was not confronted in organic chemistry with the problem of correlating non-acidic oxides with properties, and therefore he had no reason to segregate salifiable from non-salifiable oxides. In all other respects, the system attributed to each organic binary radical the same degrees of oxidation available to inorganic radicals.

Though the collection of animal and vegetable acids that Lavoisier listed in this same chapter already exceeded the possibilities of this spare system of nomenclature, Lavoisier was confident that his proposal would suffice:

> Il est probable que cette variété de langage sera suffisante pour indiquer toutes les variétés que nous présente la nature, & qu'à mesure que les acides végétaux seront bien connus, ils se rangeront naturellement & pour ainsi dire d'eux-mêmes dans le cadre que nous venons de présenter.[1]

The thought never occured to Lavoisier that the diversity and wealth of organic chemistry would ever outstrip his simple outline of names. Until chemists clearly perceived the generic nature of most organic materials, it remained possible to contemplate a nomenclature of small scope. The notion that most organic substances would be resolved into a profusion of specific compounds was urged on the chemical profession only by Berzelius.

[1] *Ibid.*, p. 127.

PART II
The Berzelian Theory of Salts

Introduction

1. Berzelius's achievements

The genesis and maturation of Berzelius's system is the subject of Part II of this study. Portrayed here as a theory of salts, the system acquired its main outlines by 1813 through Berzelius's studies of inorganic chemistry and received its mature articulation by 1820 through its deployment in mineralogy and organic chemistry. By 1820, Berzelius was forty-one years old, and he had run nearly half the course of his scientific career. Thereafter, his intellectual, professional, and personal life all changed significantly, as Söderbaum points out. The intellectual turning point was marked chiefly by the publication in 1819 of Berzelius's *Essai sur la théorie des proportions chimiques et sur l'influence chimique de l'éléctricité,* which presented a nearly exhaustive account of his full system. Accordingly, the book is a principal target of this investigation.[1]

In the initial stages of its construction (through 1813), the components of the system emerged from efforts to solve the problems of the experimental inorganic chemist. The later deployments of the system in mineralogy and organic chemistry, fields lying outside the one that had fathered the system, posed special problems that stimulated the maturation of the system which culminated in the *Proportions chimiques.* The goal that inspired both the initial construction of the system and its articulation in subsequent deployments was that of specifying compounds. In addressing this issue, Berzelius made several major departures from eighteenth-century precedents. In no case is his theory properly viewed as a mere synthesis of already available theoretical material or as an integration, as explicit doctrine, of prevailing implicit assumption.

Berzelius's concern for specificity is evident in his first experiments with the Voltaic pile (1803) and it remained the constant theme of all his subsequent investigations. This attitude underlay most of Berzelius's differences with La-

[1] Paris. Henceforth cited as *Proportions chimiques.* This book was a slightly revised version of volume III (1818) of Berzelius's textbook, *Lärbok i Kemien,* 3 vols. (Stockholm: 1808—1818), together with tables of atomic weights, formulae and formula weights, and analytical results, originally published separately in 1818 as a supplement to the third volume of the textbook: *Tabell som utvisar vigten af större delen vid den oorganiska kemiens studium märkvärdiga enkla och sammansatta kroppars atomer, jemte deras sammansättning, räknad i procent: Bihang till tredje delen af Lärboken i Kemien* (Stockholm: 1818). Much information about the bibliographic details, literary references, and the like appears in Colin A. Russell, "Introduction to the reprint edition [of the *Proportions chimiques*]," in Berzelius, *Essai sur la théorie des proportions chimiques et sur l'influence chimique de l'électricité* (1819), The Sources of Science, No. 99 (New York: Johnson Reprint Corporation, 1972), pp. [v]—[xlix].

voisier. Though Lavoisier had offered a list of elements, he still focussed on the generic property of acidity; though he presented a list of bases, he focussed on their salification by acids. In sharp contrast, Berzelius, pursuing the implications of his studies with the Voltaic pile, eliminated the generic principle of acidity and focussed on the elementary radicals as the seats of acidity or basicity, the latter now also a positive property. As specific properties, not generic ones, acidity and basicity lay in the specific components, not the generic principles. Similarly, in his theory of salts, the saline components, generally oxides of qualitatively opposing radicals, lay not in a hierarchical, taxonomic relationship, but on the same ontological plane. A salt was a compound of two ontologically equivalent, qualitatively opposing species. The system was a double-entry two-component theory. However, more than qualitative characters defined the specificity of acids and bases. The quantity of their oxygen as well as their electrochemical nature distinguished them from one another. Similarly, the specificity of salts depended not only on the identity of their constituent acids and bases, but on the proportions of these constituents. Electrochemistry determined qualitative specificity; stoichiometry determined quantitative specificity.

Electrochemistry served to generalize the concepts of acidity and basicity. Acidity, as usually understood, was extreme electronegativity; basicity, extreme electropositivity. By calling less markedly reactive oxides weakly electronegative or electropositive, Berzelius was asserting that they could serve as acids or bases, even though they lacked the usual characteristics of acids and bases, and he was maintaining that the compounds of these weakly reactive components could be considered saline, even though they shared few of the properties common to salts of strong acids and bases. Accordingly, electrochemical characterization not only distinguished substances from one another qualitatively; it also labelled substances lacking in saline character—e.g., the chlorides of sulfur—as saline and thus subject to Berzelius's stoichiometric laws for the composition of salts.

Chapters IV and V are devoted to Berzelius's initial construction of the theory of salts. Attention is centered on Berzelius's growing ability to invoke novelties in electrochemistry and stoichiometry in the service of a chemistry of specific substances. It is shown how he was able to account for virtually all laboratory chemicals, even the most complex then known, including some not typically saline. The role of his theoretical apparatus in gaining access to the specific is examined in detail. Aspects of his work that do not bear directly on this issue, however, even if otherwise prominent or interesting, are either ignored or deemphasized. These include the details of Berzelius's microscopic electrochemical ontology of matter; the consequences of the electrochemical theory for the caloric theory of heat; the detailed fate, in Berzelius's hands, of eighteenth-century affinity theory; and, except as necessary to sustain the central themes of the narrative, Berzelius's role in developing the chemical atomic theory.[2]

Berzelius deployed his theory of salts in mineralogy and organic chemistry in hope of specifying compounds that were much more complex than those of the inorganic laboratory. In mineralogy, he allied himself with Haüy as a partisan of definite proportions, but he intended to show that stoichiometry (without reference to morphology) could succeed in specifying minerals. Sustained by his powerful new techniques, he made a bold attempt to remove mineralogy from the sphere of natural history and to bring it decisively under the sway of chemistry. The special problems posed by this attempt, some of which were raised by Berzelius's Wernerian critics, compelled further articulations of his chemistry, especially the mature expression of his doctrine of dualistic combination. Though his invasion of mineralogy could not have been fully successful in the absence of knowledge of isomorphism, his work underlay the clearer victory that was attained after Mitscherlich's discovery. The challenge that mineralogy posed to his chemistry, then, was strong, but not strong enough to repulse his advance. The mineralogical deployment reveals particularly clearly the workings of his theory of salts, that is, of his electrochemical characterizations and his stoichiometric laws, as an instrument for gaining access to the specific level.

The deployment in organic chemistry was less immediately fruitful. Here, too, Berzelius distinguished himself sharply from Lavoisier by formulating the problem of the specific level in organic chemistry and making among the earliest efforts to address it as such. He hoped that stoichiometry could be made to serve the same functions here as it had elsewhere, and he explicitly sought to discover the stoichiometric laws that he expected must govern the composition of organic compounds. He did find stoichiometric regularities, but he realized that, rather than delimiting possible combinations, they permitted any combinations consistent with multiple proportions. Precisely in this circumstance Berzelius saw the major obstacle to attaining the specific level in organic chemistry. There would be no chemical conquest of organic compounds as there had been of minerals. Anxious to circumscribe this area in which his inorganic laws were of no avail and seeking to understand the ontological foundations of unlimited combinations, Berzelius articulated a sharp division between the organic and the inorganic, one that proved characteristic of chemical vitalism in the early nineteenth century. In organic chemistry, qualitative characterizations would have to suffice, even though these permitted access primarily to the generic rather than the specific level. Berzelius did not proscribe a scientific, as opposed to a natural-historical organic chemistry, but he anticipated no early progress in attaining it.

The dominant position that Berzelius's system acquired in the minds of chemists in the first half of the nineteenth century lay in its ability to render

[2] Several of the themes here left out of consideration are admirably treated by Anders Lundgren in his *Berzelius och den kemiska atomteorin*, Acta Universitatis Upsaliensis. Studier i idé- och lärdomshistoria, Vol. II, utgivna av institutionen för idé- och lärdomshistoria vid Uppsala Universitet (Uppsala: 1979); henceforth cited as Lundgren, *Berzelius och atomteorin*.

tractable, both theoretically and practically, these two sciences of complex compounds. Even if, by 1819, Berzelius had not claimed much territory in the region of organic chemistry, he had defined the state of the science. Moreover, when the older radical theory seemed to provide a clear path into a dense thicket, the theory was established on coneptual underpinnings provided by Berzelius's system. Chapter VI examines the mineralogical and organic deployments of the system and exhibits them as efforts to attain the specific level. The further articulations compelled by this work, including the mature expression of Berzelius's dualism and the elaboration of the organic/inorganic boundary, are shown to have flowed from the special problems posed by these disciplines. By analyzing the mature elaboration of Berzelius's chemistry, which guided subsequent research by nineteenth-century practitioners, the concluding chapters of this study set the stage for later analysis of Berzelius's role in the development of the new science of organic chemistry.

The story of Berzelius's labors is not easily told. Most of the difficulties stem from two peculiarities of his thought. A synthetic thinker, Berzelius regurlarly employed any experimental result, any idea, of whatever magnitude, in a variety of contexts, often quite removed from the original one. It is difficult to speak of any narrowly circumscribed set of concerns without constant need to invoke the disparate and apparently unrelated ones with which Berzelius connected them. Moreover, many of Berzelius's ideas were scarcely articulate and protean. Appearing first in one context, they reemerged elsewhere, apparently altered, and in both contexts their full significance may remain obscure. The historian is confronted with the task of analyzing often minimally articulated materials and of observing their changes from one application to another—a difficult and risky affair. Yet such efforts are essential to understanding Berzelius's science.

The analysis that follows breaks down the development of Berzelius's tightly integrated but constantly evolving ideas into four broad and partially overlapping themes:

1) Theoretical and experimental developments to 1819 on electrochemistry, acidity and basicity, and the role of oxygen.
2) Experimental developments through 1811: The *Versuch* and the rule of oxides.
3) Experimental developments through 1819: The tripartite research program.
4) Deployments in organic chemistry and mineralogy: hierarchical dualism and the separation of the organic and the inorganic.

Though none of these themes can be isolated completely from the others, this selection attempts to disentangle the threads and minimize the need for backtracking. By pursuing the four themes, one is led from the early experiments on electrochemistry in 1803 through the articulation of the full system as expounded in the *Proportions chimiques* of 1819.

2. The setting for Berzelius's achievements[3]

When Alessandro Volta announced the invention of the pile in 1800, Jöns Jacob Berzelius was nearly twenty-one years old. Then a student under Johan Afzelius (1753—1837) at Uppsala, where he had begun his studies in 1796, Berzelius immediately began to experiment with the new instrument, first jointly with his half-brother Christoffer Eckmarck, also a student at Uppsala, later in connection with his medical studies. Having obtained a summer appointment in 1800 under Sven Anders Hedin (1750—1821) at the mineral baths at Medevi, Berzelius there investigated the effect of the pile—using one of his own construction—on chronic diseases. The study provided him with the subject of his dissertation *pro gradu medico,* defended in 1802 and leading to the M.D. in 1804. As a preliminary to undertaking his own studies with the pile, examined below, Berzelius published in 1802 a lengthy (145 pages) *Afhandling om Galvanismen,* a review of all prior research concerning the effect of the pile on salts and minerals. He thus poised himself to embark on a series of experiments that had a decisive influence on the subsequent course of his entire scientific career.[4]

In the same year as the review of galvanism, Berzelius obtained a post as unpaid assistant to Anders Sparrman (1748—1820), professor of medicine and pharmacy at the School of Surgery at Stockholm. Then part of the Collegium Medicum, the school was reorganized in 1810 as an independent medical school, *Karolinska Mediko-Kirurgiska Institutet.* In conformity with local tradition, a naturalist, not a chemist, was named to succeed Sparrman when his chair became vacant in 1805, and so Berzelius, who then supported himself as physician to the poor, was passed over. After the new incumbent's unexpected early death, however, Berzelius was appointed to the chair in 1807. He was given a laboratory in the former institution's new quarters and in *Karolinska Institutet* when it occupied new premises in 1816. Elected to the Swedish Academy of Sciences in 1808, Ber-

[3] The fundamental biographical source for Berzelius is H.G. Söderbaum, *Jac. Berzelius: Levnadsteckning,* 3 vols. (Uppsala: 1929—1931). Briefer treatments are Söderbaum's *Berzelius' Werden und Wachsen, 1779—1821,* Kahlbaum's Monographien aus der Geschichte der Chemie, Vol. III (Leipzig: 1899) and J. Erik Jorpes, *Jac. Berzelius: his life and work* (Berkeley: 1970). Also useful is Berzelius's *Autobiographical notes,* ed. H.G. Söderbaum, tr. Olof Larsell (Baltimore: 1934); Swedish, *Jac. Berzelius: Själfbiografiska anteckningar, utgivna af Kongl. Svenska Vetenskapsakademien genom H.G. Söderbaum* (Stockholm: 1901). A detailed list of sources appears in Partington, *History,* IV, pp. 144—147, text and notes. Söderbaum's earlier book is henceforth cited as Söderbaum, *Werden und Wachsen.*

[4] For the intellectual context that Uppsala provided Berzelius, see Lundgren, *Berzelius och atomteorin,* pp. 25—39; Anton Blanck, "Berzelius som medicine studerande," *Lychnos,* 1948—1949 (1950), 168—205; and Gunnar Eriksson, "Berzelius och atomteorin; den idéhistoriska bakgrunden," *Lychnos,* 1965—1966 (1967), 1—37, esp. 15—24.

Berzelius's *Afhandling om Galvanismen* appeared in Stockholm in 1802. For a discussion of Berzelius's early work in electrochemistry and an analysis of the *Afhandling om Galvanismen,* see Lundgren, *Berzelius och atomteorin,* Ch. 2; cf. also Söderbaum, *Werden und Wachsen,* Ch. 2.

zelius became its president in 1810 (occupying the post from February to August) and its secretary in 1819. From the latter date, he occupied both personal quarters and a laboratory in the Academy building. The poor medical student who took summer employment at mineral baths, the unpaid assistant to Sparrman, and later *medicus pauperum* had acquired professional appointments, laboratory facilities to support his work, and honorific distinctions.

Berzelius's accession to the post of secretary also marked the end of the troubled initial period in his relationship with the Academy.[5] His early papers rejected for publication in the Academy's *Handlingar,* in part because they employed the new nomenclature, Berzelius had to struggle to gain recognition from the central scientific institution of his own country while his fame grew abroad. His frustration led him to start a new journal in 1806, the *Afhandlingar i Fysik, Kemi, och Mineralogi,* with the financial support of his friend, collaborator, and patron, Wilhelm Hisinger (1766—1852).[6] Though Berzelius's election to the Academy in 1808 and his presidency in 1810 indicated the start of more cordial relations, he continued to publish outside the *Handlingar,* which he considered in need of reform. Only when Berzelius became secretary, thereby gaining control of the *Handlingar,* did his own journal become unnecessary.

Hisinger's was one of two significant acquaintanceships that Berzelius made as a young man in Stockholm. The other was that of Johan Gottlieb Gahn (1745—1818).[7] Once Bergman's assistant (1760—1770), Gahn pursued a distinguished career in the mining industry of Sweden, made numerous chemical discoveries, and devised much now standard apparatus, but he published little about either. His innovations were developed with industrial applications in view, and his ideas, inventions, and influence entered chemistry largely through the work of his friends and colleagues. Perhaps the most significant scientific influence that these two figures exerted on Berzelius was mineralogical. With Hisinger in 1803 Berzelius discovered cerium in a rare mineral, Bastnäs tungsten, and henceforth he pursued mineralogical studies as a consistent, though not yet principal interest. Envisioning in 1814 a mineralogical deployment of his chemical innovations, Berzelius grew more concerned with mineralogy, visiting the mines at Falun during the summers as Gahn's guest. Like Hisinger, Gahn became both teacher and collaborator of Berzelius. From them Berzelius learned much of what he knew of mineralogy. As Gahn's willing apprentice, Berzelius mastered the

[5] For Berzelius's early relations with the Academy, see Sten Lindroth, *Kungl. Svenska Vetenskapsakademiens historia 1739—1818, utg. av Kungl. Vetenskapsakademien,* 2 vols. in 3 (Stockholm: 1967), II, pp. 70—71, 123—129, 394—403; henceforth cited as Lindroth, *Akademiens historia.*

[6] The *Afhandlingar* appeared in 6 vols. (Stockholm: 1806—1818). In 1808, Berzelius also published in the short-lived *Economiska Annaler* of the Academy (for his involvement with this journal, see Lindroth, *Akademiens historia,* II, pp. 123—129). Berzelius also published many of his papers abroad. For Hisinger, see Lindroth, *Akademiens historia,* II, pp. 390—393.

[7] For Gahn, see Olsson, *Kemiens historia,* pp. 151—152, and Partington, *History,* II, p. 201.

techniques of the blowpipe. Inducing Gahn to collaborate in writing a chapter on the subject for the second volume of his textbook (1812), he later elucidated the techniques in a treatise on the subject, *Om Blåsrörets Användande i Kemien och Mineralogien* (Stockholm: 1820).[8] A minor classic, the work capped a tradition of Swedish development of this little instrument; it was widely translated and went through many editions.

Besides their mineralogical concerns, Gahn and Hisinger shared with Berzelius his early interest in the Voltaic pile. The invention of the pile was, as Davy remarked, an "alarm bell" resounding around Europe.[9] Everywhere, those with scientific interests began investigations with the Voltaic apparatus.[1] Gahn and Hisinger were two leaders of the impromptu Galvanic Society that took form in Stockholm, with which Berzelius, a roomer in a house owned by Hisinger, readily associated himself.[2] It was with Hisinger that Berzelius collaborated in his first significant original scientific study (1803), the first of a series that provided a major ingredient of his chemical system.

[8] For this, see Olsson, *Kemiens historia,* p. 152, and Partington, *History,* III, p. 201, and IV, p. 146.

[9] Quoted by Partington, *History,* IV, p. 16.

[1] For an overview of early electrochemistry, see *ibid.,* pp. 3—28, 40—48, 166—177.

[2] For the Galvanic Society, see Lundgren, *Berzelius och atomteorin,* p. 41.

CHAPTER IV

The Construction of the Berzelian System: Theoretical and Experimental Foundations

1. Theoretical and experimental developments to 1819

a. The electrochemical sorting phenomenon[1]

Less well known than Davy's investigations, Berzelius's studies with the Voltaic pile nevertheless rivalled the Englishman's in importance. The two men made similar discorveries and, for a time, followed similar lines of reasoning. However, their thought soon diverged in numerous respects. Among the most important of these divergences concerned the theory of salts and saline components. Davy moved from one hierarchical, two-component theory, the oxygen theory of acidity, to another one, the hydrogen theory of acidity, and he offered no theory of basicity. Berzelius, however, moved to a horizontal theory of both salts and saline components, one that denied generic components and affirmed specificity.

Though the successful decomposition of the alkalies and earths gave the decisive impetus to this development, the way had been prepared by the prior studies by Davy and Berzelius of the electrochemical sorting phenomenon. Davy's studies in this connection, already well known, are largely ignored here in favor of Berzelius's. Collaborating with Hisinger in an 1803 paper—thus prior to Davy's famous 1806 Bakerian lecture—Berzelius exhibited the ability of the pile to separate salts into constituent acids and bases, which were then sorted by accumulation around the opposing poles.[2] Berzelius's and Hisinger's

[1] For a general account of Berzelius's electrochemical theories and researches, see Anders Lundgren, *Berzelius och den kemiska atomteorin*, Acta Universitatis Upsaliensis. Studier i idé- och lärdomshistoria, Vol. II, utgivna av institutionen för idé- och lärdomshistoria vid Uppsala Universitet (Uppsala: 1979), pp. 36—60, 76—80; and H.G. Söderbaum, *Berzelius' Werden und Wachsen, 1779—1821*, Kahlbaum's Monographien aus der Geschichte der Chemie, Vol. III (Leipzig: 1899), pp. 26—55. (Lundgren's book is henceforth cited as Lundgren, *Berzelius och atomteorin;* Söderbaum's as Söderbaum, *Werden und Wachsen.*) See also Colin A. Russell, "The electrochemical theory of Berzelius," *Ann. Sci.*, 19 (1965), 117—145; and *idem,* "Introduction to the reprint edition [of Berzelius's *Proportions chimiques*]," in J.J. Berzelius, *Essai sur la théorie des proportions chimiques et sur l'influence chimique de l'électricité* (1819). The Sources of Science, No. 99 (New York: Johnson Reprint Corporation, 1972), pp. [v]—[xlix], esp. pp. [xxiv]—[xliv]; henceforth cited as Russell, "Introduction." Berzelius's 1819 *Proportions chimiques* is cited fully above, p. 145, n. 1, and is cited henceforth as *Proportions chimiques*.

[2] For Davy, see Partington, *History*, IV, pp. 40—48. The 1803 paper of Berzelius and Hisinger is "Versuch, betreffend die Wirkung

conclusion, unlike their introductory paragraph, suggested the ontological equivalence of the constituents that the sorting process implied.[3] Ontological equivalence carried the further implication, made fully explicit only later with the decomposition of the alkalies and earths, that the opposing constituents belonged to groups that possessed positively enumerable and opposing properties: basicity was no longer merely the passive capacity to serve as a substrate; it was a set of properties that were commensurate with acidity because they rested on an equivalent ontological basis. The base was no longer a mere patient of the acidic agent, but both acids and bases were at once agents and patients.

Also emerging in the 1803 paper was an electrochemical generalization of the notion of acidity and of its increasingly well-recognized counterpart, basicity. Berzelius and Hisinger pointed to the separation by the pile of the saline constituents into *their* constituents—typically radicals and oxygen—and suggested what later became a fundamental tenet of Berzelius's science: that all substances, of whatever level of composition, accumulating around a given pole of the pile exhibited a certain analogy among themselves and stood in opposition to the ontologically equivalent group of substances that accumulated at the other pole. Since the pile segregated acids from bases, the other substances that it moved to the same place had to be like acids or bases, respectively. Acidity and basicity were not merely the properties exhibited by the clear-cut acids and bases, but were reflections of an underlying electrical endowment of matter, and could be attributed to any substance that accumulated around the same pole as did typical acids or bases. Therefore, the clear-cut acids and bases were representatives of extreme cases of electronegativity and electropositivity.[4]

der elektrischen Säule auf Salze und auf einige von ihren Basen," *Gehlen's Journal,* 1 (1803), 115—149; reprinted, *Gilbert's Annalen,* 27 (1807), 269—324; Swedish, "Försök med elektriska stapelns verkan på salter och på några af deras baser," *Afhandlingar,* 1 (1806), 1—38; henceforth cited, from *Gehlen's Journal,* as "Wirkung der Säule." Regarding this paper, see Lundgren, *Berzelius och atomteorin,* pp. 56—58, and Söderbaum, *Werden und Wachsen,* pp. 40—47.

[3] Berzelius and Hisinger ("Wirkung der Säule," 115—116) introduced their study as an attempt to investigate the previously reported ability of the pile to decompose salts in such a way that "... die Säure den mit ihr vereinigten Stoff, er sey Erdart, Alkali, oder Metall, fahren lasse, um eine demselben entsprechende Menge des zum positiven Leiter gebrauchten Metalls aufzulösen; dies letztere möge übrigens, wenn man in einigen Fällen Gold und Platin ausnimmt, in was immer für einem Grade mit dem Sauerstoff verwandt sein. Dies ist aber den bekannten Verwandtschaftsgesetzen schnurgerade entgegen: daher wir uns vornahmen, diesen Gegenstand einen genauern Prüfung zu unterwerfen..." Here the emphasis was clearly on the acid and its migrations, while the base continued to be seen as substrate. In the concluding section, however, the first in the authors' list of "general conclusions" describing the effects obtained with the pile clearly suggested the ontological equivalence of the separated constituents (142): "Wenn sich die electrische Säule durch eine Flüssigkeit entladet, so sondern sich die Bestandtheile dieser Flüssigkeit dergestalt von einander ab, dass sich einige von ihnen um den negativen Pol, andere um den positiven ansammeln."

[4] In the second of their "general consequences" (*ibid.,* 142), Berzelius and Hisinger asserted: "Diejenigen Stoffe, die sich zum Drathe eines und desselben Poles hin begeben, stehen unter sich in einer gewissen Analogie. Zum negativen

For Berzelius as for Davy, the implication of these ideas was an electrochemical reinterpretation of affinity theory. Though, unlike Davy, Berzelius refused to make a simple identification of electricity and affinity, he was sufficiently impressed by the electrochemical sorting phenomenon to pursue it as a leading theme of his theoretical chemistry.[5] The firm decision to do so, however, awaited another development that decisively established for Berzelius the reliability of the perception of the fundamental ontological equivalence of saline components: the revelation in 1807 and 1808 that the alkalies and earths were oxides of metals. That these substances were compounds was widely believed; that at least the earths were metallic calces had been predicted by various figures in the eighteenth century, including Pelletier, Bergman, Macquer, and Lavoisier; and that the alkalies were not simple had been a belief of Lavoisier, who accordingly excluded them from his table of simple substances. However, the implications of these views, especially of the oxide nature of the earths, were only tacit in Lavoisier's *Traité*.[6] They were explored by Davy and Berzelius.

Following Davy's decomposition of the alkalies, Berzelius, working in collaboration with Pontin and proceeding on the analogy of the earths with the

Pol gehen alle brennbare Körper, Alkalien und Erdarten; zum positiven hingegen Sauerstoff, Säuren und oxydirte Körper."

Davy expressed substantially the same conclusion in his 1806 Bakerian Lecture, "On some chemical agencies of electricity," in Davy, *Works*, V, pp. 1—56; henceforth cited as Davy, "Chemical agencies." There (pp. 28—29) Davy stated, in words that became more famous than those of his predecessors, Berzelius and Hisinger: "It will be a general expression of the facts that have been detailed [here], relating to the changes and transitions by electricity, in common philosophical language, to say, that hydrogen, the alkaline substances, the metals, and certain metallic oxides, are attracted by negatively electrified metallic surfaces, and repelled by positively electrified metallic surfaces; and contrariwise, that oxygen and acid substances are attarcted by positively electrified metallic surfaces, and repelled by negatively electrified metallic surfaces; and these attractive and repulsive forces are sufficiently energetic to destroy or suspend the usual operation of elective affinity."

[5] In the 1806 Bakerian Lecture, Davy proceeded immediately to suggest the identity of chemical affinity with electricity. He persisted in maintaining this view, for example in his *Elements of chemical philosophy* (London: 1812), republished as Davy, *Works*, IV, though he added there the qualification (pp. 119—120) that electrical effects and affinity effects were to be seen as two manifestations, one macroscopic and one microscopic, respectively, of a single power. He then made use of this viewpoint to elucidate both the new phenomena of the pile and the same sorts of issues confronting the traditional affinity theory that had occupied a variety of late-eighteenth-century figures such as Guyton and Berthollet; (e.g., in his "Chemical agencies," *Works*, V, pp. 41—42; and in the *Elements*, *Works*, IV, pp. 114ff).

Berzelius long found this identification uncongenial, and he was more cautious and deliberate in formulating his electrochemical ideas. These themes are explored by Lundgren, in his *Berzelius och atomteorin*.

[6] For Bergman's, Lavoisier's, and Pelletier's predictions, see Söderbaum, *Werden und Wachsen*, pp. 58, 78—80. For additional information about Lavoisier's prediction see Partington, *History*, III, p. 460; Hélène Metzger, "La philosophie de la matière chez Lavoisier," *Actualités scientifiques et industrielles*, no. 218 (1935), 1—48 (separate pagination), esp. 34—36; and Lavoisier, *Traité*, I, pp. 174, 179—180. For Bergman's prediction, see also Partington, *History*, III, p. 191. For Macquer's prediction, see W.A. Smeaton, "Macquer, Pierre Joseph," *DSB*, VIII, pp. 618—624, esp. p. 620.

alkalies, succeeded in decomposing, besides potash, several of the earths into metals and oxygen.[7] Their technique, involving amalgamation of the alkali or alkaline earth metal with metallic mercury serving as the negative electrode, led them to discover the mercury amalgam of ammonia.[8] Taken to be fully analogous to the amalgamation of the new metals, the discovery permitted Berzelius and Pontin to regard ammonia as entirely analogous with the alkalies and earths by

[7] For Davy's decomposition of the alkalies, see Partington, *History*, IV, pp. 45—47. Berzelius and Magnus Martin Pontin (1781—1858), "Electrisch-chemische Versuche mit der Zerlegung der Alkalien und Erden," *Gilbert's Annalen*, 36 (1810), 247—280; Swedish, "Försök med alkaliernas och jordarternas sönderdelning," *K.S.V.A. Economiska Annaler*, 5 (May, 1808), 110—130; (June, 1808), 113—118; henceforth cited from the German text as "Zerlegung." As Lundgren (*Berzelius och atomteorin*, pp. 76—78) and Söderbaum (*Werden und Wachsen*, p. 49) point out, this paper was preceded by an anonymously published review by Berzelius of Davy's work, entitled "Alkaliernas decomposition," *K.S.V.A. Economiska Annaler*, 6 (April, 1808), 114—121. There, Berzelius suggested that Davy's success pointed strongly to the experimental fulfillment of Lavoisier's prediction about the earths: "Die Aehnlichkeit der Alkalien mit den alkalischen Erden lässt mit Bestimmtheit voraussetzen, dass auch letztere eine solche metallische Basis enthalten, vielleicht von gleicher oder grösserer Brennbarkeit und von grösserem specifischem Gewicht. Die völlige Analogie in ihrem Verhalten führt uns zu demselben Schlusse hinsichtlich der Zusammensetzung der übrigen Erden. Da nun die ähnliche Beschaffenheit der eigentlichen Erden und der Metalloxyde auch den gleichen Schluss in Bezug auf die letzteren zulässt und die Erfahrung mehrerer Jahrhunderts uns dessen Richtigkeit lehrt, so haben wir schon den Anfang und das Ende dieser Gedankenkette bewiesen, und es wird stets dem Genie des unsterblichen Lavoisier zur Ehre gereichen, sich zu derselben Erkenntnis durchgearbeitet zu haben, wenn auch durch einen Schluss in umgekehrter Richtung." (Quoted by Söderbaum, *Werden und Wachsen*, p. 49). The analogy of the alkalies with the alkaline earths and the other earths implied that the latter two categories of substances shared the composition of the alkalies. Berzelius's next paper succeeded in demonstrating precisely this point and had unanticipated consequences connected with a second theme in the anonymous review.

Lundgren points out (*Berzelius och atomteorin*, pp. 76—80) that Berzelius was quite taken with some of the bold speculations about the composition of the elements in which Davy permitted himself to indulge. In particular, Berzelius was struck by Davy's suggestion in the paper on the alkalies, which he supported with some quantitative evidence, that ammonia (by virtue of its analogy with the other alkalies) contained oxygen, and that therefore the alkali metals (by analogy with ammonia) contained hydrogen. This conjecture was the potential basis for a still broader generalization of the idea of salifiable bases.

The truth of these ideas seemed to be borne out by the results of the work done in collaboration with Pontin. By the time these experiments were reported, Davy had still not succeeded definitively in decomposing the alkaline earths (for this, see Partington, *History*, IV, pp. 47—49). Berzelius and Pontin, then, were the first to demonstrate the oxide nature of the earths and were evidently pleased to have confirmed the predictions of Lavoisier, Bergman, and Pelletier, all of whom they cited ("Zerlegung," 259). Only after having learned from Berzelius of the appropriate experimental technique (described shortly) was Davy able to pursue a successful line of experiments with the decomposition of the alkalies (for this, see Partington, *History*, IV, pp. 47—48; and Söderbaum, *Werden und Wachsen*, pp. 51—52).

[8] This observation came as no small surprise ("Zerlegung," 262): "Nachdem uns Davy mit der Analyse der feuerfesten Alkalien vorangegangen war, lies sich eine mit diesen analoge Zusammensetzung der alkalischen Erdarten schon im voraus vermuthen. Wenn man aber in dem folgenden Versuche die bekannten Bestandtheile des *Ammoniaks* an den negativen Pol zusammentreten und einen Metallkörper bildet sieht; wenn wir erzählen, wie wir ein Metall unter unsern Augen haben in einer Flüssigkeit entstehen sehen, deren Bestandtheile gänzlich bekannt und ihrem Verhältnisse nach ziemlich genau bestimmt sind: so wird man vielleicht ein Wunder zu lesen meinen, und unsere Glaub-

supposing it (as had Davy earlier) to be the oxide of some metallic basis.[9] Ammonia differed from the other alkalies and from the earths only in that the basis was compound, not simple. These considerations led Berzelius to a thoroughgoing generalization of the character of the salifiable bases:

> Durch die Entdeckung der Analogie des Ammoniaks mit den Metallen wird, wie uns dünkt, die Entdeckung der metallischen Natur der feuerfesten Alkalien sehr vervollkommnet. Denn ist nun erwiesen, dass die ganze Klasse der Salz-Basen, die Alkalien nämlich, die Erdarten und die Metalloxyde sämmtlich zusammengesetzte Körper sind, und dass sie aus Sauerstoff und einem anderen Stoffe bestehen, der sich durch einen metallischen Charakter auszeichnet, der aber übrigens von sehr vershiedenem eigenthümlichen Gewichte ist, von dem Platin an bis zum Ammoniak-Metall, welche von allen gewiss das leichteste ist.[1]

Three consequences flowed from this demonstration that the typical bases, like the typical acids, were all oxides. First, the veil obscuring that branch of chemistry that Lavoisier had merely glimpsed was pulled aside. The properties and chemistry of the metallic oxides were now considered just as worthy of investigation as were the acidic oxides. This theoretical insight was made largely

würdigkeit bezweifeln, — wie wir denn anfangs selbst unsern Sinnen nicht glauben wollten, bis wiederholte und vielfältig abgeänderte Versuche jeden Zweifel entfernten." The peculiarity of nitrogen and its compounds is a constant theme of Berzelius's work and dates from this observation. Aspects of his pursuit of this issue are described in Chapter V. The observation led immediately to some conclusions that Berzelius and Pontin believed they had confirmed experimentally ("Zerlegung," 262—263): "Hr. Davy hatte in der Reihe seiner Versuche auch die Grundmischung des Ammoniaks zu bestimmen gesucht, und als Resultat angegeben, es enthalte 20 Procent Sauerstoff den verbrennlichen Bestandtheilen beigemischt, die wir aus Scheele's, und später aus Berthollet's und Davy's Versuchen, schon seit geraumer Zeit als Wasserstoff und Stickstoff kennen. Da nun die Eigenschaften der feuerfesten Alkalien denen des Ammoniaks so äussert ähnlich sind, so schlossen wir, dass *erstens* auch die metallische Basis, welche wir im Kali und im Natron gefunden haben, aus denselben brennbaren Bestandtheilen, als das Ammoniak, zusammengesetzt seyn möge, und *zweitens,* dass, wenn die Bestandtheile des Ammoniaks unter gleiche Umstände, wie die des Kali, versetzt würden, sie sich zu einem eben solchen metallischen Körper, als man aus den feuerfesten Alkalien erhält, vereinigen würden. Diese Folgerungen spornten uns an, einen so wichtigen Punkt der chemische Theorie genauer zu untersuchen, und wir haben unser Raissonnement völlig gegründet gefunden."

9 "Zerlegung," 268—269. Davy's suggestion that ammonia may be an oxide of a compound radical appeared in the 1807 Bakerian Lecture on the decomposition of the alkalies: Davy, "On some new phenomena of chemical changes produced by electricity, particularly the decomposition of the fixed alkalies, and the exhibition of the new substances which constitute their bases; and on the general nature of alkaline bodies" (read, 19 Nov. 1807), in Davy, *Works,* V, pp. 57—101, esp. pp. 91—98; henceforth cited as Davy, "Decomposition of the alkalies."

1 "Zerlegung," 269—270. Berzelius and Pontin offered this generalization despite their failure to decompose some other earths, especially alumina, yttria, and silica: "Ungeachtet dieses Missgelingens hoffen wir aber doch noch, irgend einen Weg zu entdecken, auch diese Erden in einem metallischen Körper und in Sauerstoff zu zerlegen; denn die Analogie macht dieses äusserst wahrscheinlich." Berzelius later decomposed silica by ordinary chemical means: Berzelius, "Zerlegung der Kieserlerde durch gewöhnliche chemische Mittel," *Gilbert's Annalen,* 36 (1810), 89—102; Swedish, "Försök til kiseljordens analys," *Afhandlingar,* 3 (1810), 117—128.

explicit in Berzelius's 1811 "Essai sur la nomenclature chimique"[2] and was realized experimentally in a programmatic study of oxides (and sulfides), discussed in Chapter V. In the "Nomenclature," Berzelius offered a classification of oxides, and his definition of his second class, called simply "oxides," reveals his growing distance from Lavoisier:

> Leur caractère est de former des bases salines, ou du moins de se combiner à d'autres corps oxidés sans posséder des propriétés acides. Comme le même combustible produit quelquefois deux oxides appartenant tous deux à la même classe de corps oxidés, je servirai pour les distinguer, de la méthode employée dans la première nomenclature antiphlogistique, pour distinguer les deux degrés d'acidification du même radical, c'est-à-dire en terminant le nom du combustible par *osum* quand il est moins, et par *icum* quand il est plus oxygéné. Ainsi j'appellerai l'oxide de fer noir *oxidum ferrosum*, et l'oxide de fer rouge *oxidum ferricum*. Plusieurs métaux ne donne qu'une seul base saline; j'ai donné à leurs oxides la terminaison en icum, parce qu'il n'est pas probable qu'il existe chez eux quelqu'autre degré de basification, et cette terminaison se prête mieux pour en faire des noms composés des combinaisons qui ont lieu parmi ces oxides.[3]

With his vision no longer narrowed by the blinders of two-component theories, Berzelius could take the step of extending the new chemistry in a fashion that never occurred to Lavoisier and his collaborators in the *Méthode*. For Lavoisier, the higher oxides of metals were misfits in chemistry, in that they neither served as substrates for salification nor displayed acidic properties. For Berzelius, however, the higher oxides could and did form salifiable bases and had positively enumerable properties—a chemistry—of their own. Like the lower oxides, they were bases, and they possessed positive properties opposed to those of acids. The degrees of oxidation of the metals now merited a system of analytic linguistic denomination corresponding to that accorded to the acid oxides but denied to the metallic ones in the antiphlogistic nomenclature. Lavoisier's use of the epithet denoting color having been shown inadequate, Berzelius devised a new nomenclature by extending the acid nomenclature to the bases.

The second consequence issuing from the demonstration of the oxide nature of all bases pertained to the question that now immediately arose, what must be the seat of the properties of the bases? Two answers to this question were proposed in succession, first by Davy, and then, following in his footsteps, by Berzelius. Initially each entertained the idea that oxygen conferred alkalinity as

[2] Berzelius, "Essai sur la nomenclature chimique," *J. de physique*, 73 (1811), 253—286; Swedish, "Försök till en chemisk nomenclature," *K.S.V.A. Handlingar*, [2] 33 (1812), 28—74; henceforth cited from the French version as "Nomenclature." This paper was occasioned by Berzelius's participation in efforts to revise the Swedish pharmacopoeia. See "Nomenclature," 255, and Berzelius, "Utkast till förbättringar i beredningssättet af åtskilliga praeparata chemica för den nya editionen af Svenska pharmacopoeen, uppförde efter den förbättrade chemiskt pharmaceutiska nomenclaturen; jemte en critisk uppställning af samma praeparaters namn och beredningssätt i de förnämsta nyare europeiska pharmacopoeer (Upplässt d. 12 mai 1812)," *Svenska läkaresällskapets handlingar*, 1 (1813), 11—100.

[3] "Nomenclature," 262.

well as acidity. Davy suggested this notion briefly in his 1807 paper reporting the decomposition of the alkalies.[4] Though Berzelius echoed him in the first volume of his Swedish textbook (1808),[5] neither he nor Davy long adhered to the belief that a single substance, oxygen, could at once be responsible for the genesis of diametrically opposed properties. The conclusion that the seat of acidity or basicity lay in the particular combustible radicals—that those properties derived from the specific component—was all but inevitable. Davy suggested this in his 1808 paper reporting on the decomposition of the earths,[6] and Berzelius, as is shown below, followed him in taking the critical step of assigning to the radicals rather than to oxygen the principal source of the properties of oxidized bodies. What had been the generic properties of oxides was henceforth to be seen as the specific properties of radicals. From here, however, Davy and Berzelius took different roads. Davy did not proceed to a chemistry of specific properties, but, as is well-known, adopted a new two-component theory of acidity, the hydrogen theory. Berzelius, however, inspired by the implications of Davy's and his own achievements with the new instrument of chemical research, persisted singlemindedly in articulating a chemistry of the specific properties of matter.

The third consequence flowing from the generalization of the notion of basicity was a decisive reinforcement of Berzelius's faith in the ontological equivalence of bases with acids. Both groups were oxides of radicals. As oxides, both depended for their properties not on the constant, oxygen, but on the variable, the radicals. Moreover, all radicals were of one *kind* of entity—electrochemical—and each could be distinguished from the others qualitatively by variation in *degree* of electrochemical character. In its full breadth, Berzelius's perspective later embraced this variability in terms of so many positions on a continuum, his electrochemical series of the radicals, having near its extremes the radicals of the typical acids and bases. Any pair of oxides could then be labelled as relatively acidic and basic in accordance with their relative positions in the continuum. Berzelius introduced a new electrochemical vocabulary to express this variability: acidity and basicity were cases of electronegativity and electropositivity, respectively. In Berzelius's hands the generic uniformity of matter was becoming the specific diversity of matter. The oxygen theory of acids, the

[4] Davy, "Decomposition of the alkalies," p. 97: "... ammonia will bear the same relations to the fixed alkalies, as the vegetable acids with compound bases do to the mineral ones with simple bases. Oxygen, then, may be considered as existing in and as forming an element in all the true alkalies; and the principle of acidity of the French nomenclature, might likewise now be called the principle of alkalescence."

[5] As reported by Söderbaum, *Werden und Wachsen*, pp. 80, 95—96.

[6] Davy, "Electro-chemical researches on the decomposition of the earths, with observations on the metals obtained from the alkaline earths, and on the amalgam procured from ammonia" (read, 30 June 1808), in Davy, *Works*, V, pp. 102—139, esp. p. 119.

last powerful representative of the two-component tradition, was giving way before an affirmation of equivalent and diverse, specific forms of matter.

b. The roles of oxygen and the roles of the radicals

Berzelius arrived at a clear expression of these ideas slowly. A partisan of Lavoisier's views, he detached himself from them through a gradual evolution of ideas, beginning with the collaboration with Pontin and ending in the *Proportions chimiques.* Lavoisier had elaborated a classification of oxides in which he correlated properties with degree of oxidation, and, as his expectations about metallic oxides reveal, he had minimized the role of particular radicals in favor of the generic role of oxygen. In Berzelius's views, the classification of oxides, though not excised from his chemistry, became decisively overshadowed by the electrochemical classification of radicals, while the correlation of properties with degree of oxidation underwent complete reconstruction; and, as the electrochemical classification of radicals shows, the specificity of the radicals, not the degree of oxidation, became the chief determinant of properties.

This development rested on a conceptual distinction that Berzelius gradually worked out. By the time of the *Proportions chimiques,* Berzelius had coupled his insight that *among radicals* acidity was determined by the nature of the radicals with the recognition that *among the oxides of a given radical* variation in acid character did indeed depend upon degree of oxidation. To achieve this clarity required distinguishing sharply the two variables—nature of the radical and degree of oxidation—that determined the character of the oxides and establishing the details of the correlation between composition and properties. The first of these requirements is considered here briefly, for Berzelius met it in the theoretical context here under discussion. The second requirement, developed largely in an experimental context, is examined below, in Chapter V, section 9.

In the 1811 "Nomenclature," Berzelius attempted an accommodation of Lavoisier's classification of oxides with his own insights.[7] His classification invoked both qualitative characters and degree of oxidation, just as Lavoisier's had. However, Berzelius did not insist, as Lavoisier had done, that each degree of oxidation was available to each radical and was constitutive of a genus (from oxide, through acids in *-ous* and *-ic,* to acid in *oxygenated -ic*). Rather, he believed that the number of oxides produced by any radical was a fact to be discovered. The oxides found by experiment were then sorted into categories. Though the order of the categories largely corresponded to increasing oxidation (suboxide—displaying no marked acidity or basicity—oxide, acid, superoxide), the classes contained more than one oxidation stage. Oxides, for example, included oxides in *-ous* and *-ic,* and acids included acids in *-ous* and *-ic.* Accordingly, the classifica-

[7] "Nomenclature," 261—264.

tion exhibited both degree of oxidation and qualitative characters. However, the initial criterion of classification evidently was the qualitative character of the oxides, not of their radicals, and not the degree of oxidation. The classification was therefore a reflection of experience, not a theoretical statement about the determination of properties, as it had been for Lavoisier. It would not become such a theoretical statement until Berzelius realized more fully the implications of placing the seat of acidity or basicity in the radicals and thus distinguishing the role of the radicals from the role of oxidation in determining properties.

Illustrating Berzelius's difficulty is his class of "superoxides," corresponding to Lavoisier's acid in *oxygenated -ic.* Though he had removed this category from his 1808 *Lärbok,* because it contained only one compound, as Söderbaum notes, he reinstated it in the 1811 "Nomenclature." The category was described as containing higher oxides of positive or basic radicals (such as the alkali metals) that exhibited diminution of basicity with high oxidation (though Lavoisier's category contained acids, or at least one acid, oxymuriatic, that had suffered diminution of acidity with high oxidation).[8] However, among the members of the category were two compounds of the supposed muriatic radical, a negative, not a positive substance. One of the compounds was chlorine, which Berzelius now proposed to name *superoxidum muriatosum,* and the other was Davy's euchlorine, which Berzelius proposed to call *superoxidum muriaticum.* Thus, the category contained compounds that were similar to the suboxides in possessing no markedly acidic or basic properties but were too highly oxidized to be classed among the suboxides.[9] Despite the description of the category as derived from

[8] For the removal of superoxides from the classification in the 1808 *Lärbok,* see Söderbaum, *Werden und Wachsen,* pp. 79—80. In the "Nomenclature" (262), Berzelius described the reinstated class thus: "La quatrième classe contient tous les corps oxidés qui se sont combinés avec une si grande quantité d'oxigène, qu'ils cessent d'être des bases salines, sans qu'ils soient devenus des acides. Je les appellerai superoxida. Leur caractère principal est qu'ils ne se combinent point avec d'autres corps oxidés, sans se défaire de la partie de leur oxygène qui les a convertis en suroxides. Tels sont les suroxides de kalium et de natrium, dont nous devons la découverte à MM. Gay-Lussac et Thenard; l'oxide rouge et l'oxide brun de plomb, etc."

[9] The picture was further muddied by Berzelius's placement of two other ostensible oxides of the muriatic radical, one a lower oxide than the superoxides and one a *higher* oxide than both of them, in the class of acids. These were muriatic acid, a name Berzelius retained, and superoxygenated muriatic acid (chloric acid), which Berzelius proposed to call "oxymuriatic acid." In this connection, see Berzelius's notes 22 and 26 in the "Nomenclature," 284. As of the "Nomenclature," Berzelius believed that four oxides of the muriatic radical existed. They are summarized in the table below, where M represents the muriatic radical, O represents oxygen, and the subscripts represent proportions of oxygen. They are in the ratio a:b:c:d::1: 1 1/2:2:4.

Berzelius's calculations of the composition of the first, second, and fourth oxides are taken

Modern name and formula	Composition acc. to B.	B's name, 1811	Previous name
hydrogen chloride, HCl	MO_a	acidum muriaticum	muriatic acid
chlorine Cl_2	MO_b	superoxidum muriatosum	oxymuriatic acid
chlorine dioxide ClO_2	MO_c	superoxidum muriaticum	Davy's euchlorine
chlorine acid, $HClO_3$	MO_d	acidum oximuriaticum	superoxymuriatic acid

the higher oxidation of basic radicals (that is, as relying initially on the properties of the radicals), the inclusion of the muriatic compounds shows that this category, too, was based on the properties of its member oxides and was a reflection of experience, not theory.

Underlying Berzelius's difficulty in distinguishing the role of the radicals from that of oxygen was his tendency to apply equivalent and parallel principles of classification to both radicals and oxides. In the "Nomenclature," Berzelius introduced his new vocabulary, distinguishing electronegative and electropositive substances—whether radicals or oxides[1]—and then he offered two classifications, one of radicals and one of oxides. Each classification amounted to an electrochemical series consisting of several groups referred to the point of electrochemical indifference or neutrality separating the electronegative and electropositive bodies.[2] In offering these two classifications, Berzelius did not yet realize that the classification of radicals reflected the influence of only one variable, the

up below (p. 197, n. 8 and accompanying text). The calculation of the composition of euchlorine is in Berzelius, "Auszug aus einem Briefe des Hrn. Prof. Berzelius an den Prof. Gilbert," *Gilbert's Annalen,* 38 (1811), 227—232. The letter is dated 20 May 1811. The calculation appears on 229.

[1] "Nomenclature," 257—258. Having first distinguished the impoderables—electricity, light, caloric, and magnetism—from the ponderables, Berzelius asserted in regard to the latter: "A l'égard de leurs rapports avec les électricités, nous les partageons en substances électropositives et substances électronégatives.

"J'appelle substances électropositives, celles, qui, dans les décompositions opérées par la pile électrique, se rangent autour de pôle positif, et éléctronégatives, celles qui se rassamblent autour de pôle négatif."

[2] For the radicals, Berzelius divided them (*Ibid.*, pp. 257—258) into two categories, oxygen, the sole absolutely electronegative body, and the combustibles. He divided the latter into metalloids (e.g., sulfur, phosphorus, carbon, and the muriatic, fluoric, and boracic radicals) and metals. The former, Berzelius declared "sont tous des corps éléctropositifs, formant des acides avec l'oxygène, et leur ordre se change par degrés presque insensibles, en celui de métaux électropositifs." The whole collection of radicals formed a series, Berzelius observed (258—259): "Dans la série des corps simples que je vais exposer, je commencerai par l'oxigène comme étant le seul qui soit absolument électropositif; viendront ensuite les métalloïdes qui sont électropositifs relativement à la plupart des autres substances; puis les métaux commençant par ceux qui sont électropositifs au plus haut degré, et qui par conséquent, forment des acides; puis ceux qui affectent une espèce d'indifférence électrochimique, en ne produisant ni des acides, ni des bases; ensuite ceux qui ne donnent que des bases. Cette série des corps simples commence donc par le corps le plus électronégatif et finit par ceux qui possède au plus haut degré les propriétés électrochimiques opposées." Accordingly, the series had several regions, distinguished by the formation from their radicals of acidic, basic, or indifferent oxides.

For the oxides (261—264). Berzelius offered four classes of oxides (plus a fifth class, containing the oxides of the supposed metal ammonium): suboxides, oxides, acids, and superoxides. The suboxides possessed neither acidic nor basic properties, because they contained "si peu d'oxigène" (261). The oxides served as bases but exhibited no acidic character. The acids, of course, were the markedly acidic oxides and the superoxides were so oxidized that they ceased to be bases but did not become acids.

Berzelius here was using the sign convention, which he soon abandoned, of labelling "electropositive" whatever was attracted to the positive pole of the pile and *vice versa.* Oxygen, accordingly, was electropositive; the metals, electronegative. His later, reversed sign convention, is the one that he and the profession continued to use, and it is employed in the present analysis.

nature of the radicals, while that of the oxides reflected the influence of two variables, the nature of the radicals and the degree of oxidation.[3] He came to perceive this distinction as he reflected further upon the problem of arriving at a correct serial order for the radicals.

In a paper of the following year, written in part as a "Rechtfertigung" of the "Nomenclature," Berzelius showed himself closer to his mature position.[4] There, he offered a classification of radicals and no classification of oxides.[5] In this connection, he emphasized two points that he had only suggested in the "Nomenclature:" that the serial order of the radicals could be determined only by comparing the relative positivity or negativity of pairs of substances[6] and that this comparison must rest on examination of oxides.[7] The difficulty of arriving at a properly ordered series in this manner became clear to Berzelius when he asked himself how to accommodate the production, by one radical, of several oxides with different electrochemical characters. In the "Nomenclature," this question did not arise. In the "Rechtfertigung," the classification of radicals explicitly accommodated the existence of several differently characterized oxides of the same radical but stood independently of different amounts of oxygen.[8] Implicit in the approach was the realization that the order of the radicals could be determined from the behavior of their oxides by regarding the presence of oxygen as a constant. This point became explicit in the *Proportions chimiques.*

There, Berzelius attained his mature position and his maximal departure from Lavoisier. He presented his electrochemical series of radicals accompanied by two observations: that the order of the radicals generally reflected that of their highest

[3]E.g., in the classification of oxides described in the preceding note, both oxides and superoxides were near the point of indifference, but they came to reside there for different reasons (proximity of the radicals to the point of indifference among radicals in the case of oxides, high oxidation of basic radicals in the case of superoxides).

[4]Berzelius, "Versuch, die chemischen Ansichten, welche die systematische Aufstellung der Körper, in meinem Versuch einer Verbesserung der chemischen Nomenclatur begründen, zu rechtfertigen," *Schweigger's Journal,* 6 (1812), 119—114, 144—176, 284—322; 7 (1813), 43—78; Swedish, "Försök till ett rättfärdigande af de theoretisk chemiska åsigter, på hvilka den systematiska upställning i mitt försök till en förbättring af den chemiska nomenclaturen grundar sig," *K.S.V.A. Handlingar,* [2] 33 (1812), 166—198, 233—244; 34 (1813), 39—62, 175—215; henceforth cited from the German text as "Rechtfertigung." Theoretical material forms only a portion of the paper; the rest consists of reports of a series of analyses of the composition and the acidic or basic properties of many radicals. This work, conducted by Berzelius and his students, was intended to support and illustrate the classification of oxides. It forms a major part of Berzelius's programmatic study of oxides and sulfides, discussed in Chapter V.

[5]"Rechtfertigung," 128—130.

[6]*Ibid.,* 127—128. Cf. "Nomenclature," 257.

[7]*Ibid.,* 128n: "Wie wesentlich es ist, die Körper nach ihren Eigenschaften im oxydirten Zustand einzutheilen, erhellt schon daraus, dass die Oxyde die electrochemische Tendenz der Radicale behalten, da diese dagegen durch die Verbindung der Radicale unter einander aufgehoben wird." Cf. "Nomenclature," 257.

[8]E.g., the "abwechselnde" radicals, "Rechtfertigung," 129. Cf. the passage from the "Nomenclature," quoted above, p. 161, in n. 2, describing the classification of radicals.

oxides and that the electrochemical character of oxides generally reflected that of their radicals:

> Les propriétés électro-chimiques des corps oxidés dépendent presque toujours exclusivement ... de leur élément électropositif, c'est-à-dire, de leur radical; l'oxide est d'ordinaire électro-négatif à l'égard des autres oxides, lorsque son radical est négatif à l'égard de leur radical, et même à l'inverse. Par exemple, l'acide sulfurique est électro-négatif à l'égard de tous les oxides métalliques, par la raison que le soufre est négatif par rapport à tous les métaux. Les oxides de potassium et de zinc sont au contraire électro-positifs à l'égard de tous les corps oxidés envers les radicaux desquels le potassium et le zinc sont positifs. Ce fait, dont nous ne pouvons expliquer la cause, rectifie une idée inexacte sur le principe de l'acidité, que dans la théorie antiphlogistique l'on crut être l'oxigène. Nous trouvons maintenant qu'il réside dans le radical des acides, et que l'oxigène y joue un role si indifférent qu'il entre également dans les plus fortes bases salifiables, c'est-à-dire les oxides électro-positifs, et dans les plus forts acides, ou les oxides électro-négatifs. Quelquefois pourtant il arrive qu'un oxide positif acquiert, par une plus haute oxidation, des propriétés moins électropositives, qui le rapprochent des électro-négatifs, comme, par exemple, l'oxide d'étain: mais, dans les plus fortes bases, telles que la potasse et la soude, une addition d'oxigène peut bien détruire la réaction positive, sans néanmoins en produire une négative; et c'est ainsi que se forment les superoxides des fortes bases salifiables.[9]

The radicals finally achieved their full status. Their principal properties were regarded as specific to them and were correlated with their electrical behavior, which served as the basis for classifying them. Their oxygen compounds, insofar

[9] *Proportions chimiques,* 90—91. These insights confronted Berzelius with a theoretical lacuna. Though the electrochemical interpretation of acidity and basicity could account for the influence of the nature of the radical on the properties of an oxide, after the fall of Lavoisier's theory of acidity there was no explanation of the dependence of acidity on degree of oxidation. The best Berzelius could offer was a vague, mechanical sort of explanation (*Proportions chimiques,* pp. 115—116): "... nous avons tout sujet de croire que les atomes ne sont unis qu'un à un dans les corps qui ne manifestent que de faibles affinités; tels que le gaz oxide de carbone, les oxidules de cuivre, de mercure, d'or, etc; en sorte que l'on pourrait bien présumer que tous les corps composés d'un atome d'oxigène, ont, d'une manière plus ou moins prononcée, le caractère des sub-oxides. D'ailleurs il paraît maintenant, que les atomes des plus forts acides et des plus fortes bases contiennent plus d'un atome d'oxigène. Le nombre des atomes simples dans un atome composé devant nécessairement influer sur la forme de l'atome composé, et par conséquent sur ses propriétés, il est permis de croire que des oxides qui contiennent le même nombre d'atomes d'oxigène, ont au moins de commun quelques propriétés générales qui les distinguent d'avec ceux qui en ont plus ou moins, et à l'aide desquelles, au défaut de toute autre donnée, on peut établir ces conjectures avec quelque probabilité. Ainsi, par exemple, lorsque nous avons sujet de présumer que les oxides qui ne contient qu'un atome d'oxigène sont ceux qui ont les plus faibles affinités, il se présente toute une série de bases salifiables plus prononcées, dans lesquelles le nombre des particules d'oxigène, comme nous le verrons plus bas, doit être plus grande que dans les précédents; elles ont les plus fortes affinités, et il est probable, pour cette raison, que toutes les plus fortes bases contiennent deux atomes d'oxigène. Celles qui en contiennent trois sont au contraire plus faibles, et beaucoup d'entre elles peuvent même être électro-négatives à l'égard de quelques oxides électro-positifs. Cette conjecture est d'autant plus probable, que le même radical a quelquefois des oxides qui présentent ces différences de composition et de caractères. Il paraît que les acids contiennent de préférence trois atomes d'oxigène, et c'est le cas du plus grand nombre de ces corps; quelques-uns en contiennent 2, 4, 5, 6 et 8 comme nous en verrous la preuve dans l'examen particulier de chaque acide."

as they displayed varying degrees of acidity or basicity, were seen as generally reflecting their specific properties, though the latter occasionally presented some modification with high levels of oxidation. The evolution from the "Nomenclature" was completed, with elimination of most vestiges of Lavoisier's two-component theory.

Berzelius's attainment of clear insights about the role of radicals in the determination of the electrochemical properties of acidity and basicity did not shift the focus of his attention back from oxides to their radicals. Berzelius's chemistry, like Lavoisier's, was and remained a chemistry of oxides. For Berzelius, the characteristic properties of the radicals were displayed in their oxides but suffered extinction in the direct combinations of the radicals with one another. The chemistry of the non-oxygenated compounds was an area of small compass in the purview of both men. Berzelius's successful analysis of the role of the radicals apart from their degrees of oxidation only facilitated the task of investigating their relative acidity or basicity in connection with degree of oxidation. The bulk of Berzelius's experimental work following the early electrochemical studies was dedicated to the investigation of the electrochemically defined properties of oxides and of their composition, as well as to the composition of salts, the compounds of these electrochemically characterized acids and bases.

This labor provided the experimental counterpart to the theoretical ideas just analyzed here about acidity and basicity. The experimental analyses accompanying the theoretical discussion in the "Rechtfertigung" were offered with that intention. More significantly, this work permitted Berzelius to reconstruct Lavoisier's correlation of degree of oxidation with changes in the manifestation of properties. Having a generalized notion of the relative acidity and basicity of the radicals in terms of positions on a continuum, Berzelius could elaborate a correspondingly generalized correlation of composition and properties of the oxides. He later established the existence of several patterns of this correlation and showed that the patterns were common to many radicals, even though they were located in different portions of the series. The classification of oxides receded in favor of an exhibition of these shared patterns, though the names of the various categories were retained to provide labels—electrochemical characterizations that carried implications for composition—for the various oxides participating in the patterns. Berzelius succeeded in this, however, only by further elaboration of his quantitative chemistry, and in particular by establishing connections between the composition of oxides—the constituents of salts—and that of the salts themselves. Accordingly, the emergence of the central tenets of Berzelius's quantitative chemistry is found in the experimental work on oxides and salts. The origins of this work lie in Berzelius's "Versuch, die bestimmten und einfachen Verhältnisse aufzufinden, nach welchen die Bestandtheile der unorganischen Natur mit einander verbunden sind."

2. Experimental developments to 1811: The *Versuch* and the rule of oxides

a. Introduction

The importance of the *Versuch* as a document in the history of qualitative chemistry is less apparent—though no less real—than its significance in the history of quantitative science. Its fame rests chiefly on its well-deserved reputation as a milestone in the rise of exact techniques in quantitative analysis and as a pioneering effort to establish experimentally the fundamental laws of chemical combining proportions. The simultaneous presence in the *Versuch* of both qualitative and quantitative themes faithfully reflects its parentage. It was born from the union of two chief concerns occupying Berzelius: the electrochemical phenomena with their consequences for a chemistry of qualitatively diverse matter and pre-Daltonian developments in stoichiometry, which had begun to gain increasing attention once the battle for the new French chemistry had been fought and won. Berzelius wished to combine his qualitative chemistry with a system, or at least with techniques, for determining the combining porportions of the various kinds of matter.

From Richter, whose even then obscure works Berzelius had read as a prelude to preparing an up-to-date Swedish chemistry text,[1] he obtained two laws governing the composition of salts: the law of neutrality, which Richter had stated in Part I (1791) of his "Ueber die neueren Gegenstände der Chymie" and the law of equivalent proportions, which apppeared in the following year in general form in Part I of Richter's "Anfangsgründe der Stöchyometrie." Richter reached the former, a special case of the latter, by reflection on the phenomenon of the maintenance of neutrality upon mixing two solutions, each containing a different neutral salt.[2] If the combining proportions of the constituents of each of the two salts that were to be mixed were known, the proportions of the constituents of the salts obtained from the double decomposition that ensued upon mixing could be calculated directly. The 1792 generalization of this conclusion was the law of equivalent proportions, which, in more modern terms, states that the amounts of the bases that saturate a given amount of an acid always stand

[1] For the origins of Berzelius's textbook, see Söderbaum, *Werden und Wachsen*, p. 76; *Proportion chimiques*, p. 16; and Berzelius, *Autobiographical notes*, ed. H.G. Söderbaum, tr. Olof Larsell (Baltimore: 1934), p. 65; henceforth cited as *Autobiography*. For Swedish version, see above, p. 149, n. 1.

[2] On Richter and his accomplishments, see Partington, *History*, IV, pp. 674—688; *idem*. "Jeremias Benjamin Richter and the law of equivalent proportions," *Ann. Sci.*, 7 (1951), 173—198; 9 (1953), 289—314; and Ida Freund, *The study of chemical composition: an account of its method and historical development* (1904; rpt., New York: 1968), pp. 171—186. Freund's book is cited henceforth as Freund, *Study*. Berzelius was originally attracted to Richter's ideas because of the possibility they offered of using proportions to measure affinities. For this, see Lundgren, *Berzelius och atomteorin*, pp. 72, 75—76, 91, 97.

in the same proportion to one another; and that the same holds of the amounts of the acids that saturate a given base. That this reversibility implied the equivalence of acids and bases likely enchanced its attractiveness for Berzelius. In the first volume of his *Lärbok* (1808) Berzelius summarized the law in relation to available data on the quantities of acids and bases that saturated one another.[3] Berzelius's early analytical work was motivated in part by his effort to exploit Richter's law of equivalent proportions by calculating the composition of a large number of salts from a small body of analytical data. The failure of available analyses to conform to Richter's law induced Berzelius to perform his own analyses of salts.[4] Berzelius accepted Richter's law despite its lack of significant experimental documentation and gave the law a normative function by insisting that analytical data conform to it.

During his earliest analytical studies, Berzelius independently discovered another law that he later found had been stated by Richter in Part VIII (1797) of the "Neuere Gegenstände" as another deduction from the law of neutrality. In this study, the law is called the "law of the constancy of basic oxygen," or the "law of basic oxygen." Linking first and second order compounds,[5] the law stated simply that the oxygen in the bases of all the neutral salts of an acid was the same (e.g., in lead sulfate $PbO.SO_3$, barium sulfate $BaO.SO_3$, and ferrous sulfate $FeO.SO_3$, the metallic oxide bases all contained the same quantity of oxygen). The law had been stated earlier by Bergman in phlogistic terms in the 1782 *De diversa phlogisti quantitate in metallis* as a deduction from the phenomenon of metallic displacement and later by Gay-Lussac from observation of the same phenomenon in antiphlogistic terms.[6] Berzelius's independent discovery is considered below in subsection 2.*c*.

Berzelius's interest in quantitative studies had thus already been sparked by the time of Davy's decomposition of the alkalies. Davy's and Berzelius's ex-

[3] Lundgren, *Berzelius och atomteorin*, p. 75; Söderbaum, *Werden und Wachsen*, pp. 140—141.

[4] Berzelius stated in several passages that he had tried to use available analyses, especially those of Rose and Bucholz, but that these data could not be brought into conformity with Richter's laws. Accordingly, he decided to do his own analytical work. See *Autobiography*, pp. 64—65, and Berzelius, "Ueber das Verhältniss der Sauerstoffmengen etc. zu einander, welche die Körper auf verschiedenen Oxydationsstufen aufnahmen, etc. (Aus einem Schreiben an den Akademiker Gehlen vom 14. Aug. 1810)," *Schweigger's Journal*, 1 (1811), 257—262, esp. 258; henceforth cited as "Sauerstoffmengen."

[5] Berzelius introduced his vocabulary of "orders" only in 1813, as is pointed out below, p. 287, n. 4 and accompanying text. However, because of its convenience, this terminology is employed in parts of this study devoted to the period prior to Berzelius's first use of it. First order compounds were unions of the elements with one another, usually oxygen with radicals or sulfur with other radicals. Second order compounds were unions of first order compounds with one another. Third order compounds were unions of second order compounds with one another or with compounds of lower orders; and so on.

Though Berzelius never used the terms "higher order compounds" or "higher order composition," they are used in this study to refer to combination above the first order.

[6] Partington, *History*, III, p. 678 for Bergman, Richter, and Gay-Lussac.

periments in the electrolytic decomposition of hitherto undercomposed substances offered new and previously unexpected material—the composition of the alkalies and earths—for submission to quantitative study and calculation. Ammonia in particular assumed in this connection a special importance for Berzelius. Regarding it as an oxide of a metallic substance made manifest in the mercury amalgam, he hoped to determine its oxygen content. Because of Berzelius's repeated failures to isolate the substance and because of its rapid supposed oxidation even when amalgamated, the oxygen content could not be determined directly, but only by calculation. Already pursuing Richter's ideas for calculation of second order composition, Berzelius now needed to determine first order composition, and he became attentive to Richter's ideas in this connection as well. The foundation of such an enterprise was the possession of exact analyses of a variety of metallic oxides, at least one acid, and the salts formed by these constituents. From the union of Berzelius's prior quantitative study of salts with his new concern for oxides was born the *Versuch.*[7]

The dual background of the *Versuch,* then, had directed Berzelius's attention to two levels of composition that he later described as being of the first and second order. Both the law of basic oxygen and the newly decomposed bases attracted Berzelius to the problem of determining the composition of the oxides of radicals. Investigation of the composition of first order oxides—and sulfides—was a major part of Berzelius's life-long effort to establish the combining weights and oxidation and sulfuration stages of all elementary radicals and provided the material for his reconstruction of Lavoisier's classification of oxides. Richter's work had called Berzelius's attention to the composition of salts as well as to that of oxides. The laws of neutrality, of equivalent proportions, and, again, of the constancy of basic oxygen, all applied to salts. Berzelius was just as interested in determining the composition of salts—second order compounds of first order oxide or sulfide constituents—as in determining the composition of the first order constituents themselves. Unlike Dalton, Berzelius always pursued his quantitative studies simultaneously with respect to both first and second order composition, and, later on, still higher order composition. A central technique in this approach was the use and discovery of laws that established a direct connection between levels of composition. The first of these was the law of the

[7] For Berzelius's desire to calculate the oxygen content of ammonia in the light of his interest in Richter's innovations, see "Sauerstoffmengen," 257, and Berzelius, "Vermischte Notizen (aus den Briefen an den Herausgeber [Gehlen] vom 19ten April 1808, 2ten September und 8ten October 1809)," *Gehlen's Journal,* 9 (1810), 585—589, esp. 588; for a preliminary report of the result of the calculation, see 586; henceforth cited as "Vermischte Notizen." For the union of the already ongoing analytical enterprise with the interest in the composition of ammonia, as well as of the other alkalies and earths, see *Proportions chimiques,* pp. 16—17. For attempts to isolate the amalgam, see "Zerlegung," 262—296 and *Versuch,* pp. 69—70. The *Versuch* is cited fully in the next note.

constancy of basic oxygen. The second, which, like the law of basic oxygen, connected first and second order composition, was the rule of oxides. Berzelius discovered it and made it the keystone of his chemistry. He later found or proposed other laws governing even more complex compounds. By dictating links between levels, the laws helped Berzelius determine one level of composition from another, or else both together, by weighing the consequences on one level of an assumption made about the other and *vice versa.*

b. The *Versuch*

Berzelius's *Versuch* consisted of a number of parts written and published at different times. Though some knowledge of the chronology of the drafting of its various components is needed to support parts of the discussion below, for the moment a brief look at its organization is sufficient. In its best known form, published in Gilbert's *Annalen der Physik* and later republished as number 35 of *Ostwald's Klassiker,* the *Versuch* consisted of five main sections—the first half, the second half, and three separate continuations or *Fortsetzungen*—and a few smaller additional sections, all appearing under the common title. The first and second halves originally appeared together in Berzelius's and Hisinger's *Afhandlingar* in 1810 and then in German translation, with some additions detailed later, in *Gilbert's Annalen.* The minor additions and the entire content of the *Fortsetzungen* were written in German expressly for *Gilbert's Annalen,* appearing in that journal in 1811 and 1812. The *Ostwald's Klassiker* edition (1892) was a republication of the entire German text of the *Versuch* as it appeared in the *Annalen der Physik,* but without Gilbert's editorial notes.[8]

[8] The first and second halves of the *Versuch* had been completed by October, 1809, when Berzelius reported ("Vermischte Notizen," 587) that the third volume of the *Afhandlingar* in which they appeared was in press. The Swedish text of the two halves is Berzelius, "Försök rörande de bestämde proportioner, hvari den oorganiska naturens beståndsdelar finnas förenade," *Afhandlingar,* 3 (1810), 162—276; German, "Versuch, die bestimmten und einfachen Verhältnisse aufzufinden, nach welchen die Bestandtheile der unorganischen Natur mit einander verbunden sind," *Gilbert's Annalen,* 37 (1811), 249—324, 415—465; the material on 324—334 and 465—472 consists of additions that are detailed below (pp. 180—181, nn. 8-5 and accompanying text). The *Forsetzungen* appeared under the same title as the first and second halves; first *Fortsetzung, Gilbert's Annalen,* 38 (1811), 161—226; second, 40 (1812), 162—208; third, *ibid.,* 235—320. Swedish versions of the *Fortsetzungen* appeared later: "Försök rörande de bestämde proportioner, hvari den oorganiska naturens beståndsdelar finnas förenade," *Afhandlingar,* 5 (1818), 94—128; "Försök öfver vattnets förhållande till syror, saltbaser och salter; öfver basiska och dubbel salters sammansättning, samt öfver salpetersyrliga salter," *ibid.,* 132—189. The last section of the third *Fortsetzung* was a summary and conclusion for the entire *Versuch:* "Allgemeine Uebersicht der Resultate meiner Versuche, welche in diesen verschiedenen Abhandlungen beschrieben sind," *Gilbert's Annalen,* 40 (1812), 320—330; Swedish, "Allmän öfversigt af hittils vunna resultat af undersökningarna om bestämda proportioner," *Afhandlingar,* 5 (1818), 190—197. The entire German text of the *Versuch,* including both halves, minor additions, *Fortsetzungen,* and conclusion, was republished from *Gilbert's Annalen,* without Gilbert's editorial notes, as *Versuch, die bestimmten und einfachen Verhältnisse aufzufinden, nach welchen die Bestandtheile der*

Pursuant to the inspiration from Richter, the first half contains a sequence of analyses of a series of metallic oxides—the bases of metallic salts—and of the two oxyacids of sulfur; and it establishes the first of Berzelius's rules of combining proportions. The second half presents an account of Berzelius's further experimental studies of the electrolytic decomposition of the alkalies and contains both direct and calculated analyses of these compounds. In the introductory paragraph of the second half, Berzelius informed his readers that this experimental work was a continuation of the study conducted in collaboration with Pontin and chronoligically preceded the analyses of the first half.[9] The latter, therefore, contains Berzelius's earliest sustained analytical investigation into chemical proportions, begun immediately following the achievement of the decomposition of the alkalies and earths. The analyses were conducted with the care and foresight, dependent upon an impressive command of inorganic reaction chemistry, that gained for Berzelius the reputation of a consummate experimentalist. Indeed, it was as an experimentalist that Berzelius first entered upon the stage of European chemistry, though he is remembered as much, if not more, for his theoretical achievements. A preeminently experimental study, the *Versuch* inaugurated his rise to prominence. Berzelius recounted that work with the express intention of minimizing the theoretical discussion, and he arranged the conclusions, which ultimately did acquire great theoretical importance, in a fashion that all but compelled assent to the laws they were intended to document.[1]

The avowed intention of the first half was to establish two laws of combining proportions: the law of multiple proportions, which Berzelius indicated he had seen, after he had already begun his own analytical work, in an account of Wollaston's famous 1808 paper, "On super-acid and sub-acid salts;" and the law of equivalent proportions, which Berzelius credited to Richter.[2] Cited in the

unorganischen Natur mit einander verbunden sind, ed. Wilhelm Ostwald, Ostwald's Klassiker der exakten Wissenschaften, no. 35 (Leipzig: 1892). Unless otherwise noted, all citations from the *Versuch* in this study are taken from the *Ostwald's Klassiker* edition, which is cited as *Versuch.* For summary overviews of some of the content of the *Versuch,* see Partington, *History,* IV, pp. 154—157; and Freund, *Study,* pp. 162—166, 179—184.

[9] *Versuch,* p. 54.

[1] Berzelius was aware of the power of his presentation and explicitly minimized theoretical material to permit his analytical work to speak for itself (*Versuch,* p. 6): "Ich werde meine Versuche in der Ordnung mittheilen, in welcher sie den Gegenstand am besten aufklären, und werde mich des Theoretisirens gänzlich enthalten. Inwiefern die Resultate der Versuche die Theorie bestätigen, wird jedem von selbst in die Augen fallen, und die Ideen, auf welche sie führen, werden gewiss, ohne mein Zuthun, bei jedem aufmerksamen Leser sich einfinden." Berzelius's insistence on the experimental foundations for chemical proportions (as opposed to the speculative foundations provided by Dalton) is a prominent theme of Lundgren, *Berzelius och atomteorin.*

[2] William Hyde Wollaston, "On superacid and subacid salts," *Phil. Trans.,* 98 (1808), pp. 96—102; also published in *Nicholson's Journal,* 21 (1808), pp. 164—169. Berzelius (*Versuch,* p. 4) referred to the latter. For Wollaston as a source for Berzelius's knowledge of Dalton's atomic theory and the laws of stoichiometry, see Lundgren, *Berzelius och atomteorin,* pp. 88—89.

opening paragraphs, not as the focus of the work, but as an incidental discovery in which he was preceded by Richter, was the law of the constancy of basic oxygen.[3] Despite this casual mention, Berzelius accorded it no small significance, calling it, in a summary of the first half presented elsewhere, "meine Hauptsatz."[4] Its importance to him lay particularly in its role in sustaining the calculations of the composition of the alkalies in the second half, in its relation to an analytical technique deployed in the first half, and, historically, in its status as the first step to the rule of oxides. Hardly had the law of multiple proportions entered chemistry, than it was compelled to share center stage, in Berzelius's mind, with another law, one governing higher order composition.

c. The constancy of basic oxygen

Berzelius's route to the law of basic oxygen, though roundabout in comparison with Richter's straightforward deduction, is important for several reasons: it rested on an innovation in analysis, it contributed a prominent conceptual theme to his chemistry, and it helped lead him to the rule of oxides, a cornerstone of his chemical edifice. The details of his route cannot be reconstructed accurately, however, because the few passages in which Berzelius described his earliest analytical labors, of which the principle was one outcome, are brief, cryptic, and partially inconsistent with one another. Still, the intellectual and practical connections that Berzelius traced can be signalled, even if the order in which he followed them cannot be determined definitively.

While pursuing some of Richter's ideas on the combining proportions of salts, Berzelius encountered the law of multiple proportions—whether by independent discovery or by noticing it in the literature is unclear.[5] Having previously analyzed the three oxides of lead and listed their composition in per-

Berzelius's statement of the two laws read (*Versuch*, p. 5): "Man wird aus dem Folgenden sehen, dass, wenn zwei Körper, A und B, sich in verschiedenen Verhältnissen mit einander verbinden, dieses immer nach folgenden fest bestehenden Proportionen geschieht: 1A mit 1B (Zusammensetzung im Minimum); 1A mit 1 1/2 B (oder vielleicht richtiger 2A mit 3B); 1A mit 2B; 1A mit 4B. Unter meinen Versuchen findet man aber kein einziges Beispiel von 1A mit 3B.

"Man wird ferner ersehen, dass, wenn zwei Körper, A und B, beide zu zwei anderen, C und D, Verwandtschaft haben, die Menge von C, wodurch A gesättigt wird, sich zu der Menge von D, wodurch A gesättigt wird, genau so, wie die Mengen von C und von D, wodurch B gesättigt wird, zu einander verhalten. Wenn z.B. 100 Th. Blei im Minimum 15.6 Th. Schwefel und 7.8 Th. Sauerstoff, — ferner 100 Th. Eisen, der Analyse zu Folge, die ich weiter unten angeben werde, im Minimum 58,8 Th. Schwefel aufnehmen, so lässt sich die Mischung des Eisenoxyduls durch die einfache Proportion 15,6 : 7,8 = 58,8 : 29,4 Th. Sauerstoff verbunden sein. Die Versuche, welche ich hier mittheilen werde, bekräftigen dieses. Es lassen sich auf diese Weise alle binären Zusammensetzungen berechnen. Der verdienstvolle Richter hat schon längst bewiesen, dass sich die Mischung der Salze durch eine ähnliche Rechnung müsse bestimmen lassen."

[3] *Versuch*, p. 4.

[4] "Sauerstoffmengen," 259.

[5] Independent rediscovery, "Sauerstoffmengen," 257; from the literature, *Versuch*, p. 4, and *Proportions chimiques*, p. 17.

centages, he now recalculated the results to reveal the oxygen in each combined with a given weight of metal. In apparent conformity with multiple proportions, the quantities of oxygen proved to be in the ratio of 1 : 11/2 : 2 (they are actually in the ratio of 1 : 11/3 : 11/2).[6] He next established that sulfides, too, obeyed the law of multiple proportions by showing that the quantities of sulfur in the two sulfides of iron, for a given weight of metal, were as 1 : 2 (this is correct).[7] Looking for a similar pattern in the sulfides of lead, he could find only one but noticed that the weight of the sulfur in the sulfide, taken to be the sulfide *in minimo,* was twice that of the oxygen in the lowest oxide.[8] This observation motivated the line of reasoning that led Berzelius to the law of the constancy of basic oxygen.

Though Berzelius did not identify clearly the next step in his thinking, the ideas described in the remainder of this subsection and in the next suggest what that step was. Berzelius inquired whether the same quantity of sulfur that was combined with a given weight of metallic lead in lead sulfide—a weight of sulfur equal to twice the weight of oxygen in the sulfide *in minimo*—also was combined with lead in a compound formed from the oxidized radicals, lead oxide and sulfuric acid. The compound of the oxidized radicals was lead sulfate. Berzelius was asking, then, whether a second order compound, lead sulfate, contained the two radicals lead and sulfur in the same proportions as in its first order analogue.

Answering this question required a knowledge of the composition of sulfuric acid.[9] However, prior analyses of the sulfur oxyacids by Rose and Bucholz were suspect in Berzelius's mind, apparently because their results did not conform to Richter's laws.[1] Hoping to avoid the sources of error that undermined Rose's and Bucholz's results, Berzelius devised a new analytical technique. His method was to oxidize the sulfide to the sulfate and partition the oxygen consumed between the lead and the sulfur. The technique permitted Berzelius to determine simultaneously the composition of sulfuric acid and the answer to the question motivating the analysis, that is, was the proportion of lead and sulfur in the lead sulfide preserved in lead sulfate.

The expectation that the proportion was preserved soon acquired the status of a principle in Berzelius's chemistry. In this study it is called the "principle of the preservation of combining proportions" and is examined for its own sake

[6] *Autobiography,* n. 110 on p. 172.

[7] "Sauerstoffmengen," 258—259; *Versuch,* p. 31 (with reference to Proust).

[8] "Sauerstoffmengen," 259; *Versuch,* p. 9.

[9] For Berzelius's analysis of sulfuric acid, see *Versuch,* pp. 10—18; of sulfurous acid, pp. 18—20. Berzelius's interest in the composition of the sulfur oxyacids was enhanced by two additional concerns. First, the oxidation of sulfur provided more evidence for the law of multiple proportions (pp. 19—20). Second, Davy's speculations about the possibly compound nature of sulfur provided material that perhaps could be resolved by analysis (pp. 10—11).

[1] *Ibid.,* p. 4; *Autobiography,* p. 65.

in the next subsection. Oxidizing lead sulfide with *aqua regia,* Berzelius satisfied himself that neither the lead oxide nor the sulfuric acid was present in excess in the resultant sulfate, thereby proving the preservation of the combining proportions of the two radicals, lead and sulfur, through the oxidation (that is, PbS became $PbO.SO_3$, the proportion of sulfur to lead being the same in both compounds). Since sulfate of lead was a compound of the lowest oxide of lead with sulfuric acid and since the weight of the oxygen of the oxide had been shown to be half the weight of the sulfur in the sulfide, Berzelius concluded that in lead sulfate the oxygen in the metallic oxide weighed exactly half as much as the sulfur in the sulfuric acid.[2] From this in turn Berzelius deduced that a given weight of any oxide saturated by an amount of sulfuric acid contained half as much oxygen by weight as the sulfuric acid did sulfur. That is, the amount of oxygen in the base of any neutral sulfate was always the same. This was a statement of the law of the constancy of basic oxygen that specified the weight of basic oxygen for a salt containing a given amount of acid.[3]

For sulfates, that weight could be determined by completing the analysis of sulfuric acid. Berzelius determined that 100 parts by weight of sulfuric acid contained 40.58 parts sulfur, and therefore, that the oxygen of the base saturated by 100 parts of sulfuric acid was 20.29 parts.[4] By similar reasoning he determined that any base saturated by 100 parts of muriatic acid must contain 30.93 parts of oxygen.[5] Generalizing, Berzelius held that all saline combinations obeyed this law.[6]

[2] "Sauerstoffmengen," 259; *Versuch,* pp. 11—12. See also Freund, *Study,* p. 181.

[3] *Versuch,* p. 12.

[4] *Ibid.,* p. 13.

[5] *Ibid.,* pp. 27—28. As presented, the calculation of the composition of the copper muriate contained two errors. Berzelius claimed (*Versuch,* p. 27; *Gilbert's Annalen,* 37 [1811], 291) to have analyzed a 6 gram sample of the muriate of the lower oxide of copper. His results followed only if the sample weighed 5 grams. This was evidently an error of typography or translation, for 5 grams was the weight reported in the Swedish version, *Afhandlingar,* 3 (1810), 197. Moreover, Berzelius's value for the amount of muriatic acid in the muriate of copper was in error. He obtained this value by dissolving the muriate, precipitating it with silver nitrate, weighing the precipitate of muriate of silver, and multiplying the weight by the percentage of muriatic acid in muriate of silver (18.7), already determined. The weight of muriatic acid he reported was 1.321 g., but, by his own analysis of muriate of silver, it should have been 1.331 g. Though 1.321 g. was the value reported in the original Swedish as well as in the two German versions, Berzelius's values for the percentage composition of the muriate of copper follow only if this correction is made.

[6] *Versuch,* p. 47. His reasoning was summarized in "Sauerstoffmengen," 259—260. Though the preceding paragraphs have focussed on the sulfates, Berzelius claimed (*Versuch,* p. 4) that he first obtained this version of the law with the muriates (salts that were also analyzed by Bucholz and Rose, whose results Berzelius had had difficulty reconciling with Richter's laws). For his discussion of Rose's and Bucholz's analyses of muriatic acid, see *Versuch,* pp. 24—27. For his discussion of Bucholz's, Rose's, Klaproth's, and Richter's analyses of sulfuric acid, see *Versuch,* pp. 13—18. In still another location, Berzelius claimed to have arrived at the law of basic oxygen by the same route as had Gay-Lussac: Berzelius, "Einige prüfende Versuche über die Theorie der Volta'schen Säule, und Nachricht seinen [Berzelius's] Untersuchungen über die Metallisirung der Alkalien, und über die Zahlenverhältnisse, wonach Sauerstoff und Schwefel sich mit den Metallen verbinden . . .

The law of the constancy of basic oxygen was the breakthrough that Berzelius had been seeking. It permitted him to calculate the composition of the alkalies and earths, and it did so by linking the composition of these oxides with the composition of their salts. The basic oxygen of a metallic salt of an acid was the same as that of an alkali or alkaline earth salt of the acid, and the basic oxygen was the oxygen content of the base, that is, the alkali or earth. Berzelius offered these calculations of the oxygen content of the alkalies and earths in the second half of the *Versuch*. For example, after detailing there a variety of unsuccessful attempts to isolate the supposed compound basis of ammonia, Berzelius went on to calculate the proportion of oxygen in the ammonia by application of the law of basic oxygen to a substance, sal ammoniac, that, ironically, later would be shown to contain no oxygen. He determined that this salt was composed of[7]

muriatic acid	49.55	100.00
ammonia	31.95	64.48
water	18.5	
	100.00	

By the law of basic oxygen, the 64.48 parts by weight of ammonia must contain 30.93 parts of oxygen, or, ammonia was composed of[8]

basis	52.714	100.00
oxygen	47.286	89.623
	100.000	

d. The principle of the preservation of combining proportions

The organization and content of much of the first half of the *Versuch* was determined by Berzelius's discovery of the principle of the preservation of combining proportions. Repeatedly, he compared the composition of the oxides of metals with the composition of their sulfide *in minimo*, produced their salts from the latter by direct oxidation, and compared the proportions of the metallic radicals with sulfur in their various compounds. He did not yet attempt any

In einem Briefe an den Prof. Gilbert. Stockholm, den 20. Apr. 1810," *Gilbert's Annalen*, 35 (1810), 269—277, esp. 273. In still another location, Berzelius presented himself as having merely verified Bergman's statement of the law: Berzelius, "Ueber die bestimmten und einfachen Verhältnissen nach welchen die Bestandtheile der unorganischen Natur verbunden sind. Summarischer Bericht über diesen Gegenstand an den königl. Acad. in Stockholm erstattet," *Schweigger's Journal*, 2 (1811), 297—326, esp. 307; Swedish, "Om de bestämde proportioner, hvari den oorganiska naturens beståndsdelar finnas förenade, summariskt resultat af de försök som häröfver blifvit anställde," *K.S.V.A. Handlingar.* [2] 32 (1811), 169—197, esp. 178.

[7] *Versuch*, p. 73. Berzelius extended the law to cover any compound in which an acid entered in definite amount. The other constituent could be considered a base. One prominent example is Berzelius's calculation (*Versuch*, pp. 78—79) of the amount of oxygen contained in oxymuriatic acid (chlorine) in excess of that contained in muriatic acid, for a given weight of the muriatic radical. The excess oxygen was treated as a base in combination with the acid.

[8] *Ibid.*, p. 73.

general comparison of degrees of sulfuration with degrees of oxidation or examine the preservation of the proportions of higher sulfides through oxidation into salts of higher oxides. In the *Versuch* he compared only the proportion of metal and sulfur in the sulfide *in minimo* with the saline compounds of the two radicals, including the salts, both neutral and basic, of the higher oxides. Only after much additional analytical work, performed in conjunction with the rule of oxides, did Berzelius attain full clarity about the relation of degrees of sulfuration and oxidation and of these with salts of higher oxides. This clarification is considered below in Chapter V, section 5. Here only Berzelius's early deployments of the principle, its relation to his saline theory, and his treatment of anomalous cases are examined.

Among Berzelius's more striking applications of the principle appeared in his analysis of sulfurous acid. He prepared a sulfite of barytes and oxidized it, as he had oxidized the sulfide of lead, and produced a sulfate of barytes. By showing, again, that there was no excess of either barytes or acid, he proved that the proportions of the radicals in the sulfite had been preserved through the oxidation to the sulfate. Because the quantity of sulfur was the same in the sulfate as in the sulfite from which it was obtained by oxidation, the quantity of oxygen combined with the sulfur in the sulfite, that is, the oxygen content of the sulfurous acid in the salt, could be determined by subtracting the sulfur content of sulfate of barytes (the S in $BaO.SO_3$) from the sulfurous acid content of sulfite of barytes (the SO_2 in $BaO.SO_2$). The sulfur content of the sulfate was known from the previously determined composition of sulfuric acid and the quantity of sulfuric acid in the sulfate, which Berzelius obtained by a separate analysis. The sulfurous acid content of the sulfite of barytes was likewise obtained separately. From these data, Berzelius showed that for a given weight of sulfur, the proportions of oxygen in the two sulfur oxyacids were as 1 :1 1/2.[9]

Pursuing the example of the oxidation of lead sulfide, Berzelius found a similar pattern in the compounds of copper.[1] He showed that the proportions of oxygen in the two oxides of copper were as 1 : 2, for a given weight of metal, and the sulfide, again taken to be *in minimo,* contained twice as much sulfur by weight as the lower oxide did oxygen.[2] The sulfate of the higher oxide of copper, he demonstrated, contained twice the weight of sulfur, for a given weight of copper, as did the sulfide.[3] He then sought explicitly a sulfate of copper, that contained the two combustible radicals in the same proportion that they exhibited in the sulfide. The neutral sulfate of the higher oxide did not, and the neutral sulfate of the lower oxide was unknown. Berzelius turned, therefore,

[9] *Ibid.,* pp. 18—19.
[1] *Ibid.,* pp. 19—20.
[2] *Versuch,* pp. 20—24. Cf. Berzelius's proof in the second half (*Versuch,* pp. 81—82) that hydrogen sulfide contained twice the weight of sulfur as water did oxygen, for a given weight of hydrogen.
[3] *Ibid.,* p. 26.

to the basic sulfate of the higher oxide (*basische schwefelsaure Kupfer*), apparently hoping that the excess base would compensate for the higher oxidation of the metal and thereby restore the proportions of metal and sulfur in the sulfide. This hope was not fulfilled. Berzelius showed that the basic salt of the higher oxide contained four times rather than two times the base of the neutral salt, and only the latter exhibited the proportions of the sulfide. He turned back, then, to calculate, by two separate procedures, the proportions of the unknown sulfate of the lower oxide and concluded that the salt indeed would have contained the two radicals in the same proportion as the sulfide.[4] That the salts of the higher oxide might have corresponded to a higher sulfide did not yet seem to have occurred to Berzelius.

Iron compounds, rather than those of copper, presented Berzelius with what he clearly recognized as an anomalous case. The anomaly merits discussion because, in his efforts to reduce it, Berzelius developed a conceptual technique that later proved to be of central importance to his work. He approached the study of the iron salts specifically to examine the applicability of the principle of the preservation of combining proportions: "Wir haben gesehen, dass sich Schwefelblei, und, aller Wahrscheinlichkeit nach, auch Schwefelkupfer, zu neutralen Salzen oxydiren lassen," he remarked. "Die Frage is jetzt: Gilt das nämliche auch von dem Schwefeleisen?"[5] The analyses conducted to answer the question produced the anomalous results that in the iron salts there was no integral relation between the proportions of the two radicals, iron and sulfur, in the lower sulfide and in the sulfate of the lower oxide, on the one hand, and in the neutral and basic sulfates of the higher oxide, on the other. Berzelius demonstrated that the lower sulfide and the neutral sulfate of the lower oxide both contained the two radicals in the same proportion.[6] The neutral sulfate of the higher oxide, however, contained, for a given weight of metal, 1 1/2 times the sulfur present in the lower sulfide, and the basic sulfate contained one-quarter of that amount.[7] Not only did the salts of the higher oxide fail to display a preservation of the original combining proportious of the sulfide, they failed even to display an integral multiple of those proportions.

These failures contrasted sharply in Berzelius's mind with the proportions exhibited by the copper compounds. As just noted, Berzelius had shown the copper compounds to display proportions of sulfur to metal that were integral multiples or submultiples of the proportions in the sulfide (Cu_2S). Berzelius had shown the calculated composition of the sulfate of the lower oxide to display the same proportion of sulfur to metal as the sulfide. He had pointed out that the neutral sulfate of the higher oxide contained twice the amount of sulfur for a given weight of metal as did the sulfide and that the basic sulfate of the higher

[4] *Ibid.*, pp. 26—27.
[5] *Ibid.*, p. 35.
[6] *Ibid.*, p. 36.
[7] *Ibid.*, pp. 38—39.

oxide contained four times that in the neutral sulfate. In summary, the proportions of metal, for a given weight of sulfur in the four copper compounds, as determined by Berzelius's analyses, were as 2 : 2 : 4 : 1. The corresponding figures for the iron compounds were 1 : 1 : 2/3 : 8/3.

Berzelius attempted to develop an explanation for this anomaly by suggesting that the intervention or intermediation (*Dazwischenkunft* or *Hinzukunft; Sw: mellankomst*) of a third body, in this case oxygen in combination with a pair of radicals, altered the original combining proportions subsisting between the radicals. For example, the basic salt of the higher oxide of iron

> giebt auf 100 Th[eile] Eisen 22 Th. Schwefel, welches zwar 1/4 von der Menge des Schwefels ist, die wir in dem neutralen Eisenoxydsalze [that is, the neutral salts of the higher oxide] gefunden haben, aber doch in keinem einfachen Verhältnisse zu der Schwefelmenge steht, mit welcher sich Eisen, ohne Dazwischenkunft eines dritten Körpers vereinigen lässt; ein Beweis, dass sich die Natur in vielfach zusammengesetzten Körpern, unter gewissen Bedingungen, von den Verhältnissen entfernt, welche sie für einfachere Zusammensetzungen beobachtet.[8]

Intimately connected with the anomaly in Berzelius's mind was that the proportions of oxygen in the two oxides of iron, again, unlike the corresponding copper compounds, were as 1 : 1 1/2, for a given weight of metal. It was under this sort of circumstance, the occurence of nonintegral proportions in the oxides of a radical, that intermediation could give rise to a violation of the principle of the preservation of the combining proportions of two radicals. The passage just quoted went on to argue that

> Wir sehen hier endlich, dass, wenn gleich in dem neutralen Eisenoxydulsalze [that is, the salts of the lower oxide] das Verhältniss der Bestandtheile durch den Schwefel und das Eisen [in the lower sulfide] bestimmt wird, es in den basischen Eisenoxydsalzen durchaus nicht weiter von diesen, sondern von der Säure und der Basis abhängig ist; beide verbinden sich in ihnen in dem Verhältnisse, dass die Säure in dem neutralen Salze [of the higher oxide] ihre bestimmte Sauerstoffmenge [its basic oxygen], und in dem basischen Salze das Vierfache dieser Menge aufnimmt; [but this amount is different from that involved in the salts of the lower oxides;] Umstände, welche sämmtlich darin gegründet sind, dass das Eisenoxyd nicht das doppelte Quantum, sondern nur das 1 1/2-fache des Sauerstoffs im Oxydul enthält. Hierdurch erzwingt die Verwandtschaft des Sauerstoffs zum Eisen des Oxyds ein ganz anderes Verhältnis für den Schwefel und das Eisen der Eisenoxydsalze, als worin sie sich ursprünglich [that is, in the lower sulfide and in the salts of the lower oxide] zu vereinigen streben. In den Kupferoxydsalzen finden dagegen zwischen dem Schwefel und dem Kupfer einerlei Verhältnisse statt, weil das Oxyd die doppelte Menge Sauerstoff als das Oxydul enthält.[9]

This passage displays Berzelius's discontent with combining proportions in the multiple of 1 1/2 and indicates that his dissatisfaction with them seems to have derived, at least in the first instance, not from any notions about a particulate ontology of matter, but from his belief that they were the source of the

[8] *Ibid.*, p. 38.

[9] *Ibid.*, pp. 38—39.

violation of the principle of the preservation of combining proportions. He attempted here to clarify this connection by a comparison of the oxygen and sulfur compounds of copper with those of iron. The oxygen compounds of copper contained proportions of oxygen in the integral multiple by 2, and its sulfur compounds did not violate the principle; while the oxygen compounds of iron were in proportions in the multiple by 1 1/2 and the sulfur compounds of iron did violate it. Berzelius here argued that the affinity of oxygen for iron manifested in the new, nonintegral proportion of oxygen altered the proportion in which iron and sulfur combined, as if the higher oxide produced a new set of "original combining proportions" that either were preserved intact or appeared as integral multiples in the compounds of the higher oxide. An alternative statement, one that emphasized the connection of the principle with the law of basic oxygen, was that the nonintegral dose of oxygen changed the amount of basic oxygen taken up by the acid, so that the combining proportions of the radicals followed this new proportion of the acid and the base and not the original proportion of the unoxidized radicals. This view was articulated in the first sentence of the preceding passage. The term "Dazwischenkunft" was the label that Berzelius coined to describe this effect.

The only means Berzelius could imagine for eliminating the proportions in the multiple by 1 1/2 was to suspect the existence of unknown lower oxides. Nevertheless, his experiments and calculations indicated, he asserted, that at least for iron, if not for sulfur, there was no such lower oxide; unless one existed in some poorly understood form, like the state of iron in the coloring matter of blood.[1] Though this suggestion now lay fallow, Berzelius later used it, in conjunction with a modified form of the idea of intermediation, to seek for hidden lower oxides. For the moment, however, Berzelius was prepared to rest content with the idea of intermediation as a technique for eliminating the anomaly to the principle of the preservation of combining proportions. It was the best he could do, and it was enough to allow him to retain the principle, which he found too attractive to relinquish.[2]

The last point that need be made about the principle is that it was readily integrated with Berzelius's belief that in the radicals lay the seat of the acidity or basicity of their oxides. Nascent in the *Versuch,*[3] this connection reached systematic expression in the "Rechtfertigung:"

Aus der electrochemischen Ansicht folgt noch eine Berichtigung unserer Ideen vom Säure-Princip. Lavoisier, der gefunden hatte, dass der Schwefel, der Phosphor u.a., in

[1] *Ibid.*, p. 47.

[2] Berzelius's later analyses of the iron compounds, reported in the third *Fortsetzung* (*Versuch,* pp. 188—192), removed the anomalies; for this see below. Chapter V, subsection 7 *c*. Until then, Berzelius adhered to the principle despire the anomalous cases, sometimes speculating on how further discoveries might reduce the anomalies. For example, see *Versuch,* pp. 78—79.

[3] *Versuch,* in the first half, p. 19; in the conclusion of the second half, p. 82; and in the first *Fortsetzung,* pp. 96—97.

Verbindung mit Sauerstoff, Säuren darstellen, glaubte dass der Sauerstoff die saure Natur dieser Verbindungen hauptsächlich bestimme. Er hatte insofern darin Recht, als der Sauerstoff der einzige absolut electropositive Körper ist und als die Säuren electropositive Zusammensetzungen sind; da man aber seitdem Körper entdeckte, welche alle Charaktere der Säuren besitzen, ohne Sauerstoff zu enthalten, so wurde die Richtigkeit dieser Ansicht und Benennung des Sauerstoffs in Zweifel gezogen; und durch die Beobachtung, dass eine grössere Anzahl Körper mit dem Sauerstoff Salzbasen als Säuren darstellen, lag es klar am Tage, dass die sauren Eigenschaften oxydirter Körper, eben so wenig als die basischen oder alkalischen, vom Sauerstoff herrühren konnten. Die electrochemische Ansicht, nach welcher das brennbare Radical einer Säure gegen das einer Salzbase (beide im noch nicht oxydirten Zustand) electropositiv ist, in Verbindung mit der Erfahrung, dass sich Körper von entgegengesetzter electrochemischer Natur im nämlichen Verhältniss verbinden, sie mögen oxydirt oder noch im Zustande von Verbrennlichkeit seyn, erklärt uns hinlänglich wie der Schwefel sich mit dem Kalium in den nämlichen bestimmten Verhältnissen im Schwefelkalium, im Schwefelkali, im hydrothionsauren, schwefligsauren und schwefelsauren Kali verbinden kann, und wie der Schwefel in allen diesen Verbindungen die Rolle einer Säure spielt, weil alle diese Zusammensetzungen ihren Grund in dem ursprünglichen electrochemischen Verhalten des Schwefels zu dem Kalium haben. Es ist klar, dass was hier von dem Schwefel und dem Kalium geltend ist, auch von allen andern im electrochemischen Gegensatz stehenden Körpern gelten müsse, und dass es mithin die Natur des brennbaren Radicals, nicht des Sauerstoffes ist, durch welche die Charaktere der oxydirten Körper als Säuren oder als Basen bestimmt werden.[4]

In this passage, Berzelius asserted at least three broad points, though not in the order of their apparent historical chronology. Because the alkalies and earths had been shown to be oxides just as were the acids, the origin of their acidic or basic properties had to lie in their radicals. Second, the radicals, as the sources of acidic or basic character, played the role of acid or base in their direct binary combinations with one another; similarly, the combinations of the radicals with other elements, above all with oxygen and prominently also with sulfur, may enter into other compounds as acidic or basic constituents, as was the case with salts. Direct combinations of radicals with one another, therefore, were like salts. The possibility of finding acids not containing oxygen was a consequence of the presence of unoxidized acidic radicals in compounds. The case of "hydrothionic acid" (the weakly acidic compound, hydrogen sulfide), to which considerable discussion was devoted in the second half and first *Fortsetzung* of the *Versuch,* provided an example.[5] Third, not only were the acidic or basic properties of the radicals preserved in their combinations with oxygen and sulfur, and in the compounds of the oxides and sulfides; the original combining proportions in a compound of a pair of unoxidized radicals were also preserved in the more complex compounds containing the radicals. The combining proportions in which oxides united with one another, then, were specific properties, dependent, like the electrochemical character of oxides, upon the radicals. Direct compounds

[4] "Rechtfertigung," 138—139.

[5] *Versuch,* pp. 79—82, 96—97.

of unoxidized radicals were therefore salts *in potentia,* both qualitatively and quantitatively. In the passage just quoted, Berzelius cited a sequence of combinations in which the radicals potassium and sulfur appeared always in the same proportion as in the sulfide of potassium, and in which the metal was the seat of basicity and the sulfur the seat of acidity. It was in the *Versuch,* however, that the first example of preservation of combining proportions appeared: the proportion of lead and sulfur was the same in the sulfide and in the sulfate.

Though Berzelius was unable to sustain this claim that the combining proportions were predictable functions of the radical—only a later chemistry, having the theory of valence, succeeded in this—Berzelius's later comparative studies of the relations of oxidation and sulfuration, in conjunction with the rule of oxides, allowed him to clarify both the principle and the rule. To the rule of oxides, then, this discussion now turns.

e. The rule of oxides and the "Spätere Zugabe"

Berzelius's rule of oxides was the keystone of his chemistry. Like the law of basic oxygen, the rule linked the composition of salts with the composition of their constituents. In particular, it established a relationship between the quantities of oxygen in each of the two constituents of a salt. In the *Versuch,* it reads:

> In den neutralen Salzen ist die Menge des Sauerstoffs, welchen die Säure enthält, ein Vielfaches der Menge des Sauerstoffs in der Basis nach einer ganzen Zahl. Diese Regel lässt sich ein wenig allgemeiner und, wie ich glaube, nicht minder richtig folgendermaassen ausdrücken: Wenn zwei oxydirte Körper einander sättigen, so enthalten sie stets den Sauerstoff nach solchen Verhältnisse, dass die Menge desselben in dem Körper, welcher in der Kette der electrischen Säule zum positiven Pole hingeht, ein Vielfaches nach einer ganzen Zahl von der Sauerstoffmenge in dem anderen Körper ist, der zum negativen Pole hinstrebt.[6]

In the first instance, the rule was a statement regarding the composition of neutral salts and was then generalized to cover all compounds formed by the union of two oxidized bodies.

Berzelius attained the rule some time between the completion of the two halves of the *Versuch* and the drafting of the first *Fortsetzung,* in which its earliest systematic elaborations were detailed. The rule was announced in a letter that Gilbert had received and published after he had received the first and second halves of the *Versuch* but before he had published them.[7] The genesis of the rule may be reconstructed by examining further materials that Berzelius

[6] *Versuch,* in the first *Fortsetzung,* p. 88.

[7] Berzelius, "Schreiben des Herrn Prof. Berzelius, Mitgl. der schwed. Akad. d. Wiss. an den Prof. Gilbert, über einige Gegenstände, welche zwischen Davy und den HH. Gay-Lussac und Thénard [*sic*] streitig sind, und über ein zweites neues Gesetz, welches er im Verfolge seiner Untersuchungen aufgefunden hat. Stockholm, d. 1. Oct. 1810," *Gilbert's Annalen,* 37 (1811), 208—220; rule, 217. This letter was dated almost exactly one year after Berzelius had completed the first and second halves of the *Versuch;* in this connection, cf. above, p. 168, n. 8.

sent to Gilbert, who published them as appendages to the first and second halves. In these passages, Berzelius offered a series of speculations accompanying the emergence of the rule. Virtually all of Berzelius's subsequent investigations, from these addenda to the *Proportions chimiques,* may be seen as the pursuit of themes first hinted in these few pages. This material therefore merits careful analysis. The emergence of the rule is examined in this subsection, and the speculative material accompanying it is investigated in the next two subsections.

The first of the addenda, chronologically, was appended, as section XXI, to the second half, and is entitled "Spätere Zugabe, die organischen Körper betreffend. Aus der Handschrift des Verfassers."[8] In a note to this section, Berzelius indicated that he was then attempting to carry out what he called "normal analyses," that is, new and more exact analyses of those substances such as the sulfides, oxides, and sulfates of lead, the muriates, sulfides, and sulfates of silver, and the salts of barytes, that had served as the foundations for various calculations of the *Versuch,* particularly in the first half.[9] His intention was to establish an unquestionably exact analytical foundation for further calculations. Berzelius indicated in the "Zugabe" that he would make these results known shortly, but Gilbert added in a note that these had already appeared in the appendage to the first half.[1] The appendage to the second half therefore antedated the appendage to the first.

The latter, labelled section XIII, was entitled "Ein späterer Zusatz aus der Handschrift des Verfassers. Neue Untersuchungen über das Schwefelblei, die Bleioxyd und die Schwefelsäure."[2] In a note to this title, Gilbert confirmed that the content of section XIII was taken from a paper written in German especially for his *Annalen* and containing Berzelius's description of his efforts to conduct normal analyses.[3] Gilbert explained that he had affixed this set of analyses to the end of the first half because of their close relation with the analyses already detailed there, a relation that is noted below. This was not the entire set of normal analyses, however. The remaining portion of the analyses was prefixed to the first *Fortsetzung* and published as its first section, entitled, "Berichtigung der Analyse des salzsauren Silbers, und einige davon abhängenden Analysen."[4]

[8] *Gilbert's Annalen,* 37 (1811), 465—472; *Versuch (Ostwald's Klassiker),* pp. 82—87. This addendum is called "Spätere Zugabe," or "Zugabe" in the text and is cited in the notes, as are all passages from the *Ostwald's Klassiker* edition, as *Versuch,* followed by page number. Citations from *Gilbert's Annalen,* necessary only for notes not included in the *Ostwald's Klassiker* edition, appear as "Zugabe," followed by page number

[9] *Versuch,* p. 84n.

[1] "Zugabe," 468n.

[2] *Gilbert's Annalen,* 37 (1811), 324—334; *Versuch (Ostwald's Klassiker),* pp. 47—53. Only the subtitle, "Neue Untersuchungen...," is given in the *Ostwald's Klassiker* edition. This addendum is called "Zusatz" in the text and is cited in the notes from the *Ostwald's Klassiker* edition as *Versuch,* followed by page number. Citations from *Gilbert's Annalen* appear in the notes as "Zusatz," followed by page number.

[3] "Zusatz," 324n.

[4] *Gilbert's Annalen,* 38 (1811), 163—174; *Versuch,* pp. 88—95.

In a note there Gilbert confirmed that these analyses, together with those appended to section XIII of the first half, constituted the full set of normal analyses.[5]

It was not merely out of a general concern to construct exact foundations for his further studies that Berzelius undertook the extraordinarily laborious task of repeating a large number of his analyses. More significant was Berzelius's wish to follow up his own suggestion that proportions in fractional multiples might be shown to be merely apparent by finding hidden oxides in still lower proportions. He hoped to succeed in this by pursuing his earlier idea, that iron in the coloring matter of blood might be present in a lower oxidation stage than observed in inorganic nature. Thus, he turned to organic matter as a possible repository of lower oxides. The "Zugabe" was the locus in which Berzelius recounted his attempt to follow this route; therefore, the passage was one "die organische Körper betreffend." Its opening paragraph testifies to this motivation:

[5] *Gilbert's Annalen*, 38 (1811), 161n. The reader may find it helpful to refer to the following table of contents of the *Versuch*. For each entry, there are two sets of page numbers. The first set, without parentheses, refers to the volumes of *Gilbert's Annalen* cited in the table. The second set of page numbers, in parentheses, refers to the *Ostwald's Klassiker* edition of the *Versuch*. Below the table of contents is a chronological list that shows the order in which Berzelius prepared his manuscripts and sent them to Gilbert and indicates how Gilbert assembled them to form the *Versuch* as published in his *Annalen* and as republished in *Ostwald's Klassiker*. Discussion of the last three items in the chronological list appears below, p. 202, n. 6.

Versuch: Table of Contents

First half: *Gilbert's Annalen*, 37 (1811)	249-334 (3-53)
Introduction	249-253 (3-6)
Sections I—XII	254-324 (6-47)
Section XIII ("Zusatz")	324-334 (47-53)
Second half: *Gilbert's Annalen*, 37 (1811)	415-472 (54-87)
Introduction	415-416 (54)
Sections XIV—XX	416-465 (55-82)
Section XII ("Spätere Zugabe")	465-472 (82-87)
First *Fortsetzung: Gilbert's Annalen*, 38 (1811)	161-226 (87-127)
Introduction	161-163 (87-88)
Section I (containing some "normal analyses")	163-174 (88-95)
Sections II—VIII	174-223 (96-125)
Conclusion	223-226 (125-127)
Second *Fortsetzung: Gilbert's Annalen*, 40 (1812)	162-208 (127-154)
Introduction	162-163 (127)
Section I (on nitric acid)	163-176 (127-135)
Section II (on basic nitrates)	176-185 (135-140)
Section III (on nitrites)	186-208 (140-154)
Third *Fortsetzung: Gilbert's Annalen*, 40 (1812)	235-330 (154-210)
Introduction	235-245 (154-160)
Sections I—III	246-320 (160-205)
Section IV (general conclusion, entire *Versuch*)	320-330 (205-210)

Ingredients of the *Versuch:*
chronological list
and location as published by Gilbert

1. First and second halves, as translated from the Swedish, including
   A. First half, through section XII
   B. Second half, through section XX
2. "Spätere Zugabe," published as second half, section XXI
3. "Normal analyses," published as
   A. "Zusatz," (first half, section XIII)
   B. First *Fortsetzung*, section I
4. Berzelius's first continuation, published as first *Fortsetzung*, sections II—VIII
5. Berzelius's "Zugabe" to his first continuation, published as second *Fortsetzung*, section I.
6. Berzelius's second continuation, published as
   A. Second *Fortsetzung*, sections II—III
   B. Third *Fortsetzung*

Ich habe in dieser Abhandlung die Vermuthung geäussert, gewisse Körper mochten wohl niedrigerer Oxydationsgrade fähig sein, als die, welche wir bisher für die Verbindungen mit Sauerstoff im Minimum angesehen habe, z.B. Schwefel und Eisen. Was besonders letztere betrifft, habe ich oben ... die Vermuthung geäussert, das Eisen möge sich wohl in den organisirten Körpern in einem solchen geringern Oxydationsgrade finden. Die Producte der organischen Natur wollen sich, dem ersten Anschein nach, nicht in die Gesetze fügen, auf die ich in der Zusammensetzung der unorganisirten Körper geführt worden bin: es war daher meine Absicht, in einer folgender Abhandlung die Revision der organischen Stoffe vorzunehmen, und dabei mit den einfachsten, d.h. mit Oelen und vegetabilischen Säuren, den Anfang zu machen. Je mehr ich mich aber mit diesen Versuchen beschäftigt habe, desto mehr bin ich überzeugt worden, dass die Data aus der Chemie für unorganischen Körper noch nicht zahlreich und, mit sehr wenigen Ausnahmen, noch nicht genau genug sind. Ich werde daher in dem Folgenden nur zeigen, wie man durch Untersuchungen der unorganischen Natur allmählich zu den Resultaten gelangen kann, die bei der organisirten zu erwarten sind.[6]

Though organic matter night indeed contain the anticipated lower oxides, Berzelius's attempts to find them led him to recognize that analytical forays into organic chemistry were premature. A more solid foundation in inorganic chemistry was necessary, he realized, and accordingly he undertook the normal analyses. Berzelius was hesitant to apply even this new analytical material to organic chemistry, however. Instead, as this passage indicates, he remained with inorganic chemistry, attempting to illustrate there the patterns that he expected to be prevalent in organic chemistry: the existence of hidden lower oxides that would eliminate fractional multiples as merely apparent. The example he chose was sulfur, one of the two radicals that had provided him with his early examples of proportions in the multiple by 1 1/2. By carrying through the attempt to find lower oxides of sulfur, Berzelius hit upon the first example of the rule of oxides.

Berzelius's original determination of the composition of sulfuric acid by oxidation of lead sulfide depended on the analysis of the lowest oxide of lead, of the sulfide of lead, and the oxygen content of lead sulfate.[7] These data had enabled Berzelius to partition the oxygen of the sulfate between the oxide and the acid. In doing so he did not notice—and this is not surprising—that the numbers he obtained for the oxygen of the saline constituents were approximately in the ratio of 3 : 1.[8] Subsequently, however, he did come to recognize this fact, which was his first example of the general phenomenon described by the rule of oxides. What induced him to see it was his comparison of the results of two separate additional calculations of the composition of sulfuric acid. The first was a re-

[6] *Versuch*, pp. 82—83.

[7] See above, p. 172, nn. 2—6 and accompanying text; for his original calculation of the composition of sulfurous acid, see above, p. 174, n. 9 and accompanying text. Berzelius had found that 10 g. of lead sulfide produced 12.65 g. of lead sulfate. Of the 2.65 g. of oxygen added, 0.67478 g. belonged to the lead oxide and 1.9752 g. belonged to the sulfuric acid. The 10 g. of lead sulfide contained 8.651 g. of lead, by Berzelius's analysis. Therefore the sulfuric acid contained 1.349 g. of sulfur and 1.9752 g. of oxygen.

[8] 1.9752 : 0.67478 :: 2.92 : 1.

petition of the calculation just noted of the composition of sulfuric acid based on the lead compounds, but using the new results of the attempted normal analyses. Because this portion of the normal analyses and the calculations based on them recapitulated those reported in the concluding section of the first half, Gilbert published them in appendage to the latter. The second was a calculation based on Berzelius's calculated composition of the anticipated lower oxide of sulfur.

The latter computation appeared in the "Spätere Zugabe." The technique by which he accomplished it is arresting because it reflected the kind of argumentation possible in a quantitative chemistry constructed without benefit of the idea of atomic weights. Relying on the gas density analysis of carbonic acid gas (carbon dioxide) by Allen and Pepys and on Gay-Lussac's analysis of carbonic oxide (carbon monoxide), Berzelius calculated the proportions by weight of carbon and oxygen in each compound and showed that the latter contained half the oxygen of the former. He next cited analyses by Thomson of two gases composed of carbon and hydrogen. The first gas, obtained by the distillation of peat, Berzelius showed to contain twice the carbon of the second, olefiant gas, for a given weight of hydrogen. Berzelius regarded the second gas as carbon saturated *in minimo* with hydrogen. He now had determined the amounts of both oxygen and hydrogen that saturated carbon *in minimo.* Next, he cited the results of his own analysis of hydrogen sulfide, revealing the amount of hydrogen necessary to saturate sulfur *in minimo.* Berzelius thus had obtained three terms, related to one another, he would soon reveal, as three of four proportionals, and he proceeded to calculate the fourth. The latter represented the amount of oxygen necessary to saturate sulfur *in minimo,* that is, the proportions of oxygen in the anticipated lower oxide. He offered no justification for this procedure, but it clearly must be taken as an application of the law of equivalent proportions. The unspoken assumption was that the law could be applied to substances capable of combining with others in more than one proportion or degree of saturation so long as each of the four terms of the proportion referred to the same degree of saturation. If E_e represents the weight of element E necessary to saturate element e *in minimo,* then Berzelius's calculation may be expressed as

$$\frac{H_c}{O_c} = \frac{H_s}{O_s},$$

where O_s, the oxygen necessary to saturate a given weight of sulfur *in minimo,* is the unknown. Accordingly, it was necessary to conceive this oxide of sulfur as an oxidation stage equivalent to that represented by carbonic oxide, and indeed, in a reference to this calculation appearing in the "Zusatz," Berzelius labelled it "eine dem Kohlenstoffoxyde entsprechende Oxydationsstufe."[9]

By this technique Berzelius determined that sulfur in its lower oxidation stage

[9] *Versuch,* p. 52.

had to contain 50 parts by weight of oxygen to 100 parts of sulfur. This, he noted, was exactly half the weight of oxygen for a given weight of sulfur that his calculation based on the normal analyses indicated sulfurous acid contained and one-third that existing in sulfuric acid. Comparing these results with the data for the lead salts, Berzelius noticed that this lower proportion of oxygen happened to equal the proportion of oxygen in the lead oxide of the lead salts. He found that 10.0 parts of lead produced a lead sulfate in which there was 0.77 parts of oxygen in combination with the lead and 1.542 parts of sulfur present as sulfuric acid. That amount of sulfur, in its lower oxide, took 0.77 parts of oxygen, or one-third the amount in the sulfuric acid. The sulfuric acid present in the lead salt, then, contained three times the oxygen of the lead oxide.

> Aus diesen Ansicht fliesst noch eine andere Methode, die Zusammensetzung der Schwefelsäure zu berechnen. Wir sehen, dass das Bleioxydul, welches eine gegebene Menge von Schwefelsäure sättigt, genau 1/3 so viel Sauerstoff als die Säure enthält, und es muss genau halb so viel Sauerstoff als die schweflige Säure, wovon es gesättigt wird, in sich schliessen, weil die schwefligsauren Salze, wenn sie Sauerstoff aufnehmen und in Schwefelsäure übergehen, den Neutralitätszustand nicht verändern. [That is, after oxidation of the sulfite, the resultant sulfate was found to be neutral, indicating that neither the acidic nor the basic component was present in excess, and that, accordingly, the original combining proportion of the radicals, lead and sulfur, was preserved.] Da nun aber 279 Th. Bleioxydul 100 Th. Schwefelsäure sättigen und diese 279 Th. Bleioxydul 19,95 Th. Sauerstoff in sich schliessen, so muss hiernach die Schwefelsäure in 100 Th. 59,85 Th. Sauerstoff enthalten; welches nur um 0,001 des Gewichts der Säure von der vorigen Bestimmung [that is, the calculation based on the oxidation of lead sulfide] abweicht.[1]

Here was the first example of a calculation based on the existence of an integral proportion subsisiting between the oxygen of the acid and that of the base. The generalization of this observation was the rule of oxides. First announced in the letter to Gilbert of 1 October 1810, the rule was given a systematic development and some applications in the first *Fortsetzung,* from which the above quotation of the rule (p. 179 accompanied by n. 6) was taken. The rule of oxides became the basis of Berzelius's ever widening applications of the theory of salts, of which he produced a steady stream, from those in the remaining portions of the *Versuch,* through a series of analytical papers, culminating in the *Mineralogy* and in the articulation of the full theory of proportions in the *Proportions chimiques.*

f. The speculations of the "Spätere Zugabe"

These applications, however, did not rely on the rule alone. Rather, Berzelius pursued them in connection with several speculative themes appearing in the "Zugabe." Unfortunately, Berzelius did not allow himself the freedom to expand upon these speculations, perhaps in conformity with his intention of minimizing the theoretical content of the *Versuch.* The "Zugabe" covered but a

[1] *Ibid.*, p. 53.

few pages and offered a number of tantalizing suggestions without elaborating them or disentangling them from one another. The historian is confronted, then, with two risky tasks: analyzing a text, the "Zugabe," at once barely articulate and critically important; and exhibiting the dependence of much later material upon this source. The speculative content of the "Zugabe" is analyzed in this subsection and in the next one. The dependence of later work on these ideas is a constant theme of the next chapter.

Having calculated the composition of the suspected lower oxide of sulfur, Berzelius hoped to demonstrate its existence. He offered no indication that he expected it to be found as an isolable substance. Rather he thought it might exist as an ingredient in a compound, just as he had already suggested might be the case with a suspected lower oxide of iron in the coloring matter of blood. The model for such a combination was clearly the saline: perhaps the lower oxide of sulfur formed salts with other oxidized bodies, just as did the higher oxides of sulfur, sulfurous and sulfuric acids. Berzelius believed he had found compounds containing not only this lower oxide, but also another still lower one, in the substances (later understood to be chlorides of sulfur) formed from the combination of sulfur with muriatic acid. Thomson had discovered one of these in 1803, calling it "sulphuretted muriatic acid" (the monochloride, S_2Cl_2), which was later analyzed by Bucholz.[2] The other compound (the dischloride, SCl_2) was analyzed by A. B. Berthollet in 1807.[3] In accordance with his own calculation of the supposed oxygen content of oxymuriatic acid, Berzelius determined that Berthollet's compound, for a given weight of sulfur, contained half the quantity of oxygen in sulfurous acid and that Bucholz's compound contained half as much oxygen as Berthollet's.[4] He then suggested that the two compounds could be conceived as the muriates of these two lower oxides of sulfur, that is, that they were composed on the model of salts.[5] In this case, however, the oxides of sulfur played the basic and not the acidic role.

[2] Thomas Thomson, "On the compounds of sulphur and oxygen," *Nicholson's Journal,* 6 (1803), 92—109; for sulphuretted muriatic acid, see pp. 104—109. Christian Friedrich Bucholz, "Versuche über Thomson's schwefelhaltige Salzsäure (Verbindung von Schwefel, Sauerstoff, und Salzsäure); zur näheren Kenntniss dieser merkwürdigen Verbindung," *Gehlen's Journal,* 9 (1810), 172—188. For Bucholz, see Partington, *History,* III, pp. 581—583.

[3] A.B. Berthollet, "Mémoire sur le combinaison du soufre avec l'oxigène et l'acide muriatique," *Mém. soc. Arcueil,* 1 (1807), 161—179.

[4] *Versuch,* p. 85.

[5] *Ibid.,* pp. 85—86: "Nehmen wir nun an, bei diesen Versuchen [of Berthollet and Bucholz], bei deren ersten [Berthollet's] man die grösste mögliche Menge Sauerstoff, bei deren zweiten man die grösste Portion Schwefel in die Verbindung zu bringen gesucht hatte, sie in beiden irgend eine kleine Unrichtigkeit bei der Sättigung oder der Wägung vorgefallen, und es hätten z.B. im Berthollet'schen Versuche 100 Th. Schwefel in der That 214 Th. oxygenirte Salzsäure (statt 204 Th.), in dem Bucholz'schen Versuche dagegen 100 Th. Schwefel 107 Th. dieser Säure (statt 111 Th.) aufgenommen, so würden wir hier zwei Verbindungen haben, von denen die letztere ein salzsaures Schwefeloxydul, die erstere ein salzsaures Schwefeloxyd wäre, worin der Schwefel mit der doppelten Menge Sauerstoff und Säure als in jenem vereinigt wäre, dem ähnlich, was wir bei den Metallsalzen schon gesehen haben."

To attain this view, Berzelius had considered and rejected several alternatives:

> Ein solches Schwefeloxyd [the suspected lower oxide containing one-third the oxygen of sulfuric acid] kennen wir nicht, es sei denn in der von Thomson entdeckten schwefelhaltigen Salzsäure vorhanden. Ich habe die Versuche sorgfältig erwogen, welche Thomson, Berthollet und Bucholz mit diesem Körper angestellt haben, und sie scheinen mir übereinstimmend weder oxygenirte Salzsäure, noch schweflige Säure in ihm zu erkennen zu geben. Es lässt sich also nur annehmen, aller Schwefel sei mit dem Sauerstoffe vereinigt, dieser neue Körper sei also nicht anders als eine Verbindung der Salzsäure mit Schwefeloxyd. Vermengt man ihn mit Wasser, so folgt die Salzsäure einer nähern Verwandtschaft, und das abgesonderte Schwefeloxyd erleidet einer Zerlegung in Schwefel und schweflige Säure, der Zerlegung der Kupferoxyduls ähnlich, wenn man es mit verschiedenen Säuren behandeln.[6]

Berzelius was here invoking a portion of his detailed knowledge of inorganic reaction chemistry to support the analogy between the new muriatic compounds and traditional salts. It is known today that the hydrolysis of the monochloride proceeds by a series of reactions:

$$\begin{aligned} S_2Cl_2 + 2HOH &= 2HCl + S_2(OH)_2 \\ S_2(OH)_2 &= H_2S + SO_2 \\ H_2S + S_2(OH)_2 &= 3S + 2HOH \\ SO_2 + S_2(OH)_2 &= \text{polythionic acids} \end{aligned}$$

Berzelius could not have observed the polythionic acids, but he was clearly aware of the production of sulfurous acid (SO_2) and of elemental sulfur and recognized that the muriatic acid emerged from the reaction intact: it "folgt einer nähern Verwandtschaft," while the supposed oxide of sulfur underwent a decomposition (disproportionation) into elemental sulfur and sulfurous acid. He here compared the decomposition of the oxide to the behavior of cuprous oxide when undergoing acidic decomposition, a reaction that may also be written in modern terms as a disproportionation:

$$Cu_2O + H_2SO_4 = CuSO_4 + Cu + HOH.$$

Just as the cuprous oxide was both oxidized to a higher oxide, in this case forming a salt with the acid, and reduced to the elemental state, so the lower oxide of sulfur was oxidized to a higher oxide, sulfurous acid, and was reduced to the elemental state. Cuprous oxide was a salifiable base, and the analogy of its chemistry with that of the supposed *Schwefeloxyd* permitted Berzelius to assimilate the latter to the category of base without having to observe the oxide in the isolated state. Furthermore, Berzelius seems to have been untroubled by the failure of the chlorides of sulfur to display any of the properties of the typical salts: the dischloride, for example, is an orange-yellow liquid, fuming in moist air and having a repulsive and suffocating odor. In extenuation of this saline view, Berzelius went on to cite the examples of certain compounds of more clearly saline character but not displaying typically neutral properties,

[6] *Ibid.*, pp. 84—85.

namely the arsenious acid compounds of muriatic acid and chromium (*-ic*) oxide.[7] However, no listing of analogies could veil Berzelius's readiness to extend the model of saline combination to compounds that no other chemists were prepared to regard as salts, nor could analogies mitigate the need to assume the nonisolability of one of the saline components, the sulfur oxide.

As a theoretical underpinning for the saline role of ostensible, nonisolable components, Berzelius invoked a concept that he had employed earlier, though in a somewhat different form, of combination by "intermediation" (*Dazwischenkunft* or *Hinzukunft*):

> Ist diese Ansicht richtig, so ist der niedrigste Oxydationsgrad des Schwefels, wie man ihn in Bucholz's Versuch findet, 100 Th. Schwefel gegen 25 Th. Sauerstoff, und die folgenden Oxydationsgrade sind Multipla durch 2,4,6 mit Uebergehung der ungeraden Zahlen 3 und 5. Ich erinnere hierbei ausdrücklich, dass ich die Resultate dieses Raisonnements durchaus nur als Veranlassung zu weiteren Forschungen will angesehen haben, und für diese werden solche Mischungsverhältnisse, die niemals selbständig und ohne Hinzukunft eines anderen Körpers existiren, gewiss nicht die am wenigsten wichtigsten sein.
>
> Vielleicht werden nun durch Verbindungen dieser Art Multipla nach 1 1/2 veranlasst, so dass diese immer einen niedrigeren Oxydationsgrad voraussetzen, von welchem sie Multipla nach 6 oder 12 sind. Und dann liesse sich viellecht künftig beweisen, dass diese Gradationen immer Steigerungen mit geraden Zahlen, mit 2, 4, 6, 8 und vielleicht darüber sind.[8]

This use of the idea of intermediation differed somewhat from its earlier deployment in the first and second halves of the *Versuch.* There, it was a case of the alteration of the proportions of two combustible radicals in combination with one another by the intermediation of oxygen. Here, it was an example of the alteration in the proportions of a radical with oxygen occasioned by the presence of another body such as another oxidized radical. In particular, an oxide of a radical with a lower proportion of oxygen to radical than was observed in the ordinary isolated oxides of the radical might have existed as an ingredient of a compound with other bodies by virtue of the intermediation of those bodies. The connection with the occurence of proportions in apparent nonintegral multiples was retained from the earlier application of the idea, but the nature of that connection had also changed. In regard to the iron compounds, Berzelius had suggested that an oxide with a fractional multiple in the proportion of oxygen to radical was itself an example of an intermediating body that produced an alteration of the proportions in which that radical combined with another. Here it was suggested *not* that a nonintegral dose of oxygen constituted the intermediating substance and as such was the causal agent, but that a nonintegral dose in the proportion of oxygen to radical was an appearance resulting from the occultation of lower oxides by intermediating bodies. Fractional multiples would disappear when lower oxides, now hidden because incapable of

[7] *Ibid.*, p. 86.

[8] *Ibid.*

an independent existence, became known. In the case of the oxides of sulfur, for example, the apparent multiple of 1 1/2 in the proportion of oxygen to sulfur could be replaced, once the two lower oxides were taken into account, by a series of proportions in the integral ratios of 1 : 2 : 4 : 6.[9]

g. Further speculations: the "Zugabe" and organic matter

It was chiefly with regard to the composition of organic matter, the issue that had motivated the "Zugabe," that Berzelius believed this pattern would prove significant. The case of the lower oxides of sulfur was offered only as a relatively uncommon inorganic example—though later he considered it widespread—of a pattern that he expected to prove commonplace in organic chemistry. Berzelius's analytical materials from inorganic chemistry were sufficiently well-developed to permit this illustration of the anticipated nature of organic combinations, to which his methods were not yet applicable. Apparently, he believed that the complexity of organic compounds probably resulted from numerous cases of intermediation, each one preserving intact ingredients that were analogous to saline components and that consisted of compounds in low proportions, incapable of an independent existence. On this belief seemed to rest Berzelius's suggestion that in the coloring matter of blood a lower oxide of iron might be found. He offered an example of the possible application of these views to other organic compounds as well:

Nehmen wir einen vegetabilischen Körper zum Beispiel, so finden wir in demselben Kohlenstoff, Wasserstoff und Sauerstoff, den letztern aber in so geringern Verhältnisse, dass es selten dem niedrigsten bekannten Oxydationsgrade eines der beiden andern entspricht; sie müssen also noch geringerer Oxydationsgrade fähig sein. Dieses gibt uns z.B. Veranlassung, zu untersuchen, ob etwa gasförmige Kohlenstoffoxyd [carbon monoxide] und Kohlensäure [carbon dioxide] nicht, was den Sauerstoff betrifft, nur Multipla des niedrigsten Oxydationsgrades nach 2 und 4 sind, so dass, bei diesem untersten Grade, 100 Th. Kohlenstoff mit 62.9 Th. Sauerstoff verbunden waren. Eben so lässt sich aus den gefundenen Verhältnissen zwischen Kohlenstoff und Wasserstoff [the proportions in which hydrogen saturated carbon in the two gases, one from the distillation of peat and the other olefiant gas] erwarten, dass sich 100 Th. Wasserstoff im Minimum durch 74,584

[9] So taken with this analysis was Berzelius that it continued to guide his researches through the rest of the *Versuch* and beyond, as is shown in the remaining chapters, and it dictated the nature of his restatement of the law of multiple proportions in the final section of the *Versuch*. The law was there presented as the burden of the first in a series of four general categories of the results of his study (*ibid.*, p. 205): "Wenn zwei Körper, welche wir jetzt für einfach halten, sich in mehreren Verhältnissen vereinigen können, so sind diese Verhältnisse, wenn die Menge des negativ-elektrischen Körpers unverändert bleibt, Multipla nach 1 1/2, 2, 4, u.s.f. von dem kleinsten Verhältnisse, in welchem der positiv-electrische Körper mit dem negativ-electrischen verbunden sein kann.

"Es deuten aber mehrere Erscheinungen darauf, dass das Multiplum nach 1 1/2 nur scheinbar ist, und daraus folgt, dass die Verbindung, nach welcher er als Multiplum berechnet worden, nicht das wahre Minimum der Verbindung darstellt; sondern dass andere, niedrigere Verhältnisse vorhanden sind, nach welchen es ein Multiplum nach 6, 12, 18, u.s.w. sein kann."

Th. Kohlenstoff werden sättigen lassen, und die beobachteten Verhältnisse Multipla nach 4 und 8 sind. Wir vermissen also bei diesen Zusammensetzungen, ausser dem niedrigsten Verbindungsgrade, noch die Multipla durch 2 und 6, die wir in der Mischung organisirter Körper zu suchen veranlasst sind.[1]

This saline view of organic compounds was a departure from the precedent of the radical theory of Guyton and Lavoisier. Organic compounds were taken to be binary combinations of nonisolable oxides of simple radicals, rather than oxides of compound radicals. Doubtless Berzelius hoped that in the binary view lay a conceptual technique capable of rendering the complexities of organic composition accessible to treatment by the methods that he was developing for inorganic chemistry. If, remaining with Berzelius's own example, one could conceive the coloring matter of blood as a compound of oxidized elementary radicals, that is, of oxidized saline components, then the composition of the components and the proportions in which they combined with one another could be treated in terms of Berzelius's saline theory. But why so ready a departure from Lavoisier's precedent in extending the radical theory to organic chemistry?

Berzelius indicated that he had originally adopted the older view, but that exploratory experiments with the Voltaic pile had failed to confirm it:

Ich habe mit der electrischen Säule verschiedentlich versucht, das brennbare Radical der Gewächssäuren von dem Sauerstoff zu scheiden, es hat aber nie gelungen wollen. Ich wurde auf diese Versuch gebracht durch die Reduction des Ammoniaks, von dessen [metallischen] Basis ich damals glaubte, sie sei aus Wasserstoff und Stickstoff zusammengesetzt, und verhalte sich so zu den Metallen, wie sich das zusammengesetzte brennbare Radical der Gewächssäuren zu dem Schwefel und dem Phosphor verhalten wird.[2]

The striking phenomenon of the amalgamation of ammonia by means of the pile had, if anything, strengthened in Berzelius's mind the notion of the existence of compound radicals, but his attempt to extend the idea to organic matter had failed. If the pile—so confident was Berzelius of its reliability as an indicator of chemical ontology—could not separate an organic radical as it had separated the radical of ammonia, then likely organic radicals did not exist. Only inorganic radicals, elementary ones, existed. Combining, by virtue of intermediation, at low levels of oxidation, they formed organic compounds that, at bottom, were salts.

This early failure to isolate organic radicals by electrolysis explains why the author of the famous system of "electrochemical dualism" never seemed to bolster his views of organic chemistry with electrolytic studies comparable to those he performed in inorganic chemistry. This step had to await that famous partisan of Berzelius's science, Hermann Kolbe. Early in his career, still mindful of the novelty and power of the Voltaic pile, Berzelius was frustrated in his at-

[1] *Ibid.*, p. 86.

[2] *Ibid.*, p. 87. The change in Berzelius's view of ammonia suggested by this passage is examined in the next chapter.

tempt to deploy the pile in organic chemistry. His experiments could not sustain the idea of compound organic radicals, leading him to offer instead two other possibilities:

Vielleicht sind sie [the supposed compound radicals] alle Verbindungen, die für sich nicht existiren können [; but do exist as radicals in oxides by virtue of the intermediation of the oxygen with which they are combined. Or else, v]ermuthlich ist auch der Wasserstoff niedrigerer Oxydationsgrade, als im Wasser, fähig, die darin Divisionen mit 8, 6, 4 oder 2 entsprechen müssen. Tritt bei der Zerlegung solcher Körper der Fall ein, dass z.B. Kohlenstoff und Wasserstoff mit dem Sauerstoff nach Verhältnissen zusammen sind, welche mit den eigentlich zu erwartenden Zahlen nicht übereinstimmen, so versuche man, den Sauerstoff unter beide zu vertheilen; und gelangt man auf diese Art zu den vermutheten [niedrigeren] Verhältnissen, so ist es erlaubt, den Körper aus zwei Oxyden zusammengesetzt zu betrachten.[3]

Either the complex radicals were themselves incapable of an independent existence apart from oxygen, or else the compounds were binary unions of oxides that displayed lower proportions of oxygen than observed in isolated compounds and that were themselves incapable of an independent existence. Berzelius clearly chose the latter view. The most common organic compounds were salt-like compounds of nonisolable lower oxides of simple radicals. No matter that their properties were far removed from the typically saline: the muriates of sulfur were no more salt-like, yet the model seemed successfully to account for them.

From this standpoint, Berzelius regarded organic compounds as composed of supposed lower oxides of carbon and hydrogen. In the last quoted passage, for example, he suggested the existence of lower oxides of hydrogen; in an earlier portion of the same passage (quoted above, p. 189, accompanied by n. 1), he made a similar suggestion in regard to carbon. The propriety of this view was supported by the kind of argument possible in the absence of the idea of atomic weights. The expression of the occurrence of multiple proportions was ordinarily given by Berzelius as a statement in which the proportions of one element, such as oxygen, were reduced to the lowest common denominator of a given weight of the other, such as a radical, the latter expressed as 10 or more usually 100 parts. As new compounds, especially organic compounds, were discovered and analyzed, many were found to display lower proportions of one ingredient in relation to a given weight of another. With the appearance of lower and lower proportions, the anticipated minimal combining proportion of one ingredient with the other grew progressively smaller. Berzelius therefore could consider these compounds as examples of the lower combining proportions that the idea of intermediation suggested, and he was willing to entertain the belief that isolable compounds displayed very high multiples of these minimal proportions. His restatement of the law of multiple proportions in the concluding pages of the *Versuch,* in which proportions in multiples as high as eighteen are considered,

[3] *Versuch,* p. 87.

is indicative of this tendency.[4] Later sections of the *Versuch* revealed his dealing with very much higher multiples.

h. A glance ahead: the central thrust of Berzelius's chemistry

Though the "Zugabe" was penned with an eye to organic composition, Berzelius did not pursue organic chemistry directly in the light of these speculations. Rather, he first pursued inorganic chemistry, where the ideas of oxides in minimal proportions incapable of independent existence, intermediation, high multiples of minimal proportions, and the binary character of oxides first bore fruit. Many themes and conclusions of Berzelius's inorganic chemistry, notably his views of mineral acids, are shown in the next chapter to have depended on elaborations and applications of these notions. Only as Berzelius neared the end of the studies that constituted the *Versuch* did he again turn to organic chemistry, where the speculations of the "Zugabe" exerted their influence primarily through the medium of the inorganic precedents.

Throughout the course of Berzelius's research, the guiding principle was that of saline combination, of binary compounds of ontologically equivalent and phenomenologically opposing oxides, in proportions governed by the rule of oxides or its variants. Berzelius never considered more than fleetingly any other archetype of composition. The successful reduction of the muriates of sulfur to this model and the subordination of all substances viewed in terms of that example to the rule of oxides or its variants permitted Berzelius to extend his saline view to substances that otherwise might have proven conceptually inaccessible to the qualitative ideas then available to him and quantitatively refractory to the stoichiometric techniques at his disposal.

In this archetype and its deployments lay the fundamentals of the much-belabored dualism of Berzelius's chemistry and the mainspring of its expansive force. In dealing with complex compounds, Berzelius attained the level of the specific by showing quantitatively that these substances were composed of less complex constituents, united in accordance with precisely articulated laws. It was the electrochemical theory—specifically, the electrochemical generalization of the idea of acidity and basicity—that permitted the subsumption of the constituents to the categories of acidic or basic saline components and therefore of the compound to the category of salt. As such, a complex compound was subject to the quantitative laws of the saline theory, the application of which established the quantitatively defined specificity of the compound. Berzelius's studies of the saline role of water, his quantitative definition of neutrality for both simple and double salts, the quantitative specification, on that foundation, of basic salts, all of which are discussed in the next chapter, provide striking examples of Berzelius's successful categorization of certain substances, especially

[4] Quoted above, p. 188, in n. 9.

the complex, as specific examples of simple, double, basic, or neutral salts by application of stoichiometric principles to qualitatively identified ingredients. As Berzelius penetrated ever more deeply into the hierarchy of increasing complexity and diversity of chemical composition, the quantitative theory of qualitatively opposing saline constients was the instrument enabling him to organize and rationalize a wealth of chemical material that otherwise might have remained intractable. Only when Berzelius reached the most complex compounds—naturally occurring minerals and the wealth of organic compounds just becoming known—did the approach demand further theoretical articulation. The further articulations in mineralogical and organic deployments are considered in Chapter VI.

CHAPTER V

The Elaboration of the Theory of Salts

Experimental Developments through 1819

The Tripartite Research Program from the "Spätere Zugabe" to the Proportions chimiques

1. Introduction

Berzelius's discovery of the rule of oxides intensified his commitment to the research concerns he had been pursuing up to the "Zugabe" and invested his subsequent investigations with a programmatic character. The program consisted of three distinct but related branches: the quantitative analysis of simple binary compounds, especially the oxides and sulfides of radicals; the qualitative characterization of oxides and sulfides in terms of the generalized notions of acidity and basicity; and the quantitative analysis of salts, that is, compounds containing oxides and sulfides as proximate constituents. The simultaneous pursuit of the first and third branches was a manifestation of the intimate connection, in Berzelius's mind, between first order and higher order composition (cf. above, Chapter IV, subsection 2.*a*.); that of the first and second was a reflection of Berzelius's concern for the correlation of degree of oxidation (or sulfuration) with properties (cf. above, Chapter IV, subsection 1.*b*.2)).

The characteristics of this tripartite research program became clearly marked in the first *Fortsetzung*, where Berzelius presented analyses that confirmed the rule of oxides and constituted its first deployments. Especially his study of arsenic compounds, which demanded simultaneous analysis of a series of oxides and their salts, formed a model for much of his subsequent analytical work, which Berzelius (and his students) prosecuted relentlessly. Using the rule in conjunction with qualitative characterizations, Berzelius developed the general applicability of the saline model to many categories of compounds not previously regarded as saline; and, exploiting it to exhibit the patterns connecting the composition of oxides and that of their salts, he was able to reconstruct Lavoisier's correlation of composition and properties. The results of this work were displayed in a series of reports, beginning in the *Fortsetzungen* of the *Versuch* and culminating in the *Proportions chimiques.*[1] The tripartite program is the subject of this chapter.

[1] The most prominent of the reports, besides the *Fortsetzungen*, were two lengthy papers, each devoted to analyses of compounds of many radicals. The first of these papers was the "Versuch,

The *Proportions chimiques* represented more than the outcome of the tripartite program, however. It resulted from the confluence with the currents originating in the *Versuch* of three additional streams: the atomic theory, mineralogy, and organic chemistry. The atomic theory occupied a significant place in Berzelius's work after 1812, when he finally obtained from Dalton himself a copy of the latter's *New System of Chemical Philosophy.*[2] Despite his disappointments with Dalton's work, Berzelius gradually grew committed to atomism during the years preceding the *Proportions chimiques,* which marked his acceptance of a corpuscular ontology as the foundation for the phenomena of chemical proportions. The focus of Lundgren's *Berzelius och den kemiska atomteorin,* Berzelius's acceptance of the atomic theory is pursued here only as required to exhibit the development of the saline theory. In this chapter, atomism emerges primarily in Berzelius's translations of his rules governing proportions into rules for the determination of formulae and in his continuing efforts to eliminate proportions in fractional multiples.

Atomism also appears in the next chapter in connection with Berzelius's mineralogical interests and his research in organic chemistry. Because minerals and organic substances were more complex than the laboratory chemicals of inorganic chemistry, Berzelius felt he would need to invoke all the elaborate rules and principles that he had devised in his inorganic work. However, Dalton's theory provided no sure guides in this relatively uncharted region, and the deficiencies, as Berzelius perceived them, of Dalton's rules of simplicity threatened to undermine the doctrine of chemical proportions in its application to mineralogy and organic chemistry. These perceptions, and their influence on Berzelius's work in mineralogy and organic chemistry, are explored in the next chapter.

2. Summary and overview of the tripartite research program

a. First and second order composition

Viewed with hindsight, from the standpoint of the succession of Berzelius's atomic weight tables, the program is a prominent part of the traditional histories

die chemischen Ansichten, welche die systematische Aufstellung der Körper, in meinem Verbesserung der chemischen Nomenclatur begründen, zu rechtfertigen," *Schweigger's Journal,* 6 (1812), 119—144, 144—176, 284—322; 7 (1813), 43—78; for Swedish version, see p. 162, n. 4; henceforth cited as "Rechtfertigung." The second was "Essay on the cause of chemical proportions, and on some circumstances relating to them; together with a short and easy method of expressing them," *Ann. Phil.,* 2 (1813), 443—454; 3 (1814), 51—62, 93—106, 244—257, 353—364; no Swedish version of this paper was published; henceforth cited as "Cause of proportions." Besides these studies of many radicals, Berzelius published numerous studies each devoted to one or a few radicals. These are cited where needed.

[2] For this, see Partington, *History,* IV, p. 154.

recounting the development of the post-Daltonian atomic theory.[3] In this respect, the program is already well-known, and only two points need be noted about it here. First, a systematic effort to discover and analyze all the oxides and sulfides of radicals was entirely novel. The immediate followers of Lavoisier, notably Fourcroy, had attempted to discover new *salts,* hoping to fill in the blanks in the chart of the *Méthode* exhibiting all possible saline combinations, in the belief that all saline constituents, at least the inorganic ones, were known or soon to be known.[4] However, a programmatic study of saline constituents themselves was new. Moreover, though many contemporaries, such as Gay-Lussac, were prepared to study the compounds of an occasional radical, Berzelius was almost alone in undertaking a study of all radicals. In this he had only one fellow traveller, Thomas Thomson, and Berzelius's work vastly excelled in both quantity and quality that of his Scottish contemporary.[5]

Second, as it was pursued from the first *Fortsetzung,* the program extended and completed the research recounted in the two halves of the *Versuch.* In the earlier work, Berzelius had attended almost exclusively to the radicals of the bases and the salts formed by the bases with but a few acids. To analyze the bases, particurlarly in the light of the metallization of ammonia, was a major motive for the *Versuch,* and the exploitation of Richter's methods to this end demanded analysis of few acids. Therefore, Berzelius confined his attention to sulfuric acid (and, to a lesser extent, sulfurous acid), while studying the sulfates (and sulfites) of numerous bases. With these materials, Berzelius exhibited and sustained the new laws of equivalent and multiple proportions, and he developed some methods for calculating the composition of salts. Only the discovery of the rule of oxides shifted his attention to the acids.

The shift occurred because the rule linked the composition of the acidic component with that of the basic component. A quantitative study of acids was itself a groundbreaking effort, for, as Berzelius observed in his letter to Gilbert announcing the rule, the composition of very few acids was then known[6]—a telling assertion, coming as it did more than twenty years after Lavoisier's *Traité,* which had placed so much stress on correlating the properties of acids with their composition. In addition, the rule provided more powerful means for determining the composition of salts than either the principle of the preservation of com-

[3] The best known example is Hermann Kopp, *Die Entwickelung der Chemie in der Neueren Zeit* (Munich: 1873).

[4] Lavoisier, *et al., Méthode,* pl. facing p. 100. For Fourcroy's studies of salts, see Partington, *History,* III, pp. 542—543.

[5] Thomson's culminated in his *An attempt to establish the first principles of chemistry by experiment,* 2 vols. (London: 1825).

[6] Berzelius, "Schreiben des Herrn Prof. Berzelius, Mitgl. der schwed. Akad. d. Wiss. an den Prof. Gilbert, über einige gegenstände, welche zwischen Davy und den HH. Gay-Lussac and Thénard [*sic*] streitig sind, und über ein zweites neues Gesetz, welches er im Verfolge seiner Untersuchungen aufgefunden hat. Stockholm, d. 1. Oct. 1810," *Gilbert's Annalen,* 37 (1811), 208—220; rule, remark on acids, 217; henceforth cited as "Schreiben."

bining proportions or the law of the constancy of basic oxygen. Accordingly, Berzelius determined the oxidation stages of acidic radicals not only for their own sake, but also to aid in determining the composition of salts.

The study of the acidic oxides was rapidly integrated with the ongoing study of the basic ones. The dividing line between the two, already blurred by the electrochemical generalization of the ideas of acidity and basicity, grew still less determinate as Berzelius increasingly documented what Lavoisier had suspected: that the higher oxides of even metallic radicals might be acidic. From the first *Fortsetzung,* where Berzelius presented a series of analyses to document the rule of oxides, the fusion of the study of the acidic with that of the basic radicals became increasingly apparent. The analyses presented there of the compounds of arsenic, a semi-metal, provided the bridge between the acidic and the basic radicals and constituted the model for Berzelius's approach to later studies of all radicals. Each radical, regardless of its position in the electrochemical series, received similar treatment, with similar ends in view.

b. The reconstruction of the correlation

The analytical enterprise appeared uniform in its outcomes as well as in its procedures. Berzelius showed that many radicals from widely different positions in his electrochemical series shared the property of forming acids at high levels of oxidation, exhibited a similar progression in the proportions of oxidation from one stage to the next, and formed salts in agreement with the rule of oxides (or its variants). Though, as noted shortly, the evidence that Berzelius first collected in seeking a correlation was seriously deficient, the very deficiencies helped motivate much of Berzelius's subsequent research. His optimism that the anomalies he found at first could be reduced rested on his faith in the power of two tools that Lavoisier had lacked: the law of multiple proportions and the rule of oxides. Invoking these laws, Berzelius devised a reconstructed correlation that expressed widely shared patterns in the progression of oxidation of radicals and in the corresponding conformity of their salts to the rule of oxides. Thus, unlike Lavoisier's, Berzelius's correlation involved second order composition as well as that of the first order and exhibited shared patterns on both levels.

The effort to attain the new correlation dominated all of Berzelius's analytical work from the first *Fortsetzung* through the *Proportions chimiques,* and indeed, beyond it to the revised atomic weight tables of 1826.[7] Though the details of the reconstruction are not analyzed until the end of this chapter, some of its themes need be signalled here to provide a framework for the earlier parts of the chapter. The confirmation and first deployments of the rule exhibited in the first *Fortsetzung* rapidly put Berzelius into a position of conflict. Though clearly

[7] For revisions of Berzelius's patterns of oxidation and salt formation, on which the 1826 atomic weights were based, see Partington, *History,* IV, p. 212.

expecting a correlation to emerge from the analytical data, Berzelius found little confirmation in his early results. He could confirm the rule of oxides for a number of acids, including metallic ones, but he found that few others shared stoichiometric regularities. For example, he found that carbonic acid contained either twice or four times the oxygen of the base in its salts and that phosphoric acid contained twice the oxygen of the base; that arsenious acid contained three times and arsenic acid twice the basic oxygen. The salts of the oxides of the supposed muriatic radical, the composition of which Berzelius had calculated in one of the first applications of the rule, appeared similarly subject to it: muriates contained twice the basic oxygen; superoxymuriates, eight times.[8] Few analogies, however, could be found among the proportions of radical and oxygen exhibited by the various acids, or in the integral multipliers that linked the oxygen content of the basic to the acidic component of the salts of these acids. Though an oc-

[8] Berzelius, *Versuch, die bestimmten und einfachen Verhältnissen aufzufinden, nach welchen die Bestandtheile der unorganischen Natur verbunden sind* (Leipzig: 1892), pp. 109—125. This work is cited in full, above, p. 168, n. 8, and its content is summarized on p. 181, n. 5. Further bibliographic information is given below, p. 202, in n. 6; henceforth cited as *Versuch*. In his first application of the rule, Berzelius determined the composition of three acidic oxides of the supposed muriatic radical, substances that had thus far resisted analysis. In the second half of the *Versuch*, using the law of the constancy of basic oxygen, Berzelius had been able to calculate the quantity of oxygen that oxymutiatic acid (chlorine) contained over and above that in muriatic acid; (see above, p. 173, n. 7). The oxygen content of muriatic acid itself, however, had remained unknown; it could not be determined by direct analysis, because the acid had resisted all attempts to decompose it into its radical and oxygen.

By the thermal decomposition of superoxymuriate of potash (potassium chlorate), Berzelius found that oxygen was given off in a quantity equal to six times the oxygen of the potash in the remaining muriate of potash. By the rule of oxides, however, the oxygen of the potash must be an integral multiple of that of the muriatic acid. The only multiplier conformable with Berzelius's data was two. Therefore, the superoxymuriatic acid contained eight times the oxygen of the potash. Berzelius indicated that the amount of oxygen in oxymuriatic acid in excess of that in muriatic acid, for a given weight of radical, was known from the experiments of Davy and of Gay-Lussac and Thenard to be equal to the oxygen contained in the bases of the salts of muriatic acid. Thus, for a given weight of base, the oxygen content of oxymuriatic acid was to that of muriatic acid as 1:1 1/2. The quantities of oxygen in the three known oxidation stages of the muriatic radical were as 1:1 1/2:4. For these calculations, see *Versuch*, pp. 122—125. These results provided material for the classification of oxides in the 1811 "Nomenclature" (cf. above, p. 160, n. 9 and accompanying text). In the first *Fortsetzung*, Berzelius looked for, but could not find, an oxide containing twice the oxygen of muriatic acid. This was later provided by Davy's euchlorine.

The experiments of Davy appear to be those in *Works*, V, pp. 191—197, 222—224; of Gay-Lussac and Thenard, in part of their "Extrait des mémoires lus à l'Institut national, depuis le 7 mars 1808 jusqu'au 27 février 1809," *Mém. soc. Arcueil*, 2 (1809), pp. 295—358; henceforth cited as Gay-Lussac and Thenard, "Mémoires lus." The experiments of interest to Berzelius appear on pp. 339—342: "De la nature et des propriétés de l'acide muriatique et de l'acide muriatique oxygéné. (lu le 9 janvier 1809)." An abbreviated account, "Extrait d'un mémoire sur les acides muriatique et muriatique oxygéné," appears in *N. Bull. Soc. Philom.*, 1 (1807—1809), 301—303, and in *J. d. mines*, 25 (1809), 70—72. Some details of this work became particularly important to Berzelius and are explored below (pp. 216—217, nn. 3—5 and accompanying text).

casional shared characteristic was observed such as the sesquialteral multiple in the oxides of arsenic, sulfur, and a few other radicals, Berzelius remarked,

Aber welche Verschiedenheiten selbst mitten unter diesen Analogien? Die schwefligsauren Salze nehmen Sauerstoff auf, ohne die innere Zusammensetzung zu verändern [that is, the two radicals still stand in the same proportion to one another], und werden dadurch schwefelsaure Salze [e.g., oxidation of $PbO.SO_2$ gives $PbO.SO_3$]. Die phosphorigsaure Salze geben einen Theil von ihrem Phosphor her, damit er Phosphorsäure werde, und die überoxygenirt-salzsauren Salzen, um in salzsaure Salze überzugehen. Die [arsenigsaure] Salze werden dagegen im Feuer nur durch Zutritt von Sauerstoff verändert, weil sehr mehr Arsenik und mehr Sauerstoff [for a given weight of base] als die arseniksauren Salze enthalten [which is the opposite of the case of the salts of the sulfur acids]. Kann aber ein Theil von der darin befindlichen arsenigen Säure sich höher oxydiren, so drängt sie einen andern Theil, der mit der verbrannten ein gleiches Gewicht hat, aus der Mischung aus. Dieses sind Erscheinungen, welche wir, ohne dass wir nöthig gehabt hätten, sie zu sehen, aus den vorhergehenden angeführten Ansichten [in regard to the composition of the acids and their salts] hätten berechnen können. — Es ist bemerkenswerth, dass keine von diesen doppelten Säuren [that is, pairs of acidic oxides of the same radical] demselben Gesetze in Ansehung der Sättigungs-Capacität der Basen unterworfen ist. Die Schwefelsäure enthält im Verhältniss gegen die von ihr gesättigte Basis die Hälfte Sauerstoff mehr, als die schweflige Säure. Die Menge von phosphoriger und von Phosphorsäure, welche einerlei Basis sättigen, enthalten dagegen gleiche Theile Sauerstoff; die arsenige Säure enthält in Beziehung auf die sie sättigende Basis die Hälfte Sauerstoff mehr als die Arseniksäure, indess die überoxygenirte Salzsäure, wie wir werden sehen, dreimal mehr Sauerstoff gegen die von ihr gesättigte Basis als die Salzsäure enthält.[9]

For Berzelius, degree of oxidation of a radical was unquestionably connected with the relative acidity and basicity of its oxides and with the composition of the salts of the acidic oxides. Unfortunately, the precise form of these relations was now anything but clear. What had been the clean correlations of oxygen content and acidic character established by Lavoisier had become thoroughly convoluted in the nascent quantitative chemistry of Berzelius.

Throughout the analytical studies following the first *Fortsetzung,* Berzelius's intent was to reduce the diversity in the observed patterns of oxidation of the radicals, subjecting them to the constraint that the salts conform, in delimitable ways, to the rule of oxides. Many of the numerous themes of Berzelius's research connecting the *Versuch* with the *Proportions chimiques* may be seen as the by-products of this central thrust. The initial study of the arsenic compounds, already noted as the archetype of the tripartite research program, acquires its full significance when seen in relation to the effort to reconstruct the correlation. It was the first and exemplary case of Berzelius's effort to reduce diversity in oxidation to uniformity. The arsenic compounds conformed to the pattern shown by the sulfur oxides, which Berzelius had developed in the "Zugabe" and had adopted as his archetype. Of the four oxides, the two acidic ones contained oxygen, for a given weight of sulfur, in the proportion of 2 : 3 and the multipliers

[9] *Versuch,* pp. 117—118.

by which the basic oxygen of their neutral salts was contained in the acids were, respectively, 2 and 3; (that is, the *-ous:-ic* proportion in the amount of oxygen was 2:3; in neutral sulfites, e.g., $PbO.SO_2$, the sulfurous acid contained twice the basic oxygen; and in sulfates, e.g., $PbO.SO_3$, three times). The attractiveness of this pattern was enhanced by the neutral salts that he had shown, in the first *Fortsetzung,* to obey the rule of oxides. Berzelius surmised, therefore, that there was a connection between progression of oxides and the composition of salts. Moreover, he had successfully reduced the sesquialteral proportion of oxygen in the two acidic oxides of sulfur to an integral one by finding the two lower oxides in the muriates of sulfur. These contained respectively one-sixth and one-third the oxygen of the sulfuric acid (or one-fourth and one-half that of sulfurous acid; written in modern notation, the formulae of the four oxides are $SO_{1/2}$, SO, SO_2, and SO_3). Berzelius hoped to bring most radicals, whether acidic or metallic, into conformity with this paradigmatic case. Arsenic, a semi-metal, formed a convenient link between the typical acidic radicals and the typical basic ones. A model for the program, it contributed from the start to Berzelius's drive to subordinate both categories to common principles, while distinguishing the radicals chiefly in terms of their electrochemical diversity.

The details of Berzelius's progress in reconstructing the correlation from the first *Fortsetzung* to the *Proportions chimiques* are exhibited below in section 9. The main outcomes of this effort and some of its offshoots are signalled here and elaborated in detail later. First, Berzelius was compelled to recognize that the uniformity in patterns of oxidation was not so broad as he would have liked. Though many radicals having more than one oxide did exhibit a pattern similar to that of sulfur, many did not. In particular, Berzelius wrote the *Proportions chimiques* mindful of three categories of radicals: those more or less in conformity with the sulfur pattern; those in conformity with the phosphorus pattern, in which the two higher oxides contained oxygen in the proportion of 3:5 and formed salts often in violation of the rule of oxides but in accordance with a variant of it; and those agreeing with neither pattern, but producing salts in agreement with the rule of oxides in its original form. Because the radicals of the first and third categories agreed with the rule, Berzelius lumped them together in the *Proportions chimiques* and distinguished them from those of the second category:

> Mais, bien que cette loi [the rule of oxides] régisse le plus grand nombre des combinaisons d'atomes composés parmi les corps oxidés, on trouve quelques exceptions, qui toutefois ne se montre pas ça et là dans les oxides en général, mais se bornent à certains acides qui ont tous cela de commun, que le radical donne deux acides, dans lesquels des différentes quantités d'oxygène sont entre elles dans le rapport de trois à cinq. Ce sont les acides de phosphore, d'arsenic et d'azote, si l'on considère ce dernier comme un substance simple. Mais, il y a, même pour ces acides, une loi qui règle leur combinaison avec d'autres oxides de telle sorte, que le nombre des atomes d'oxygène dans l'oxide est d'un ou plusieurs cinquièmes, et plus rarement d'un ou plusieurs dixièmes (c'est-à-dire, un

cinquième, un cinquième et demi, ou trois dixièmes, deux cinquièmes et trois cinquièmes) du nombre des atomes d'oxygène dans les acides en *ique,* et d'un ou deux tiers de ce même nombre dans les acides en *eux.* Si donc la loi qui assigne des limites aux combinaisons des atomes de ces acides avec les autres oxides, ne paraît pas, pour le moment, être la même que celle qui règle les combinaisons de tous les autres corps oxidés, c'est qu'elle est particulière à ces acides.[1]

A major theme of this project was a characteristic search for lower oxides, again typified by the original study of the arsenic compounds. Pursuing the analogy with sulfur, Berzelius tried to prepare and analyze a suboxide of arsenic containing one-sixth the oxygen of arsenic acid, the existence of which would have eliminated the sesquialteral multiple in the proportions of oxygen. He believed that he had found such a suboxide in the powder into which metallic arsenic degenerated in the air, and a preliminary analysis indeed revealed its oxygen content to be approximately the predicted one-sixth that of the acid. This, Berzelius observed, "... giebt einen neuen Beweiss ab, dass die Multipla nach 1 1/2 nur scheinbar sind, und wirkliche Multipla mit 6, 12 u.f. darstellen."[2] Granted that the lower oxides were not common, Berzelius sought for them, as in this example, where one might not ordinarily have looked. Though they were elusive, he believed it worth the trouble to find them, for doing so revealed the lowest combining proportion of a radical with oxygen, permitted the removal of the sesquialteral multiple, and provided another potential example of conformity with the sulfur pattern. This kind of investigation provided the entries in the class of suboxides in the 1811 "Nomenclature," which Berzelius described in these terms:

Tels sont les degrés d'oxidation de kalium et de natrium inférieure à celui qui produit l'alcali; tel est l'oxide noir de l'arsenic, dont j'ai donné la description dans un de mes Mémoires sur les proportions determinées [the first *Fortsetzung*]; telles sont encore les pellicules, grises ou noires, qui tournissent par degrés la surface métallique du plomb, du zinc, du bismuth, et qui, après quelque temps, forment une enveloppe autour des métaux, qui les dérobe à l'influence ultérieure de l'air. Malgré les soins que j'ai pris pour me procurer ces derniers sous-oxides en état de pureté, je n'y ai jamais complètement pu

[1] *Essai sur la théorie des proportions chimiques et sur l'influence chimique de l'électricité* (Paris: 1819), pp. 35—36; cited fully on p. 145, n. 1; henceforth cited as *Proportions chimiques.* At the end of this passage, Berzelius signalled this footnote (p. 36n): "C'est une circonstance assez remarquable que, si l'on suppose que le radical de ces acides contient le cinquième de l'oxigène qu'il en faut pour produire l'acide en *ique,* la plupart des anomalies disparaissent, et ces trois substances obéissent aux mêmes lois que toutes les autres. Cependent, comme l'expérience ne l'a point démontré, il faut se tenir aux faits connus; et d'ailleurs cette supposition ne réduit point tous les cas anomaux à une conformité complète avec les lois générales." Remaining anomalies were those of the muriatic acid compounds of phosphoric acid (phosphorus tri- and pentachloride), as Berzelius pointed out in a study of the salts of the phosphorus oxyacids (see below, p. 259, in n. 5). Despite them, Berzelius could not yet detach himself fully from the possibility of reducing the anomalous pattern of oxidation exhibited by these three radicals. For this, see the passage quoted at the end of this chapter, p. 260 accompanied by n. 9.

[2] *Versuch,* p. 121.

réussir. Tout ce que j'en ai pu connoître, c'est qu'ils se réduisent en partie quand on y verse un acide ou un alcali, et que l'oxigène se concentre sur la portion du métal nécessaire pour la formation d'une base saline. Outre que ces sous-oxides se forment aisément sur les amalgames des métaux précités, on peut aisément, par le calcul, prouver l'existence de quelques-uns parmi eux, par exemple, de celui du plomb. Nous savons que le *minium* [the red oxide Pb_3O_4] contient 1 1/2 fois autant d'oxigène [for a given weight of metal] que l'oxide jaune [PbO; the multiple should have been 1 1/3]; ce multiple par 1/2 présuppose un degré d'oxidation inférieure à celui de l'oxide jaune, de l'oxigène duquel celui du minium doit être un multiple par un nombre entier. Il en est de même avec le fer; nous voyons aussi que le fer métallique se couvre, après quelque temps, d'une couche grisâtre, qui a un aspect moins métallique et qui n'est point l'oxide noir de ce métal.[3]

This approach typified Berzelius's later analytical work.[4]

The electrochemically based notion that nitrogen was itself a suboxide of a nitric radical permitted the assimilation of the nitrogen oxides to the sulfur pattern and entailed conformity of the salts of the nitrogen oxyacids to the rule of oxides. Two of Berzelius's major ideas thus became mutually reinforcing: the oxide nature of nitrogen was sustained by the conformity of the salts of its oxyacids to the rule of oxides, and the rule was sustained by the reduction of the nitrogen oxides to the sulfur pattern. Berzelius's persistence un upholding the oxide nature of nitrogen must be traced to the gratifying character of this interlocking relationship.[5] Only his failure to extend the sulfur pattern to phosphorus compelled him to recognize a second pattern of oxidation and salt formation, the phosphorus pattern, and to concede later that nitrogen and arsenic conformed to it. Though in the *Proportions chimiques* he had not abandoned his cherished belief in the oxide nature of nitrogen, and therefore of phosphorus and arsenic, his faith was clearly shaken.

Because of their peculiar position in Berzelius's chemistry, the nitrogen compounds are given special attention in the following discussion of the tripartite research program. Emphasis is also given to Berzelius's studies of the oxides of the acidic radicals and of the weakly reactive metallic radicals, including elements now described as amphoteric. The metals that form the bases of the typically

[3] "Essai sur la nomenclature chimique," *J. de physique,* 73 (1811), 253—286, esp. 261; for Swedish version, see above, p. 157, n. 2.

[4] In the "Rechtfertigung" (70—78), Berzelius offered examples of what he took to be the suboxides of antimony, bismuth, lead, zinc, and suggested the existence of a suboxide of manganese. At the conclusion of the section on zinc (74), he remarked, "Seitdem wir durch die Analysen der feuerfesten Alkalien, die beiden niedrigeren Oxydationsstufen von Kali und Natrum kennen gelernt haben, und seitdem die in dieser Abhandlung angeführten Facta die Existenz von mit diesen alkalischen Oxyden gleichartigen Oxydationsstufen bei dem Antimon, Wismuth, Blei und Zink bestätigt haben, glaube ich, dass die besondere Klasse von Oxyden, welche ich *Suboxyde* nenne, hinlänglich gerechtfertigt worden ist. Ich vermuthe, dass wir mit der Zeit Suboxyde bei den mehrsten Metallen entdecken werden."

[5] Cf. H.G. Söderbaum, *Berzelius' Werden und Wachsen, 1779—1848,* Kahlbaum's Monographien aus der Geschichte der Chemie, Vol. III (Leipzig: 1899), pp. 80, 119—127; henceforth cited as Söderbaum, *Werden und Wachsen.*

neutral salts, prominent in the first and second halves of the *Versuch,* are de-emphasized. Another weakly reactive oxide, water, which also acquired a prominent and consequential place in Berzelius's research, is accorded commensurate attention below. The analysis of these studies exhibits the construction of the Berzelian theory of salts within the field of experimental inorganic chemistry, and it sets the stage for the exploration in the next chapter of the deployments in mineralogy and organic chemistry.[6]

[6]Before proceeding with the analysis of Berzelius's research, it may be helpful to indicate the organization and chronology of the remaining sections of the *Versuch.* The arrangement of the *Fortsetzungen* as published by Gilbert and reprinted in the *Ostwald's Klassiker* edition did not reflect faithfully the order of their composition. It has already been shown (pp. 180—181, text accompanying nn. 8—5) that the first section of the first *Fortsetzung* consisted of a portion of the description of the normal analyses, which were repetitions of parts of the experiments in the first and second halves. The first part of the normal analyses had been followed by the "Spätere Zugabe," published in appendage to the second half. The rule of oxides was announced in a letter to Gilbert of 1 October 1810 appearing in the *Annalen der Physik* before the publication there of the German text of the first and second halves ("Schreiben"). The letter, which also antedated the composition of the *Fortsetzungen,* adumbrated in highly condensed fashion the notion of the existence, by virtue of intermediation, of non-isolable lower oxides as constituents in compounds and went on to broach some of the developments of this idea that were fully elaborated only in the second and third *Fortsetzungen.* The letter accordingly suggests the rapidity with which Berzelius was proceeding with studies occasioned by the speculations of the "Spätere Zugabe."

The first *Fortsetzung,* prefixed by the normal analyses, appeared next (*Versuch,* pp. 87—127). It contained the rule of oxides, and it offered a discussion of the acidic character of hydrogen sulfide, an account of the anticipated oxygen content of ammonia in terms of the theory of higher multiples of minimal combining proportions, and a series of studies on particular acids to establish fully rule of oxides. The *Fortsetzung* was followed immediately in *Gilbert's Annalen* by Berzelius's letter of 20 May 1811: "Auszug aus einem Briefe des Hrn. Prof. Berzelius an den Prof. Gilbert. Stockholm, d. 20 Mai 1811," *Gilbert's Annalen,* 38 (1811), 227—232; henceforth cited as "Auszug." In its opening paragraph, the letter stated that Berzelius had since communicated to Gilbert a paper on neutral nitrates and that he would follow it with a second continuation of the *Versuch,* of which he listed the contents. Because the list includes all topics covered in the remaining portions of the *Versuch,* it is clear that *for Berzelius* the first *Fortsetzung* was followed only by the paper on neutral nitrates and a second continuation; cf. Berzelius's letter to Gahn of 13 May 1811, where he reported the completion of the studies described in the second continuation and the start of their preparation for publication (*Bref,* V,ix, p. 54). In notes to the published second *Fortsetzung* (*Gilbert's Annalen,* 40 [1812], 161n, 176n), Gilbert indicated that he had received the paper on neutral nitrates as a "Zugabe" to the first *Fortsetzung.* Gilbert joined the paper on neutral nitrates with material on basic nitrates and on nitrites from Berzelius's second continuation to form the published second *Fortsetzung.* The published third *Fortsetzung* contained the remainder of Berzelius's second continuation. In the *Ostwald's Klassiker* edition, the second *Fortsetzung* occupies pp. 127—154; the third, pp. 154—205.

The *Versuch* concluded with a few pages summarizing in highly generalized form the conclusions established by the entire study. This section was divided into four parts. The first, already excerpted (p. 188, n. 9), contained the law of multiple proportions; the second listed five laws governing various combinations of pairs of oxides; the third, laws governing the combinations of three or more oxides; and the fourth concerned direct combinations of unoxidized combustible radicals composed in accordance with the principle of the preservation of combining proportions. In the *Ostwald's Klassiker* edition, the concluding section occupies pp. 205—210.

For references to the parts of the *Versuch* as they appeared in *Gilbert's Annalen* see above,

3. Early search for lower oxides of nitrogen

Because Berzelius's generalization of the idea of basicity emerged from belief in the oxide nature of ammonia, his commitment to that belief was intense. New experimental developments, however, rendered less plausible the idea that ammonia was the oxide of a compound radical. Not surprisingly, Berzelius felt compelled to respond to these developments in the interest of preserving his symmetrical theory of equivalent saline components. The theory could be salvaged by the assertion that ammonia did contain oxygen, not as an immediate or proximate constituent, but as a component of either hydrogen and nitrogen or of nitrogen alone. Rather less tenable, as Söderbaum points out, was the notion that the oxygen content of ammonia lay solely in the hydrogen, or indeed that hydrogen could be excluded from the list of elements.[7] Accordingly, the idea that ammonia was a combination of elementary nitrogen and compound hydrogen was never suggested by any of the figures who concerned themselves with these issues, and the idea of the compound nature of hydrogen was dropped first. Thereafter, belief in the oxygen content of ammonia amounted to belief in the composite nature of nitrogen. As may be imagined, Berzelius adhered to this view for a long time. Lundgren and Söderbaum recount the overall evolution of Berzelius's views about the nature of nitrogen.[8] Only those aspects of the issue that were related to the evolution of Berzelius's saline theory are considered in this study (here and in the next section).

In great measure, Berzelius followed in Davy's footsteps. Davy's belief in the oxide nature of ammonia, like Berzelius's, rested on the analogy between ammonia and the two alkalies that he had shown to be metallic oxides. In the 1807 Bakerian Lecture on the decomposition of the alkalies, before Berzelius had apprised Davy of the amalgamation phenomenon, Davy had performed a variety of experiments indicating the presence of oxygen in ammonia.[9] In view of the widespread contemporary opinion that the volatile alkali contained only nitrogen and hydrogen, he had reanalyzed it to see whether new data could accommodate the presence of oxygen as well. The results indicated that the combined weight of the hydrogen and nitrogen obtained from ammonia was slightly short of the

p. 181, n. 5. In the same note is a chronological list that shows the order in which Berzelius sent his manuscripts to Gilbert and that indicates how Gilbert assembled them to form the published *Versuch*.

[7] Söderbaum, *Werden und Wachsen*, p. 123.

[8] Anders Lundgren, *Berzelius och den kemiska atomteorin*, Acta Universitatis Upsaliensis. Studier i idé- och lärdomshistoria, Vol. II, utgivna av institutionen för idé- och lärdomshistoria vid Uppsala Universitet (Uppsala: 1979), pp. 76—80, 102—113; Söderbaum, *Werden und Wachsen*, pp. 119—127. Lundgren's book is henceforth cited as Lundgren, *Berzelius och atomteorin*.

[9] Davy, "On some new phenomena of chemical changes produced by electricity, particularly the decomposition of the fixed alkalies, and the exhibition of the new substances which constitute their bases; and on the general nature of alkaline bodies" (read, 19 Nov. 1807), in Davy, *Works*, V, pp. 57—101, esp. pp. 91—98.

weight of the ammonia used. The loss could be attributed to oxygen. At this point Davy offered his suggestion that ammonia was likely the oxide of a compound radical.[1] The metallization of ammonia reinforced his belief in its oxide nature and induced him to formulate a variety of bold speculations about the composition of ammonia and about the character of nitrogen and hydrogen.[2] The question was, how was the oxygen retained in ammonia. Davy did not readily abandon his original idea that ammonia was an oxide of a compound radical,[3] but his repeated efforts to analyze ammonia gradually made it clear that the apparent loss of weight in the decomposition of ammonia was the result of experimental errors inevitably attendent upon the collection and measurement of gases.[4] An alternative to the idea that ammonia contained oxygen *in addition* to hydrogen and nitrogen became more attractive. Davy suggested that nitrogen and hydrogen, as well as ammonia, were oxides of a still unknown radical, which he called "ammonium."[5]

Berzelius, too, persisted in his belief that ammonia was an oxide, even as he followed Davy in abandoning the belief that it contained a compound radical.[6] As is evident from a reading of the first *Fortsetzung,* his faith in this notion, like Davy's, lay in the similarity of the properties of ammonia with those of the remaining alkalies. Though Gay-Lussac and Thenard had argued from their experimental results that the metallic substance obtained from ammonia and mercury by the pile was a compound of hydrogen and ammonia, Berzelius found their views plausible, but not compelling. If the ideas of the Frenchmen seemed to have left the question moot, Berzelius believed the analogy of ammonia with the other bases tipped the balance in favor of its oxide nature.[7] Against the op-

[1] *Ibid.,* pp. 97—98.

[2] In several successive Bakerian Lectures, in Davy, *Works,* V, pp. 122—139, 141—159, 205—216, 246—267, 276—282.

[3] He was still suggesting it, for example, in the appendix to the 1808 Bakerian Lecture (*ibid.,* p. 216).

[4] Davy's own analyses (*ibid.,* pp. 214—216, 260—263), as well as those of William Henry and A.B. Berthollet, led him to this conclusion (Davy, *Works,* V, 262—263). Henry's work was reported in his "On ammonia and an account of a new method of analysing by combustion with oxygen and other gases [a letter to Davy], *Phil. Trans.,* 99 (1809), 430—449; and Berthollet's, in his "Mémoire sur l'analyse de l'ammoniaque," *Mém. soc. Arcueil.* 2 (1809), 268—294.

[5] Davy, *Works,* V, pp. 276—277 (in the Bakerian Lecture for 1809).

[6] Berzelius signalled his acceptance of Davy's suggestion that ammonia, nitrogen, and hydrogen were all oxides of the radical "ammonium" in a letter to Gilbert ("Schreiben," 211). There, he gave a preliminary version of materials about ammonia, nitrogen, and hydrogen that he developed later in the first *Fortsetzung* and in his "Ueber die bestimmten und einfachen Verhältnissen nach welchen die Bestandtheile der unorganischen Natur verbunden sind. Summarischer Bericht über diesen Gegenstand an den königl. Acad. in Stockholm erstattet," *Schweigger's Journal,* 2 (1811), 297—326, which was written at about the same time as the first *Fortsetzung.* As indicated on its first page, this paper was a translation by Berzelius himself of the Swedish version, cited above, p. 172, n. 6. It is henceforth cited from the German text as "Summarischer Bericht."

[7] *Versuch,* pp. 100—101. The experiments of Gay-Lussac and Thenard were reported in their "Sur la production d'un amalgame par l'ammoniaque et les sels ammoniacaux, au moyen de la pile électrique," *Ann. chim.,* 73 (1810), 197—

posing contention he developed a series of clever arguments. Several of these bore directly on the evolution of the saline theory and are noted here. After he presented them, Berzelius invoked his notions about lower oxides to reconcile his belief with the abandonment of the notion that ammonia contained a compound radical.

Many of Berzelius's arguments for the oxygen content of ammonia were based on a comparison of ammonia with hydrogen sulfide, which he had discussed in the immediately preceding section of the first *Fortsetzung*.[8] Though hydrogen sulfide displayed acidic properties, Berzelius showed that only weak and, in some cases, untenable arguments could sustain belief that it was an oxide. That assumption, he asserted, was in any case unnecessary, for the acidic properties of hydrogen sulfide and its ability to form salts could be attributed to the sulfur, which was electronegative in relation to the salifiable bases. He argued, along the lines of the principle of the preservation of combining proportions, that the proportions of the sulfur and the metal of the base were the same as in the corresponding metallic sulfide, and that the basic oxygen of the salt was present in just the quantity that would be necessary to form water with the hydrogen of the acid. Such a fourfold compound could exist, he asserted, so long as the affinity of the metal in the base for the basic oxygen exceeded that of the hydrogen. Otherwise, the hydrogen in fact formed water with the oxygen of the base and left the metallic sulfide behind. With this model, Berzelius could account for the exhibition in a compound of an *unoxidized* radical of that radical's specific properties—even though, as has been seen, Berzelius believed that generally the properties of radicals were extinguished in their unoxidized combinations.

Given the ontological equivalence of all radicals, a similar argument should have applied to a basic substance that contained no oxygen. Accordingly, Berzelius was ready to measure ammonia against his model for hydrogen sulfide:

> Enthielte das Ammoniak keinen Sauerstoff, so müssten wir es für eine ähnliche Salz-Basen nehmen als den Schwefel-Wasserstoff, für eine Säure, ohne dass es doch ein oxydirter Körper wäre.[9]

Berzelius then invoked a variety of arguments to demonstrate that this analogy could not be upheld.[1] Unlike hydrogen sulfide, which owed its acidic properties to sulfur, i.e., to its negative radical, ammonia would have had no constituents to which its specific properties could be referred if it had not been the oxide of a

214. Berzelius had earlier stated his position about their argument in his "Ueber die Streitigkeiten zwischen Davy und den HH. Gay-Lussac und Thenard, die Metalle aus den Alkalien, das Ammoniak und den Stickstoff betreffend, (in einem Schreiben an den Prof. Gilbert.) Stockholm, den 1. Aug. 1810," *Gilbert's Annalen*, 36 (1810), 198—203.

8 *Versuch*, pp. 96—97.

9 *Ibid.*, p. 97.

1 *Ibid.*, pp. 97—110.

metallic basis; for ammonia was more electropositive than either hydrogen or nitrogen. The behavior of the two substances, moreover, was markedly different when they were exposed to the action of the pile: ammonia could migrate intact to one of the poles but hydrogen sulfide could not; ammonia could be made to yield a metallic substance in amalgamation with mercury but hydrogen sulfide could not. Most important, the quantitative composition of the two could not be rendered analogous:

> Wenn das Ammoniak der Analyse des Herrn Gay-Lussac zu Folge 18,475 Th. Wasserstoff enthält, so wäre es zu präsumiren, dass in den Ammoniak-Salzen das Ammoniak so viel Säure sättige, als entweder genau die Menge Sauerstoff (oder ein Vielfaches desselben nach einer ganzen Zahl) enthielte, die nöthig ist, um mit diesem Wasserstoffe Wasser zu bilden. Dieses ist aber nicht der Fall; denn wir werden sehen, dass das Ammoniak in dem erstgenannten Verhältnisse nur an die Säuren gebunden ist, welche drei Mal so viel Sauerstoff als die sie sättigende Base enthalten. In den Säuren, welche entweder 2 oder 4 Mal so viel Sauerstoff enthalten, als die Basis, von der sie gesättigt werden, enthält die neutralisirte Säure entweder 2/3 oder 11/3 so viel Sauerstoff, als erförderlich wäre, um mit dem Wasserstoffe des Ammoniaks Wasser zu bilden. Dieses sind aber ungewöhnliche Vielfache, von denen wir nirgendwo anders Beispiele haben. In dem überoxygenirten salzsauren Ammoniak, das ich im Folgenden erwähnen werde, befindet sich das Ammoniak in einem solchen Verhältnisse, dass die Säure 22/3 Mal so viel Sauerstoff, als zur Wasserbildung mit dem Wasserstoffe des Alkalis nöthig wäre, enthält. Dagegen werden wir sehen, dass das Ammoniak, wenn es für ein Metalloxyd von der oben gefundenen Mischung [as calculated in the second half of the *Versuch;* cf. above, p. 173, nn. 7—8 and accompanying text] genommen wird, genau den nämlichen Gesetzen, wie die übrigen Alkalien, Erden und Metalloxyde gehörcht.[2]

If ammonia, then, could not be reduced to the model of hydrogen sulfide, it must be considered an oxidized body. Davy's suggestions that hydrogen, nitrogen, and ammonia were all oxides of a common radical, ammonium, and that hydrogen and nitrogen, respectively, were lower and higher oxides than ammonia, had a strong appeal for Berzelius. Happily seizing on them as consistent with his own speculations about lower oxides advanced in the "Spätere Zugabe," he submitted the data involved to calculation.[3] His approach was clearly intended to account for ammonia and a whole series of related compounds as a set of simple oxides of a single radical with progressively higher multiples of a minimal combining proportion. Latent in the treatment, however, was the belief that these substances were binary compounds of oxides of the ammonium radical, that is, composed like salts.

Berzelius had already calculated the total oxygen content of ammonia in the second half of the *Versuch* and had offered a new determination among the normal analyses. The point now was to partition that amount of oxygen between the two putative oxides so that the combining proportions of hydrogen and

[2] *Ibid.*, pp. 99—100.

[3] "Schreiben," 211—216; *Versuch*, pp. 100—102.

nitrogen in ammonia and of the known higher oxides of nitrogen could all be accommodated. He clearly elaborated the foundations for his procedure:

Wenn das Ammoniakgas durch electrische Schläge zersetzt wird, so erhält man, nach den Versuchen der Herren Henry und Berthollet des Jüngern, Stickgas und Wasserstoffgas, ohne eine Spur von Sauerstoff; daraus folgt, dass der Sauerstoff des Ammoniaks in diesen beiden Gasarten enthalten sein muss. Der eine dieser beiden Körper muss also eine höhere, der andere eine niedrigere Oxydationsstufe sein; und es leidet wohl keinen Zweifel, dass der Stickstoff die höhere ist. Es muss sich also der Sauerstoff in dem Stickstoffe in einen solchen Verhältnisse befinden [according to the law of multiple proportions], dass er ein Vielfaches nach 1 1/2, 2, 3 u.s.f. von den Sauerstoffe des Ammoniaks ausmacht. Wir werden sehen, dass 1 1/2 der Multiplicator ist. Betrachtungen über die verschiedenen Oxydationsstufen des Schwefels führten mich, wie ich in der [späteren] Zugabe zu meiner Abhandlung angezeight habe, zu der Vermuthung, dass ein jedes scheinbares Vielfaches nach 1 1/2 ein wahres Vielfaches mit 6 oder 12 von einer vielleicht unbekannten niedrigeren Stufe ist, die, wenn sie für sich nicht existiren kann, wenigstens in Verbindung mit andern Körpern zu vermuthen ist. Sehen wir nun den Wasserstoff als eine solche niedrigere Stufe an, so muss er entweder 6 oder 12 mal weniger Sauerstoff als der Stickstoff enthalten. Da wir nun das Verhältniss der beiden Bestandtheile im Ammoniak kennen, so ist es leicht zu berechnen, dass der Wasserstoff nur 1/12 so viel Sauerstoff gegen 100 Th. Ammonium als der Stickstoff enthält.[4]

On this basis Berzelius calculated the composition of hydrogen, nitrogen, and ammonia as oxides of ammonium. Making use of gas density data supplied chiefly by Gay-Lussac's experiments, he went on to calculate the composition of the known higher oxides of nitrogen and that of water as ammonium oxides. He summarized these results in a table.

			Theilen Sauerstoff
		× 1 =	11,0346 den Wasserstoff.
		× 4 =	44,1384 das Ammoniumoxydul.
		× 8 =	88,2768 das Ammoniak.
100 Theile		× 12 =	132,4152 den Stickstoff.
Ammonium	11,0346	× 24 =	264,8304 das Stickstoffoxydul.
geben mit		× 36 =	397,2456 das Stickstoffoxyd.
		× 48 =	529,6608 die Salpetrige Säure.
		× 60 =	662,0760 die Salpetersäure.
		× 72 =	794,4912 das Wasser.

Es ist bemerkenswerth, dass die Sprünge vom Stickstoffe bis zum Wasser immer ein Zwölffaches ausmachen.[5]

Though procedure clearly identified ammonia as a simple oxide of ammonium, Berzelius was moving in the direction of perceiving ammonia itself as a saline compound, that is, as a combination of the two oxides of ammonium, nitrogen and hydrogen. This tendency appears to have been the result of his attempts to overcome an objection that occurred to him:

[4] *Versuch*, pp. 101—102.

[5] *Ibid.*, p. 103.

Warum giebt beim Verbinden mit Sauerstoff der Wasserstoff immer Wasser und der Stickstoff immer Salpetersäure oder Stickstoffoxyde? Oder umgekehrt: Warum erhält man durch entzeihen von Sauerstoff aus dem Wasser nur Wasserstoff und aus der Salpetersäure oder dem Salpeter immer nur Stickstoff, wenn beide Stoffe wirklich Oxyde des nämlichen Radicals sind?[6]

Why, in short, if hydrogen and nitrogen were oxides, did they behave like elements? Berzelius offered a highly speculative answer to this question, invoking electrochemical considerations, the details of which need be of no concern here.[7] For the present discussion, the important aspect of his answer was his suggestion that the two gases obtained from decomposition of ammonia were "die wahren Bestandtheile desselben," that is, that they were its proximate constituents, just as other oxides, acids and bases, were the proximate constituents of salts.[8] Ammonia, a nonsaline substance, was being assimilated to the saline category. It was a compound of two oxides of the same radical.

The existence of apparently simple oxides that actually took the form of binary compounds of oxides of the same radical formed a recurrent theme in Berzelius's later work, as is seen below. Among the most striking examples of this pattern that Berzelius later found was the black oxide of iron, a substance capable of existing in isolation and of forming the binary base of certain double salts, especially prussiates. This binary oxide of iron and related binary oxides of cobalt and nickel were shown to display a typical pattern of combining proportions and played a role in the construction of the later theory of proportions. This pattern, moreover, was extended to include binary bases composed of dissimilar oxides in double salts.

Finally, Berzelius's quantitative treatment of the nitrogen compounds reveals another example of the tendency to assimilate nonsaline substances to the saline model. Following up the investigations of Gay-Lussac and Thenard, Davy had produced, as is known, an "olive-colored matter" (potassium amide or potassamide, KNH_2) from the action of metallic potassium on ammonia gas.[9] The reaction also produced a quantity of hydrogen gas equal to that evolved by the same quantity of potassium from its action on water.[1] Using Davy's figures for the quantity of ammonia decomposed by a given weight of potassium, Berzelius showed that the evolution of the appropriate amount of hydrogen gas could be explained by regarding the olive-colored substance as a binary pair of oxides. It was a combination of *Ammoniumoxydul,* the second compound in the table, with *Kaliumoxydul,* a supposed lower oxide of the metal containing half the

[6] *Ibid.*, p. 104.

[7] *Ibid.*, pp. 104—109.

[8] *Ibid.*, p. 107.

[9] Davy, *Works,* V, pp. 143—150. Davy relied on the report of the French collaborators' work appearing in the *Moniteur,* 27 May 1808. Their work was later described in Gay-Lussac and Thenard, "Propriétés du métal de la potasse," in their "Mémoires lus," 302—311, esp. 308—311.

[1] The equations are:
$K + NH_3 = KNH_2 + 1/2\, H_2$;
$K + H_2O = KOH + 1/2\, H_2$.

oxygen, for a given weight of metal, of potash, and forming one of the entries in the class of suboxides in the 1811 "Nomenclature." The proportions of the two constituents conformed with the rule of oxides: the oxide of potassium contained three times the oxygen of the oxide of ammonium. Moreover, in accordance with the principle of the preservation of combining proportions, Berzelius regarded Davy's graphite-like residue (potassium nitride, K_3N), obtained by heating the olive-colored matter, as an alloy of potassium and ammonium in the same proportions as in the olive-colored substance.[2]

4. Saline combinations of nitrogen

Among Berzelius's arguments for the oxide nature of ammonia and therefore also of nitrogen was that preserving the basic oxygen of ammonia permitted the subsumption of the composition of its salts under the rule of oxides.[3] Once accepted, the rule could become the arbiter of this and other controversial issues. Berzelius extended this argument from the opposite standpoint, that of the *acids* of nitrogen and their salts. His study of the neutral nitrates immediately followed the first *Fortsetzung* as a "Zugabe" to the latter to bolster the views about nitrogen he had enunciated there in connection with ammonia. In his introduction to the section on neutral nitrates, Berzelius indicated that he had originally intended to present his study of these salts in connection with his analyses of animal substances; but that he chose instead to include them in the *Versuch* because they constituted further proof of the rule of oxides by providing another series of acids and their salts in conformity with it and because they supported the compound nature of nitrogen. The latter motive was proclaimed by the title of this section: "Die Salpetersäure und salpetersauren Salze, als Beweiss dass der Stickstoff nicht chemisch einfach ist."[4] The "Zugabe" on neutral nitrates was complemented by studies contained in Berzelius's second continuation on basic nitrates and on nitrites, which is examined in later parts of this chapter. These three sections had been assembled by Gilbert into the second *Fortsetzung*. The latter two of the sections fully established the existence of a series of nitrite salts of nitrous acid, provided a store of firm analyses of the salts of both acids, and contributed to the general question of the composition of basic salts, a major theme of later parts of the *Versuch*. These results were mere

[2] "Schreiben," 215—216; Söderbaum, *Werden und Wachsen*, p. 122n, indicates that Berzelius later presented these ideas in vol. II (1812) of the *Lärbok*. For Davy's graphite-like residue, see his *Works*, V, p. 150. Partington (*History*, IV, p. 156) erroneously identifies Berzelius's *Ammoniumoxydul* with Davy's olive-colored matter.

[3] See the passage quoted above, p. 206, accompanied by n. 2.

[4] The section on neutral nitrates in the second *Fortsetzung* is in *Versuch*, pp. 127—135. For completion of this work in March, 1811, see Berzelius's letter to Gahn of 27 March (*Bref*, V,ix, p. 49).

by-products, however, the chief motivation having been to place the idea of the compound nature of nitrogen and the rule of oxides in a mutually supportive relationship.

Although Berzelius gave considerable weight to the conclusions flowing from these studies, they were based on an error. As Ostwald points out, Berzelius accepted Gay-Lussac's erroneous analysis of nitric acid.[5] Using Gay-Lussac's data, Berzelius found that the nitrates did not conform to the rule of oxides, for calculation revealed that the oxygen content of nitric acid fell between four and five times that found in the base. If the oxygen content of the nitrogen in the nitric acid, as it appeared in the table of the oxides of ammonium, was reckoned into the calculation, however, then the oxygen content of the acid was found to be six times that of the base.[6] The rule of oxides, therefore, demanded that nitrogen be compound.[7]

[5] Ostwald's "Anmerkungen" to the *Versuch* occupy pp. 211—218 of the *Ostwald's Klassiker* edition. Berzelius's reliance on Gay-Lussac's erroneous analysis is pointed out in n. 30, p. 217. In question is the analysis of what is now called the anhydride of nitric acid, the modern formula of which is N_2O_5. Gay-Lussac's analysis was reported in a paper following up his famous memoir announcing and documenting the law of the combining volumes of gases. The latter is "Mémoire sur la combinaison des substances gazeuses les unes avec les autres," *Mém. soc. Arcueil*, 2 (1809), 207—234. The follow-up is: "Sur la vapeur nitreuse, et sur le gaz nitreux considéré comme moyen eudiométrique," *ibid.*, 235—251; the analysis of nitric acid appears on 237—238. Gay-Lussac's eudiometric investigations confirmed Davy's analysis, which Gay-Lussac had shown, in the combining volumes paper (215—216), to be consistent with the law of combining volumes (it corresponds to the formula NO_2). When tested with potash, the water over which the gaseous reaction was conducted produced nitrate of potash. Ostwald suggests the error resulted from the formation of NO_2, its absorption by the water, and production in solution of nitric (as well as nitrous) acid. For Gay-Lussac's work on the oxides of nitrogen, see Partington, *History*, III, pp. 80—81; and Maurice P. Crosland, *The Society of Arcueil: a view of French science at the time of Napoleon I* (London: 1967), pp. 366—367.

[6] *Versuch*, pp. 129—130. Berzelius did not here emphasize that this conformed with the sulfur pattern. He stressed this later in his elaborate efforts to reconstruct the correlation of degree of oxidation and properties. For this, see below, section 9.

[7] An interesting example of Berzelius's techniques for calculating the composition of salts containing nitrogen was his determination of the composition of ammonium nitrate (*ibid.*, pp. 130—133). In Berzelius's thinking, this typical neutral salt consisted of two oxides, ammonia and nitric acid, of the same radical, ammonium. Berzelius calculated the composition of the salt in conformity with the rule of oxides, on the assumption, justified by his quantitative investigation of the non-saline nitrogen compounds, that in it the oxygen content of the base, ammonia, was the same as that in any other nitrate (that is, that the constancy of basic oxygen held for ammonia as well as for other bases), and he advanced analytical evidence in support of the figures obtained. Because of the complexities introduced in his discussion by the use of Gay-Lussac's erroneous analysis for nitric acid, the details are not given here.

5. Correspondence of oxidation and sulfuration: clarification of the principle of the preservation of combining proportions

The principle of the preservation of combining proportions, emergent in the first half of the *Versuch* and articulated explicitly in the 1812 "Rechtfertigung," asserted the correspondence of the sulfide *in minimo* with the lowest oxide of a radical and expressed the preservation of the proportion of the basic radical to the sulfur of a sulfide in the corresponding neutral sulfate salt of the radical.[8] Though Berzelius was aware, when preparing these papers, that at least some metals had more than one sulfide, he made no attempt to seek or establish a correspondence between higher oxides and higher sulfides. Moreover, because the proportion of the radicals was preserved only in the neutral salts, Berzelius was aware that the basic salts in some fashion violated the principle. He did not specify the permissible violations or their consequences, however. He did note that in the sulfate of the higher oxide of copper, the proportion of sulfur for a given weight of metal was double that of the sulfide *in minimo,* but that the multiple for the corresponding iron compounds was fractional, a circumstance he attempted to explain with the concept of intermediation. Similarly, he pointed out that in the basic salts of copper, the proportions of the sulfur for a given weight of metal were integral multiples of the proportion of the sulfide *in minimo,* but that in the basic iron salts the multiple was fractional. It appears that Berzelius regarded integral multiples as permissible violations and fractional multiples as impermissible ones requiring explanation or elimination. However, this belief was not made explicit. In these two papers, Berzelius left unclear the relations among the lower sulfide of a radical, its higher sulfides, and its sulfate salts, neutral and basic, that violated the principle.

Berzelius clarified his thinking in regard to the neutral salts by investigating degrees of sulfuration and thereby coming to recognize their correspondence with degrees of oxidation. The attainment of this recognition and some its consequences are the subject of this section. Berzelius's studies of double salts, which enabled him to describe permissible violations, are considered below in section 8.

Berzelius's recognition of a correspondence of oxidation and sulfuration emerged in the first post-*Versuch* collection of analytical studies. Published in 1812—1813 as the bulk of the "Rechtfertigung," this report contained results of Berzelius's own labors, those of several students, and those of his friend and occasional collaborator, Hisinger. Berzelius's view here was still limited to the correspondence of the lowest oxide with the lowest sulfide, but the discovery of

[8] Cf. Chapter IV, section 2 *d.*

new sulfides later led him to an awareness of a general correlation, clearly expressed in the next collection of analytical results, the "Cause of proportions" (1813—1814).

In the "Rechtfertigung," Berzelius's studies of sulfuration were deployed as an analytical tool, invoking the correspondence of the lowest oxide and sulfide. When the lowest oxide could not be analyzed accurately or reproducibly by direct means, Berzelius used the sulfide *in minimo* as a standard against which to determine the progression of oxides.[9] Summarizing this work in a letter to Gilbert,[1] Berzelius signalled, for example, the correspondence of the minimal oxide and sulfide of antimony: "Der Schwefel-Spiesglanz [sulfide of antimony] ist so zusammengesetzt, als es seyn muss, wenn man ihn nach dem Spiesglanz-Oxyde (*oxydum stibicum* [the lowest oxide]) berechnet."[2] Of tin, however, he remarked that it "giebt uns das erste mir bekannte Beispiel eines Metalls, welche drei salzbare Oxyde, so wie drei für sich darstellbare [that is, capable of an independent existence] Schwefelungs-Stufen hat."[3] Berzelius reported that the proportions of sulfur, for a given weight of metal, were as 1 : 1 1/2 : 2.[4] Still, he considered only the correspondence of the lowest oxide and sulfide: "Das Schwefel-Zinn im Minimo ist ganz so zusammengesetzt, wie es zufolge der Berechnung nach dem Zinn-Oxydul (*Oxydum stannosum* [the lowest oxide]) seyn soll."[5] The results of Sefström, Berzelius's student, on mercury compounds, also reported in a letter to Gilbert,[6] ended Berzelius's preoccupation with the lowest sulfide. Sefström showed that cinnabar, hitherto the only known sulfide of mercury (HgS), corresponded not to the lower oxide, but to the higher; and he discovered a lower sulfide (Hg_2S) that did correspond to the lower oxide. Berzelius's efforts, throughout his analytical work, to correlate oxidation and sulfuration dated from Sefström's study.

Expectation of a general correspondence of oxidation and sulfuration led Berzelius to seek the oxide, if unknown, that corresponded to a given sulfide, and

[9] The most prominent examples in the "Rechtfertigung" were the compounds of antimony (148—155) and tin (284—311). In the "Cause of proportions" (355), Berzelius determined the composition of the oxide of silver from its sulfide.

[1] Berzelius, "Zwei Schreiben des Hrn. Prof. Berzelius, Mitgl. d. schw. Ak. d. W., an den Professor Gilbert, (Wahre Natur der Eisenoxyde und der salpetrigen Säure; Entdeckungen neuer Metall-Oxyde und Schwefel-Metalle; Kritik von Davy's Lehre von der Chlorine; eine neue und merkwürdige Erscheinung von Feuer; Druckfehler.)," *Gilbert's Annalen,* 42 (1812), 276—298. All but the last two topics listed were from a letter dated 26 March 1812 and occupy pp. 276—294. The last two topics were from a letter dated 29 May 1812, occupying 294—298. The letters henceforth are cited as "Zwei Schreiben."

[2] "Zwei Schreiben," 283. The antimony compounds are discussed in detail in "Rechtfertigung," 144—176.

[3] "Zwei Schreiben," 283.

[4] *Ibid.,* 284.

[5] *Ibid.*

[6] *Ibid.,* 287. A report of Sefström's work on mercury compounds appears in the Swedish version of the "Rechtfertigung," *K.S.V.A. Handlingar,* [2] 34 (1813), 186, and in "Cause of proportions," 355, but does not appear in the German version of the "Rechtfertigung."

vice versa. The tripartite research program accordingly came to emphasize sulfides as well as oxides. These investigations culminated in the generalization of the rule of oxides to cover compounds of sulfides and in the articulation of rules to aid in the determination of formulae. Instances of the search for sulfides corresponding to oxides appeared quite prominently in the "Cause of proportions," the next summary of Berzelius's accumulating analytical results, published serially in Thomson's *Annals of Philosophy* in 1813—1814.[7] Berzelius reported there, for example, his confirmation of Laugier's analysis of realgar (As_4S_4) and stated his belief that this sulfide contained the same weight of sulfur, for a given weight of arsenic, as arsenic acid (As_2O_5) contained of oxygen (in fact, it contains four-fifths the weight). This sulfide would have corresponded to a then unknown oxide containing half the oxygen of arsenic acid. Berzelius described the experiments by which he tried, in this case unsuccessfully, to isolate this compound, though he believed he had discovered a substance in which the suspected oxide existed as an ingredient by virtue of the intermediation of the other constituents. From these experiments, he developed arguments about the probable formulae of the oxides of arsenic.

Berzelius converted this experience with what he now recognized as a general, but not universal, correspondence of oxides and sulfides into the second of three rules presented in the *Proportions chimiques* to aid in deciding on the formulae of oxides (the first rule was to examine and compare the oxides of a given radical and the third was the rule of oxides):

> 2° [On peut] comparer les degrés de sulfuration des corps avec leur degré d'oxidation, qui ne se correspond pas toujours. On sait que l'arsenic [of the compounds of which Berzelius now had corrected analyses], par example, peut se combiner en deux proportions avec l'oxigène et avec le soufre. L'oxigène, dans ses deux oxides, est comme 3 à 5 [the tri- and pentoxides, As_2O_3 and As_2O_5]; mais le soufre, dans ses sulfures, est comme 2 à 3 [realgar is As_4S_4 and orpiment is As_2S_3; the other two sulfides, As_4S_3 and As_2S_5, were then unknown]; ainsi l'oxide à trois atomes d'oxigène correspond parfaitement au plus haut sulfure, qui contient trois atomes de soufre. On peut en conclure que le nombre des atomes de soufre et d'oxigène, dans ces compositions, est 2, 3 et 5. Dans les divers sulfures d'étain le soufre y est comme 2, 3 et 4; mais l'oxigène, dans les oxides, est comme 2 est à 4, et par conséquent, dans la même proportion que le premier et le dernier des sulfures; d'où on doit inférer que les atomes d'oxigène, dans les oxides d'étain, sont comme 2 à 4.[8]

[7] Berzelius had completed a manuscript, apparently in Swedish, of the "Cause of proportions," in June, 1813 (*Bref,* I,iii, p. 60). He sent a French version in August, 1813, to his friend Alexandre Marcet, a Genevan physician living in London, for transmission to Thomson (pp. 69—72). In November, Marcet informed Berzelius that Thomson, "déjà pleine de vous," had still not published the paper (p. 82). The first installment finally appeared in December, and the remainder followed in subsequent numbers. For the relations of the oxides and the sulfides of arsenic discussed in this paragraph, see "Cause of proportions," 97—100. Berzelius's studies of the oxides of antimony and arsenic are examined further below, in section 9.

[8] *Proportions chimiques,* pp. 117—118. Berzelius's accumulating experience with sulfides led him to restate rule of oxides to include them (pp. 34—35). A few pages later (pp. 36—37), in the conclusion of the discussion of compounds formed by oxides and sulfides, he observed that

Perception of a general correspondence permitted Berzelius to recognize that to each sulfide of a radical there corresponded a salt having the radical and sulfur in the same proportion as in the sulfide; that is, each sulfide corresponded to a particular oxidation stage of the metallic radical and this stage appeared in the corresponding salt. These perceptions were brought to fruition in the theoretical portion of the "Cause of proportions." There, Berzelius restated the rule of oxides, and, in reference to the correspondence of atoms and volumes suggested by Gay-Lussac's law of combining volumes,[9] he went on to argue that:

not only the proportions, but the atoms, followed the same law. The rule, formulated in terms of atoms, but with reference solely to oxides, appeared again in the *Propertions chimiques* (pp. 119—120) as the third of three rules for determination of the formulae of oxides. The passage in the text accompanied by this note was the second rule.

[9] Berzelius's preference for combining volumes over atomic weights may be traced to his ambivalence about Dalton's atomic theory. Though he was evidently taken with the possibilities of an atomic theory, he had reservations about the viability of a corpuscular ontology. Accordingly, he developed simultaneously two parallel theories, his "corpuscular theory"—so called to distinguish it from Dalton's "atomic theory"—and his "volume theory," based on Gay-Lussac's law of combining volumes. In the "Cause of proportions" the corpuscular theory was offered first and was followed by a series of objections to it. Then the volume theory was presented as largely equivalent in its consequences to the corpuscular theory. However, because it was based on the observed phenomena of gaseous combination, it was unburdened by ontological problems and was therefore immune to the objections to the corpuscular theory. The atomic weights offered in the paper were presented as specific gravities of unit volumes of the elements, the specific gravity of oxygen having been chosen as the unit. Accordingly, in many of the passages quoted in this study, Berzelius employed the word "volume" where, were it not for his reservations about the atomic theory, he would have used the word "atom." For details about Berzelius's use of Gay-Lussac's law, see Lundgren, *Berzelius och atomteorin*, esp. pp. 114—123, 131—135.

Though stated forcefully in the "Cause of proportions," Berzelius's rservations about the corpuscular theory had nearly vanished by the time of the *Proportions chimiques*. There, the corpuscular and volume theories were still presented in parallel (*Proportions chimiques*, pp. 21—28, 47—54), but the latter was no longer portrayed as avoiding the pitfalls of the former. Rather, they were considered as alternative representations of the phenomena of combining proportions, the only question being whether atoms and volumes represented equal quantities of matter (*Proportions chimiques*, p. 48): "Plusieurs savants ont élevé des doutes sur l'identité des atomes et des volumes; mais comme les deux théories ne sont que des manières de se représenter les éléments qui se combinent, afin de mieux comprendre les phénomènes, et que l'on n'a pas la prétention d'exposer ce qui se passe réellement dans la nature, elles sont bonnes si elles donnent les plus simples explications. Or, ce ne serait point là le mérite de celle où l'on considérerait l'atome et le volume comme des fractions l'un de l'autre." Moreover, Berzelius was sufficiently content with the atomic theory that he was no longer hesitant to speculate about what "se passe réellement dans la nature:" the *Proportions chimiques* contained the detailed electrochemical ontology of matter that Berzelius suggested as a possible foundation for both electrochemical and stoichiometric phenomena (*Proportions chimiques*, pp. 56—109).

Those of Berzelius's objections to the Daltonian atomic theory that bear on the evolution of his theory of salts and upon its deployments outside inorganic chemistry are considered in Chapter VI. For detailed analysis of Berzelius's evolving ideas about the atomic theory, see Lundgren, *Berzelius och atomteorin*.

Experiment seems to prove that if a combustible radical combine in preference with 2 or 3 volumes of oxygen, it combines likewise in preference with 2 or 3 volumes of sulfur. If a salifiable oxide be composed of 1 volume of radical and 2 or 3 volumes of oxygen; and if we neutralize this oxide by any acid whatever [possessing a radical that, like sulfur, could form a "radicalide" with the metal], it is to be supposed that the neutral combination which results ought to contain for 1 volume of the radical of the oxide as many volumes of the radical of the acid as the oxide contains volumes of oxygen [for that is also the number of volumes of radical in the "radicalide" of the metal]; and, consequently, that the number of times that the acid contains the oxygen of the oxide will be the number of volumes of oxygen combined with 1 volume of the radical of the acid; for example, we consider sulphuric acid as composed of 1 volume of the radical and 3 volumes of oxygen; because it is very probable that the quantity of sulphur and of oxygen capable of combining at an elevated temperature with a given portion of lead constitute equal volumes. But if we want to know by another method how many volumes of oxygen exist in sulphuric acid, we have only to examine the composition of some sulphate; for example, sulphate of iron (*sulphas ferrosus*). The black oxide of iron contains 1 volume of metal and 2 volumes of oxygen. It follows, from what has been said, that the black oxide of iron ought to be neutralized by a quantity of acid containing 2 volumes of sulphur for every volume of iron: so that the number of volumes of the sulphur of the acid and of the oxygen of the base shall be equal. But the acid contains three times as much oxygen as the base; consequently, it is composed of 3 volumes of oxygen and 1 volume of sulphur. If instead of the sulphate of iron, we were to make choice of the persulphate of iron (*sulphas ferricus*), it is evident that in such a case the iron is combined with 3 volumes of sulphur, so that the result is just the same.[1]

Berzelius concluded the argument by pointing out the connection between the correspondence of higher sulfides and oxides and the maintenance of neutrality:

The preceding observations explain why, when a salifiable base has combined with more oxygen, it requires always an additional volume of acid for every volume of oxygen which it has absorbed.[2]

The argument may be clarified by an anachronistic interpretation using modern formulae. In accordance with the rule of oxides, the composition of any neutral salt may be expressed as

$$R^{bs}O_n \cdot nR^{ac}O_m,$$

where R^{bs} is the radical of the base, R^{ac} is the radical of the acid, and m and n are small integers. If the salt is a sulfate, for example, it corresponds to the sulfide

$$R^{bs}R^{ac}_n,$$

where R^{ac} is S and R^{bs} is a metal. If the base of the salt, $R^{bs}O_n$, is oxidized, taking up an additional dose of oxygen, for example,

$$R^{bs}O_n + 1/2\ O_2 = R^{bs}\ O_{(n+1)},$$

[1] "Cause of proportions," 452.
[2] *Ibid.*, 453.

then, to preserve the neutrality of the salt, the base must take another dose of acid. The salt then becomes

$$R^{bs}O_{(n+1)} \cdot (n+1)R^{ac}O_m.$$

This salt corresponds to the higher sulfide,

$$R^{bs}R^{ac}_{(n+1)}.$$

From this scheme it follows that the proportion of the radicals in each sulfide is the same as in the corresponding sulfate.

If the oxidized base did not take another dose of acid, basic salts were produced in violation of the principle. Berzelius's insights about basic salts were contingent upon his prior studies of the role of water in saline combinations. To these the next section is devoted.

6. Water as a saline component

By studying the role of water in salt formation, Berzelius took major strides in documenting and elaborating his generalization of the ideas of acidity and basicity, for he demonstrated that water, *the* neutral substance, could play either role. In developing techniques for differentiation of water of crystallization from water as a saline component, Berzelius proved that other oxides, not typically acidic or basic, could take over these roles by displacing water from salts. Finally, by studying water of crystallization as a neutral component of salts, Berzelius took the first steps in his passage from second to still higher order composition. This work was recounted in the third *Fortsetzung* of the *Versuch.*

a. Basic and acidic water

Davy's denial of the oxide nature of muriatic acid was the occasion for Berzelius's study of the role of water as base. Originally adhering to the opinion of Gay-Lussac and Thenard, announced in 1809, that muriatic acid, in its ordinary state, contained one-fourth its weight in water, Davy later abandoned this belief because of his failure to decompose muriatic acid into a radical and oxygen, or oxymuriatic acid into muriatic acid and oxygen. Instead, he proposed that oxymuriatic acid was an element, chlorine, which, in combination with hydrogen, formed muriatic acid. The effects explained by Gay-Lussac and Thenard by assuming the presence of water resulted from the presence of hydrogen, which Davy now began to consider the generic component of acids.[3] Unwilling to

[3] For the work of Davy and of Gay-Lussac and Thenard on chlorine and its compounds, see Partington, *History,* IV, pp. 52—54. Passages in which muriatic acid was viewed as a compound containing water are cited above, p. 197, in n. 8. The French collaborators reached the conclusion that muriatic acid contained one-quarter of its weight as water from the observation that dried muriatic acid gas produced muriates and water from its reaction with metallic oxides. Similarly, with metallic potassium, the gas yielded the muriate and hydrogen, the latter in the same

assent to a new generic theory and to the abandonment of the muriatic radical, Berzelius responded to Davy's conclusions by sustaining the view of Gay-Lussac and Thenard, which he believed to agree nicely with the speculations of the "Spätere Zugabe."

He made his stand, then, at the same point from which Davy pushed on: the idea that muriatic acid gas contained water, in such an amount that its oxygen equalled that in the base of muriate salts. The failure, chronicled by Gay-Lussac and Thenard and by Davy,[4] to prepare dry muriatic acid gas, that is, the anhydrous acid, or to decompose muriates in the absence of water, and conversely the ability to do so in the presence of water, were taken as proof that this water was essential to the existence of the acid. In this view, the acid provided Berzelius with a prime example of an oxide incapable of independent existence and occurring only in combinations, by virtue of the intermediation of another body. The other body in question here was, like muriatic acid itself, an oxide: either water, or the base of a muriate salt. The decomposition of muriates in the presence of water was then interpreted as a displacement reaction, in which the base of the muriate was displaced from combination with the anhydrous acid by the water, which then combined with the acid to form ordinary muriatic acid.[5]

Having concluded that one of the most common and most important mineral acids in its ordinary state was a compound of two oxides, Berzelius sought to generalize his view to cover the other mineral acids. He was able to devise ingenious arguments to prove that sulfuric and nitric acids, like muriatic acid, contained water as base and could not exist independently without it, though sulfurous and nitrous acids could.[6] With these ideas Berzelius was now in a

quantity evolved by an equal amount of potassium through its action on water. They presumed, therefore, that the metal was oxidized to potash by decomposition of the water contained in the muriatic acid gas, thereby occasioning the evolution of hydrogen. The anhydrous muriatic acid then combined with the potash to produce the muriate. The decomposition of oxymuriatic acid by hydrogen into muriatic acid was taken as evidence that this acid contained just enough oxygen, above that in muriatic acid, to form water with the hydrogen, leaving anhydrous muriatic acid, which then combined with the water to form ordinary muriatic acid gas. These views permitted the conclusion that oxymuriatic acid contained just that much more oxygen than did muriatic acid as was necessary to oxidize the amount of metal that was neutralized by the amount of muriatic acid in a muriate salt.

Though for a while Davy adhered to similar views, based on similar experiments, his argreement did not last long. Struck by the observation that oxymuriatic acid could not be decomposed even by charcoal heated to whiteness, Davy doubted the presence of oxygen in that substance (Davy, *Works,* V, p. 286). He made every conceivable effort to achieve a decomposition and failed. He concluded therefore, that oxymuriatic acid was an element, chlorine, and that muriatic acid was a compound of that element with hydrogen. (See Partington, *History,* IV, pp. 54—57.)

4 In the passages cited above, p. 197, in n. 8.

5 Berzelius long offered powerful resistance to the chlorine interpretation. For an overview of the evolution of his thinking in this respect, see Söderbaum, *Werden und Wachsen,* pp. 101—119. The material analyzed in this and the succeeding two paragraphs is in *Versuch,* pp. 154—158.

6 Once again, he examined the oxides of sulfur. There were four of them, Berzelius recalled, the two acidic ones, and the two lower ones. The latter were taken to be incapable of an independent existence, and their composition had

position to explain Davy's failure to decompose oxymuriatic acid into muriatic acid and oxygen, using either dry acids or white hot charcial, without having to concede that oxymuriatic acid was simple. Though the other mineral acids could be decomposed by heat into oxygen and a lower oxidation stage, such as nitrous or sulfurous acid, the next lower oxidation stage of oxymuriatic acid was muriatic acid, a substance incapable of an independent existence. Davy's experiment, therefore, would have had to result in either a complete reduction of oxymuriatic acid into oxygen and the uncombined muriatic radical or in nothing. Since the affinity of carbon for that radical was less than that of the oxygen with which it was combined, no effect took place. Berzelius could rest content with this conclusion:

Es ist also sehr natürlich, dass eine Säure, welche für sich nicht darstellbar ist, und deren Radical eine sehr grosse Verwandtschaft zum Sauerstoff hat, nicht ohne Dazwischenkunft von Wasser oder irgend einen oxydirten Körper (welcher die Stelle der Basis vertritt), aus ihren Verbindungen getrennt werden kann.[7]

In upholding the idea of the oxygen content of oxymuriatic and muriatic acids against the view that the latter was an element, Berzelius had brought even the mineral acids into line with the saline view of matter—by granting to water the capacity of behaving as a base. In the *Versuch,* he chose to illustrate the commonplace occurrence of acids capable of existing only in combination with water or bases by analyzing three vegetable acids. He thus introduced the long influential concept that organic acids consisted of acid anhydride and water, an archetypical dualistic notion.[8] The three analyses in the *Versuch* were the early results of an extensive series of organic analyses that Berzelius published not long

been calculated in the "Zugabe." The decomposition of sulfates by heat in the absence of water, Berzelius noted, gave not sulfuric acid, but sulfurous acid and oxygen. Conducted in the presence of water, the thermal decomposition of sulfates yielded liquid sulfuric acid. Berzelius concluded, therefore, that sulfuric acid, like muriatic acid, was incapable of an independent existence. Of the four oxides of sulfur, then, Berzelius believed only one, sulfurous acid, capable of an independent existence. Berzelius offered a similar set of arguments about nitric acid, again noting that thermal decomposition of salts in the absence of water produced nitrous acid and oxygen and, in the presence of water, the liquid acid.

[7] *Versuch,* p. 157.

[8] Berzelius gave full expression to this view in the *Versuch* (p. 164): "Wir kennen also keine von den drei hier analysirten Pflanzensäuren im reinen Zustande, und es lässt sich muthmassen, dass sie, wie die vorhin erwähnten Mineralsäuren, *für sich nicht darstellbar sind.*

"Wir sehen ferner, dass sowohl die Citronensäure als die Sauerkleesäure eine Portion Wasser enthalten, welche ausgetrieben werden kann, und sich ganz wie Krystallwasser verhält, während, die Säure mit einer andern Portion Wasser verbunden bleibt, die ganz die Rolle eine Basis spielt, und sich nur durch eine stärkere Basis von ihr trennen lässt. Wir können diese Verbindungen mit Wasser füglich als Salze ansehen, in welchen das Wasser die Basis ist, und sich ganz wie die schwächern Basen unter den Metalloxyden verhält, die sich sowohl mit Säuren als mit andern Basen verbinden können. Wenn wir das in der krystallisirten Sauerkleesäure durch Verwitterung verflüchtigte Wasser, für Krystallwasser ansehen, so enthält diese zweimal so viel Sauerstoff, als der Basis anzusehende Antheil des Wassers."

after and that effectively launched the concept on its long career.[9] The idea dominated all studies of organic acids until it was overthrown by Gerhardt's discovery (1852) of the anhydrides of the organic acids.[1]

Berzelius also considered compounds in which water played the role of acid in combination with bases. There were modest precedents for Berzelius's work in this connection; he did not cite them, though he was likely familiar with them. As Partington indicates, Proust had demonstrated the existence of hydrates of the metallic oxides (now seen as hydroxides) and had himself pointed to a precedent in the investigation of the basic hydrates, Lavoisier's work on slaked lime.[2] Given that the decomposition of the alkalies and earths had since revealed the analogy of the alkaline and earthy with the basic hydrates, Berzelius was in a position to establish the existence of a general class of such compounds and to incorporate them in his saline theory by treating them as salts of basic metallic oxides in which water took the role of acid. He presented a series of analyses of the alkaline earths, of other earths, such as alumina and silica, and of one metallic oxide, that of iron. The studies suggested that these combinations were governed by a version of the rule of oxides: that the alkalies and alkaline earths entered into combination with water in such a way that the quantities of oxygen in each component of the hydrate were equal.[3]

That water was itself an oxide contributed to strengthening the analogy that Berzelius developed in these studies between water and the more usual saline components and between the aqueous compounds of acids and bases with typical salts. In the concluding section of the *Versuch,* Berzelius gave forceful expression to his views about the potential of water to assume the role of acid or base in relation to other oxides. There, he listed hydrates and compounds of acidic water together with neutral salts under the rubric of binary compounds of oxides:

> Wenn sich zwei oxydirte Körper vereinigen, so lässt sich das Verhältniss unter ihnen am leichtesten nach dem Sauerstoff bestimmen, indem der Sauerstoff des einen dem des andern entweder gleich, oder davon ein Multiplum nach einer ganzen Zahl ist.
>
> Zu diesen Verbindungen gehören
>
> a) *Salze,* oder *Verbindungen von Säuren mit Basen.* In den *neutralen* Salzen ist der Sauerstoff der Säure von dem der Basis ein Multiplum nach 2, 3, etc. bis 8...
>
> b) *Hydrate,* oder *Verbindungen des Wassers: Erstens: mit Säuren.* In diesem spielt das Wasser ganz die Rolle einer Basis; die Säure nimmt davon zur Sättigung eine Menge auf, welche genau so viel Sauerstoff als eine Salzbasis enthält, von der diese Menge Säure ge-

[9] *Ibid.,* pp. 160—165. The subsequent paper on organic analysis is Berzelius, "Experiments to determine the definite proportions in which the elements of organic nature are combined," *Ann. Phil.,* 4 (1814), 323—331, 401—409; 5 (1815), 93—101, 174—184, 260—275; Swedish, "Försök öfver de bestämda förhållanden, hvari elementen för den organiska naturen äro förenade," *Afhandlingar,* 5 (1818), 520—646.

[1] For this see Edouard Grimaux and Charles Gerhardt [fils], *Charles Gerhardt [père], sa vie, son oeuvre, sa correspondance, 1816—1856: document d'histoire de la chimie* (Paris: 1900), pp. 401—409.

[2] Partington, *History,* III, p. 646.

[3] *Versuch,* p. 167.

sättigt wird. Das mit der Säure auf diese Weise verbundene Wasser ist ganz von dem Krystalwasser einiger krystallisirenden Säuren verschieden.

Zweitens: mit Salzbasen. Das Wasser spielt hier in so fern die Rolle einer Säure, als es ihre Stelle vertritt, enthält aber nur gleiche Theile Sauerstoff mit der Basis, oder ist zuweilen ein Submultiplum vom Sauerstoff der Basis. Das Krystallwasser in einigen krystallisirenden Hydraten ist ganz von dem mit der Basis, in der Stelle einer Säure verbunden, verschieden. . .[4]

This belief in the capacity of water to serve as either acid or base was an outcome of Berzelius's electrochemical generalization of the ideas of acidity and basicity. The remaining sections of this chapter and portions of the next provide additional illustration of Berzelius's growing belief that all oxides, except perhaps the suboxides, could assume either acidic or basic roles in combinations with other oxides, even if the substances in question did not display typically acidic or basic properties. The full generally of Berzelius's ideas about acidity and basicity emerged from his work on basic and double salts and on minerals. This work provided the experimental counterpart of the theoretical generalizations embodied in the electrochemical series.

b. Water of crystallization

Berzelius's study of water of crystallization marks his departure from the realm of second order to still higher order compounds and forms the bridge across which he travelled from binary saline compounds to the most complex of laboratory chemicals, basic and double salts. Signposts for this route appear in the final, summarizing pages of the *Versuch*. There, Berzelius did not juxtapose his ideas about water of crystallization with those just cited about neutral salts and aqueous compounds of acids and bases. These last had been assembled under one heading because of their shared conformity, as binary compounds, to the rule of oxides, but salts containing water of crystallization had at least three oxides. Berzelius therefore listed his conclusions about them together with those bearing on double salts under the heading of compounds containing three or more oxides.[5] This grouping suggests that he conferred on neutral water of crystallization a role ontologically equivalent to that of acidic and basic saline components and that he saw in salts containing water of crystallization substances similar to double salts. It also indicates that he had not yet attained his later notion of hierarchical dualism. He did not yet have a clear notion of higher orders of compounds, for he merely separated binary compounds of oxides from all compounds of oxides, that is, ordinary salts from salts of three or more oxides. Though Berzelius's coining of the term "double salt" for compounds of three oxides suggests the direction in which he was soon to move, at the end of the *Versuch* a double salt for Berzelius was merely one example of compounds containing multiple oxide components.

[4] *Ibid.*, pp. 206—207.

[5] *Ibid.*, pp. 207—209.

The ability to differentiate the various conditions under which water was retained in compounds, and to do so quantitatively, was critical to a quantitative study arguing for a qualitative multiplicity of roles for water in saline compounds. Berzelius elucidated this subject directly in some preliminary remarks in the third *Fortsetzung.*[6] The themes discussed there, however, as well as some related ones that are not, were visibly emergent in earlier portions of the *Versuch,* and one of Berzelius's chief criteria for such discrimination was offered only later in the body of the *Fortsetzung.* The various passages show that Berzelius had developed three principal rules for determining the nature of aqueous components: one for distinguishing the presence of aqueous impurities in crystalline precipitates from water of crystallization; a second for deciding upon the presence or absence of crystal water in a pure precipitate; and a third for differentiating water of crystallization from basic or acidic water.

The first of these dates from at least as early as the normal analyses, when Berzelius doubtless experienced most keenly the need for pure precipitates.[7] In recounting the conduct of the normal analyses he offered a working procedure to discriminate water of crystallization from water that was merely mechanically retained in crystals as a residue of the solution from which they were precipitated. The latter appeared as water lost by decrepitation on heating. Decrepitation, in Berzelius's view, split crystals but did not destroy the crystalline character of the precipitate. Unlike water lost by decrepitation, water of crystallization must depart first at the surface, thereby providing escape passages for the interior portions of water. The point was reiterated in connection with the analysis of the nitrates, where Berzelius showed that the nitrates of barytes and of lead oxide contained no water of crystallization, because they lost water only by decrepitation.[8]

The criterion for the presence of water of crystallization was also emergent in earlier portions of the *Versuch* but was at first mostly implicit. It appeared fairly clearly in the section devoted to nitrites, where Berzelius wished to determine the quantity of water of crystallization of these salts.[9] Whether water of crystallization existed in these salts was determined from their thermal decomposition. Removal of water of crystallization demanded strong heating, which simultaneously liberated the acidic component of the salt in the form of a gas. Because the liberated acid gas dissolved in the departing crystal water, the appearance of aqueous acid in the thermal decomposition of a salt signified the presence of crystal water, while the appearance of acid gases signified its absence.

The rule for the discrimination of water of crystallization from acidic and basic water appeared twice as indirect statements in the body of the third *Fort-*

[6] *Ibid.*, pp. 158—160.
[7] *Ibid.*, p. 90n.
[8] *Ibid.*, pp. 128—129.
[9] E.g., *ibid.*, pp. 142, 143, 148, 149; a later example, p. 178.

setzung, in connection with the laws for the composition of compounds containing basic and acidic water.[1] Heating could remove water of crystallization, but only displacement could remove basic or acidic water.[2]

Berzelius presented a general discussion of the problem of determining the nature of aqueous constituents in that portion of the preliminary remarks to the third *Fortsetzung* devoted to water of crystallization.[3] There, he argued that his experimental methods would allow him to bring water of crystallization under the sway of stoichiometry. In addition, he argued that stoichiometry would allow him to overcome a further difficulty which he had been unable to surmount by experimental observation. In salts containing both water lost by decrepitation and water of crystallization, the two kinds of water were generally driven off simultaneously by heat. Accordingly, experiment could not determine definitively whether a salt that decrepitated also contained water of crystallization. Here was a role for stoichiometry, as he stated in the conclusion of his discussion of the aqueous content of precipitates:

> Ich habe diese Bemerkungen vorausgeschickt, um zu zeigen, wie es nicht selten eine schwierige Sache ist, das Krystallwasser eines Salzes mit volliger Schärfe zu finden. Dieses wird aber mit der Zeit sehr leicht sein. Denn wenn das Gesetz für den Gehalt der Salze an Krystallwasser einmal ausgemacht ist, so bedarf es dann annähernder Versuche, um unter mehreren Möglichkeiten das wahre Verhalten herauszufinden. Ich glaube in meinen hier zu erklärenden Versuchen dieses Gesetz gefunden und durch hinlängliche Beispiele bewiesen zu haben.[4]

Berzelius evidently had no doubt that a law should have existed to serve as a control on analysis of crystal water. The careful and often clever experiments he went on to recount clearly permitted him to enunciate a variant of the rule of oxides for water of crystallization. The law stated simply "... dass in den Salzen der Sauerstoff des Krystallwassers stets ein Multiplum nach einer ganzen Zahl von dem der Basis ist, oder (wie bei der Citronensäure und dem basischen

[1] In the passage quoted above, p. 218 in n. 8, for basic water. For acidic water, cf. the conclusion drawn from the exemplification of acidic water in the hydrates of chalk and magnesia (*ibid.*, p. 167): "Ich halte diese beiden Beispiele für hinreichend, die Vermuthung zu begründen, dass das nämliche bei den feuerbeständigen Alkalien und bei den übrigen alkalischen Erden statt finde. Wenn diese aber eine Menge Wasser binden, deren Sauerstoff dem des Alkali oder der Erde gleich ist, und wenn sie auch, wie bekannt ist, neue Portionen Wasser, die dem Krystallwasser entsprechen, aufnehmen können, so müssen sie das Wasser in zwei verschiedenen Zustände enthalten, ganz so, wie wir das bei den im Vorhergehenden beschriebenen Säuren [the three organic acids] gesehen haben. Die eine Portion Wasser entspricht bei diesen stärkeren Basen dem positiven Körper, oder der Säure, and lässt sich aus ihrer Verbindung mit den fixen Alkalien, dem Baryt und dem Stronti[a]n nicht durch das Glühen austreiben, sondern nur durch Hinzukommen eines andern entweder für sich, oder in Hinsicht der Temperatur positiveren Körpers."

[2] This procedure often led Berzelius to identify as water of crystallization what in modern terms would be identified as hydroxide, which corresponds to Berzelius's basic water.

[3] *Ibid.*, pp. 158—160.

[4] *Ibid.*, p. 160.

kohlensauren Kupferoxyd) ein Submultiplum nach einer ganzen Zahl von der Sauerstoffmenge der Basis ist."[5]

Berzelius's readiness to consider still other patterns marks the beginning of his attention to the problem of higher order composition. Noting that comparison of the *acidic* oxygen, rather than the basic, with that of the crystal water revealed fractional, not integral multipliers, he observed:

> Dieses hat mich auf folgende Regel geführt: "In Verbindungen von mehreren oxydirten Körpern ist der Sauerstoff desjenigen Bestandtheils, der davon am wenigsten enthält, in der Sauerstoffmenge jeder der übrigen Bestandtheile nach einer ganzen Zahl enthalten."[6]

Berzelius acknowledged that the experimental evidence was insufficient to rule out other possible patterns (for example, that there was a dominant component, the oxygen of which, though not the least of all components, might be that unit of which the oxygen of each of the other components was an integral multiple or submultiple). However, he claimed it useful to adhere to the provisionally stated rule until more evidence was accumulated.[7] This rule, that the oxygen of the oxide component having the least oxygen was the unit of which the oxygen of each of the other components was an integral multiple, is called in this study the "law of unitary oxygen." Though Berzelius offered it here merely as a preliminary suggestion, he later displayed a tenacious commitment to it as a device to delimit sharply the number of permissible inorganic compounds. This point is exhibited in the next chapter.

Though Berzelius formulated the law of unitary oxygen in a discussion of salts with water of crystallization, he considered it applicable to all compounds containing more than two oxides as the formulation just quoted indicates. In Berzelius's view, the addition of a third oxide had produced a different sort of entity from an anhydrous neutral salt. All three components were ontologically equivalent (hence the comparison of acidic with aqueous oxygen and the formulation of the unitary law), and the oxygen of any one of them, not merely of the base, might be the measure of that of the others. A salt with water of crystallization was not, therefore, at this stage, a dualistic construction of the third order, as Berzelius would later call it; but a "triple salt," in Bergman's sense, that is, one having three saline constituents.[8] Berzelius's juxtaposition, noted at the beginning of this subsection, of compounds containing water of crystallization

[5] *Ibid.*, p. 185.

[6] *Ibid.* Berzelius also presented the rule in his "Summarischer Bericht," 318—319, but he had not yet formulated it as of "Schreiben" (in which, see pp. 219—220).

[7] *Versuch*, pp. 185—186.

[8] For relevant passages in Bergman, see above, p. 50, n. 7. In Bergman's nomenclature ordinary neutral salts were called "double salts," because they contained two saline components (an acid and a base); and what have been called "double salts" since Berzelius (e.g., alum) were named "triple salts," because they contained three components (in alum, sulfuric acid, silica, and alumina).

with double salts and other compounds containing three or more oxides is consistent with this interpretation.

The notion of complex salts as compounds of multiple oxide components indeed persisted through Berzelius's study of double salts and coexisted in them with a more nearly dualistic conception. That Berzelius was tending to replace the former with the latter is suggested by his coining, in the *Versuch,* the term "double salt" to replace their Bergmanian designation as "triple." Through the end of the *Versuch,* however, the binary conception remained overshadowed by the multiple oxide model. Only in the *Mineralogy,* when Berzelius experienced difficulty in bringing still more complex compounds in line with the multiple oxide model of double salts, did he attain his nature dualistic conception. In the *Proportions chimiques,* the multiple oxide model had nearly vanished.

7. Basic salts

a. Introduction

As he had with water of crystallization, Berzelius attempted to discern rules for basic and double salts that governed and limited the composition of these substances. His work on basic salts was not heavily precedented. Rouelle had distinguished between neutral and acid salts, claiming the latter were chemical compounds, not mixtures of the neutral salt with excess acid. The excess, according to Rouelle, also had its point of saturation, a view later upheld by Bergman for acid salts and by Bayen for basic mercury salts.[9] Richter, too, had recognized that acid and basic salts existed in which an apparent excess of one component could be saturated.[1] The existence of acids that formed several salts by saturation with several different amounts of base was noted by Fourcroy in the *Méthode* (but not by Lavoisier in the *Traité*). The nomenclature adopted for them called for the addition, to the name of the ordinary neutral salt formed by a given acid and base, of the expression *acidule,* as in *tartrite acidule de potasse,* for salts in which the acid predominated, or *sursaturé,* as in *phosphate sursaturé de soude,* for salts in which the base predominated.[2] Basic salts also appeared in the Proust-Berthollet controversy.[3] Berthollet had prepared basic salts of copper oxide by precipitation of neutral salts with alkali and claimed that the acid content of the precipitate diminished as more alkali was added to the solution. Similarly, he had prepared basic sulfates of mercury and claimed that they represented cases of indefinite proportions lying between the two extremes formed by mercury with the minimum and with the maximum of oxygen. Work-

[9] For Rouelle's views and their subsequent defense by Bayen and Bergman, see Partington, *History,* III, pp. 74—75.

[1] On Richter, see below, p. 225, n. 5 and accompanying text.

[2] Lavoisier *et al., Méthode,* pp. 96—97.

[3] Partington, *History,* III, pp. 648—649.

ing in the immediate aftermath of this controversy, Berzelius was prepared to employ the methods of the *Versuch* to develop a definite quantitative treatment of these substances.

b. Definition of neutrality

Berzelius argued that a prerequisite to the study of basic salts was a definition of neutrality. The need for a definition was the more pressing because Berzelius's own work had brought to light salts of weak acids that exhibited an acidic reaction and were incapable of saturation with further base. As Berzelius remarked, "Jeder Chemiker weiss, was wir unter Neutralsalzen verstehen; dessen ungeachtet ist est so leicht nicht, durch eine gute Definition anzugeben, was *neutral* eigentlich ist."[4] The typical and therefore generally undisputed examples of neutral salts were those of the mineral acids with the alkalies or the alkaline earths. These salts exhibited minimal reactivity and exerted no effect on vegetable dyes. They were the majority of the better known salts in Lavoisier's *Traité,* where the discussion centered chiefly on defining neutral salts, not as distinct from acid or basic salts, but from the acids and the various salifiable bases. On the other hand, Richter explicitly called attention to cases in which apparently saturated salts displayed an acidic or basic reaction, and he noted that some of those could combine with an additional quantity of acid or base to form different salts. He called the typical neutral salts "absolutely neutral" and those nonneutral salts capable of forming neutral salts with another definite dose of acid or base "relatively neutral."[5]

Berzelius went beyond these precedents to accommodate the salts of weakly basic oxides. As examples, he cited the oxides of zinc and iron and the earths alumina and zirconia, all substances that had been entered into the category labelled simply "oxides" in his 1811 "Nomenclature." The acid in the salts of these bases was so loosely bound, he noted, that the salts displayed an acid reaction, though they represented the most nearly neutral compounds into which these oxides could enter.[6] Accordingly their behavior conformed neither to the qualitative norm of neutrality provided by the salts of the mineral acids with the alkalies, nor to the quantitative norm provided by Richter's "relatively" neutral salts, which could enter into more nearly neutral combinations by saturation with an additional dose of base. Berzelius offered a new definition of neutrality, one capable of assimilating the compounds of weakly basic oxides to the category of neutral salts.

He did so by turning the qualitative norm of the typical neutral salts into a quantitative one. It was the proportions of the constituents of the definitely neutral salts that would define neutrality for other salts. Any salt, regardless of its

[4] *Versuch,* p. 186.

[5] For Richter's classification of salts, see Partington, *History,* III, p. 677.

[6] *Versuch,* p. 187.

effect on vegetable dyes, was neutral if it was composed quantitatively like a definitely neutral salt:

Wir können also, glaube ich, alle diejenigen erdigen und metallischen Salze als *neutral* ansehen, in welchen die Säure eben so viel Theile Sauerstoff in der Basis mit sich vereinigt, als dieses in einer bestimmt neutralen Verbindung der nämlichen Säure mit einem Alkali oder ein alkalischen Erde geschieht. So z.B. sehe ich als *Neutralsalze* an, alle diejenigen schwefelsaure Salze, in welchen die Basis ein Drittel so viel Sauerstoff als die Säure enthält; ferner alle arseniksaure, kohlensaure, salzsaure und phosphorsaure Salze, in welchen die Basis halb so viel Sauerstoff als die Säure in sich schliesst, und so ferner. Salze, in welchen des Sauerstoffs in der Säure verhältnissmässig gegen den der Basis mehr enthalten ist, als im Neutralsalze, nenne ich *säureliche* oder *saure Salze,* und umgekehrt *basische Salze* solche, welche des Sauerstoffs in der Saure, bezogen auf den in der Basis, verhältnissmässig weniger als das Neutralsalz enthalten.[7]

The acids Berzelius cited here in illustration of his idea were precisely those that he had studied to establish the rule of oxides. The rule first emerged from Berzelius's study of sulfuric acid, and the salts of muriatic, carbonic, arsenic, and phosphoric acids were presented in the first *Fortsetzung* as the material that served to confirm the rule. The rule was there presented as a statement about neutral salts, and it was based on studies of undisputedly neutral salts of the mineral acids. Here, in the third *Fortsetzung,* the rule served as the norm defining the neutrality of salts that did not belong to the traditional category of neutral salts because they did not contain members of the traditional category of bases.

c. Composition of basic salts

Given this quantitative expression of the composition of neutral salts, Berzelius undertook a series of studies to determine how basic salts departed from it. First, he reanalyzed the basic sulfate of iron oxide, which, in the first half of the *Versuch,* had presented him with the anomaly to the principle of the preservation of combining proportions.[8] After having done the experiments recounted in the *Versuch* comparing the oxygen content of the constituents of various salts, he had come to recognize that the analysis of the iron salt contained an additional anomaly, in that the oxygen of the acid was only three-quarters that of the base and was thus neither an integral multiple nor submultiple.[9] He reanalyzed this salt, therefore, found his original analysis to be in error,[1] and arrived at new

[7] *Ibid.*, pp. 187—188.

[8] Cf. above, Chapter IV, subsection 2.*d*.

[9] *Versuch*, p. 188.

[1] To determine the source of the error in the original analysis, he repeated it (*ibid.*, p. 190) using the same procedure he had first employed (p. 38). The product, though exhibiting the same yellow color of the pure basic salt, now proved to contain ammonia. Berzelius suggested that the small quantity of nitric acid that he had used (without having mentioned it previously) to improve the yield resulted in the production of a triple compound of the oxide of iron, sulfuric acid, and ammonia. He did not think to consider that he had used ammonia to induce precipitation of the basic salt. The product was likely a mixture of the yellow basic salt and the colorless ammonium iron alum, $NH_4Fe(SO_4)_2.12H_2O$.

results that removed both anomalous aspects of this salt.[2] He found now that whereas the acid of the neutral sulfate of iron oxide contained three times the oxygen of the base, the acid in the basic sulfate contained only half the oxygen of the base. Therefore, in the basic salt a given weight of acid saturated six times as much base as it did in the neutral salt.[3] He also reanalyzed the basic sulfate of copper oxide, which he had compared with the iron compound in the first half of the *Versuch,*[4] and he added new analyses of additional examples of basic salts of sulfuric and other acids.[5] The investigation of basic nitrates and nitrites, presented in the second *Fortsetzung* as part of the effort to bring the salts of the nitrogen oxyacids under the rule of oxides, was a major part of this effort and is described below. These analyses permitted Berzelius to formulate rules governing the departures of the composition of basic salts from the proportions dictated by the rule of oxides for neutral salts:

> a) Der Sauerstoff der Säure kann in den basischen Salzen sowohl ein Submultiplum als ein Multiplum nach einer ganzen Zahl von dem der Basis sein.
>
> b) Die Menge von Basis, welche eine gegebene Menge von Säure in einem basischen Salze sättigt, kann von der Basismenge des neutralen Salzes derselben Art ein Multiplum nur nach einer solchen Zahl sein, durch welche zugleich der Sauerstoff der Säure von dem der Basis ein Submultiplum oder ein Multiplum nach einer ganzen Zahl wird. So z.B. kann in den basischen Salzen der Schwefelsäure die Basis nur ein Vielfaches nach 3, 6, 9, nicht aber nach 2, 4, 5, 7, 8 u.s.f. von der Basis des neutralen Salzes sein. Säuren dagegen, deren Sauerstoff in den neutralen Salzen ein 2- 4- 6- 8-faches von den Sauerstoff der Basen ist, können in den basischen Salzen 2, 4, 6, 8, aber nicht 3, 5, 7, 9 u.s.f. Mal so viel Basis als in den neutralen Salzen aufnehmen.[6]

Having established the departures of the basic salts from the neutral as defined by the rule of oxides, Berzelius could complete the argument begun in the "Cause of proportions" concerning the violations of the principle of the preservation of combining proportions presented by basic salts.[7] He needed only to point out what happened when the base of a neutral salt was oxidized but saturated no additional acid. This he did in the *Proportions chimiques:*

> En vertu de la loi générale [the rule of oxides], un oxide neutralisé par un acide rompt cette neutralité, s'il trouve à se combiner avec encore un ou plusieurs autres atomes d'oxigène. Il se forme alors deux combinaisons de différents degrés de saturation, toutes deux composées de telle manière que l'oxigène de l'acide est un multiple ou un sous-multiple par un nombre entier de celui de l'oxide. Si l'acide est un de ceux qui font exception à la règle générale [the acids of the phosphorus pattern], on aura le même phénomène; mais les deux combinaisons nouvelles se formeront d'après la loi particulière à ces acides.[8]

[2] *Ibid.*, p. 191.
[3] *Ibid.*, p. 189.
[4] *Ibid.*, p. 193.
[5] *Ibid.*, pp. 193—195.
[6] *Ibid.*, pp. 195—196.
[7] Cf. above, Chapter IV, section 2.*d.*, and above, section 5.
[8] *Proportions chimiques*, p. 36.

d. Basic nitrates and nitrites

Berzelius's study of the composition of the basic nitrates and nitrites affords a practical illustration of the connections be perceived between the composition of the neutral salts, dictated by the rule of oxides, and the basic salts, dictated by the two rules just quoted. He had investigated the basic salts of the nitrogen oxyacids, however, not only as part of his study of basic salts, but also to provide further evidence for the compound nature of nitrogen. The results reported in the *Versuch,* then, also prove important in the discussion below of Berzelius's reconstruction of Lavoisier's correlation.[9]

Having shown that the neutral nitrates conformed with the rule of oxides only if nitrogen was compound, Berzelius turned to the basic nitrates for further support of this idea. Though his analyses of the basic nitrates provided contradictory evidence, he persisted in his belief. He found that the basic nitrate of lead conformed to the rule of oxides on the assumption that ammonium, and not nitrogen, was the radical of the acid.[1] In the "überbasisches Salz," however, he found that the rule of oxides held only if nitrogen was taken to be the radical.[2] Berzelius declined to accept the idea that his own analyses, which this anomaly in any case induced him to repeat, were in error. Moreover, he did not yet suspect the analyses of Gay-Lussac, whose abilities as an experimentalist Berzelius justifiably held in high esteem.[3] Unwilling to drop the idea of the compound

[9]For Berzelius's study of the neutral nitrates, see above section 4. Because of its connection with the question of the nature of nitrogen, Gilbert had combined the material on the neutral nitrates with the later studies of the basic salts into the second *Fortsetzung* (cf. above, p. 202, n. 6).

[1]Neutral Pb(II) salts easily form well-defined, crystallized basic salts by addition of lead oxide or hydroxide. In modern formulae, neutral lead nitrate is $PbO.N_2O_5$; the basic salt analyzed by Berzelius contains twice the base for the same quantity of acid: $(PbO)_2.N_2O_5$. Written as compounds of the ammonium radical, Am, the two salts become $PbO.AmO_6$ (neutral) and $(PbO)_2.AmO_6$ (basic), where $Am = N_2\text{-}O$. In this conception, the acid in the neutral salt contains six times the oxygen of its base; and that of the basic salt, three times. The composition of the neutral salt is discussed in *Versuch,* pp. 129, 138—139; that of the basic, pp. 135—136. The composition of nitric acid as AmO_6 is derived from Berzelius's table, reproduced above (p. 207, accompanied by n. 5), of the oxides of ammonium.

[2]When Berzelius later returned to the analysis of the basic lead nitrates (below, section 9), he discovered that this supposed "überbasisch" salt was a mixture. Here (*Versuch,* p. 136), he concluded that, with Gay-Lussac's analysis of nitric acid, the acid and base contained approximately equal quantities of oxygen (however, the margin of error was larger than typical for Berzelius's analyses). On the assumption that ammonium was the radical, the acidic oxygen proved to be less than two but greater than one times the basic oxygen. The "überbasisch" salt is discussed in *Versuch,* pp. 136—139. In the discussion of basic nitrites (p. 150), Berzelius pointed out that in the "überbasisch" salt the quantity of base, for a given amount of acid, was a non-integral multiple, equal to 4.75, of that in the neutral salt.

[3]Berzelius expressed his opinion unreservedly in his correspondence. For example, to Alexandre Marcet he remarked in June, 1814 (*Bref,* I,iii, p. 101): "[J]e le considère comme le meilleure chimiste de Paris." In December, 1816, he observed to Marcet (p. 147): "Gay-Lussac est sans conteste le meilleure chimiste de France." In a letter to Gahn of November, 1818 (*Bref,* V,ix, p. 180), he described Gay-Lussac and Thenard as "två de största kemister af vårt tidehvarf..." Cf. also Partington, *History,* IV, p. 81.

nature of nitrogen, Berzelius concluded that in basic nitrates containing the maximum amount of base, the oxygen content of the nitrogen dropped out of the calculation as a result of undetermined causes.[4] An additional analysis, of a basic nitrate of copper, was adduced as further support of this opinion.[5]

Recognizing the controversial nature of these views, Berzelius turned to the nitrites explicitly to find further support for the compound nature of nitrogen.[6] With this study, Berzelius was the first to document definitively the existence of salts of nitrous acid and to prepare a series of both neutral and basic nitrites. These achievements rested on his having correctly interpreted a phenomenon first observed by Proust: the dissolution of metallic lead in a solution of neutral lead nitrate.[7] Berzelius recognized that the metal was oxidized at the expense of the nitric acid, that the latter was thereby reduced to nitrous acid, and that the nitrous acid then combined with the lead oxide, now present in excess, to form basic nitrite salts.[8] From his interpretation of the dissolution, Berzelius developed two arguments for the compound nature of nitrogen: first, that his analysis of the basic nitrite (but not of the "überbasisches") showed the salt to conform with his stoichiometric laws only on the assumption of the compound nature of nitrogen; second, that also on that assumption he could calculate the quantity of lead that could be dissolved in a solution of lead nitrate to convert the latter into the basic salt.[9]

To a solution of lead nitrate, Berzelius added a quantity of metallic lead equal to that present in the lead salt, and he boiled the mixture until all the lead had dissolved. On cooling, the solution deposited crystals of the basic nitrite. After analyzing the salt, Berzelius showed that its composition could be made to conform neither with the rule of oxides, nor, when the water of crystallization was taken into account, the law of unitary oxygen, on the assumption that nitrogen

[4] *Versuch*, pp. 137, 139.

[5] *Ibid.*, pp. 139—140.

[6] *Ibid.*, pp. 140—141. Cf. also p. 142.

[7] Berzelius mentioned Proust as the source of this observation, but did not provide a reference. Proust's description of the phenomenon appeared in his commentary on Fourcroy's *Connaissances chimiques:* J.L. Proust, "Remarques sur le Système des connoissances chimiques de Fourcroy," *J. de physique*, 56 (1802), 200—220, esp. 206. Proust's interpretation was that the additional lead was oxidized at the expense of the lead oxide of the neutral salt, producing a previously unobserved lower oxide of lead. This claim led Chevreul to undertake a study in search of the supposed oxide: Michel Eugène Chevreul, "Faits et observations pour servir à l'histoire des combinaisons de l'oxide de plomb jaune avec les acides nitrique et nitreux," *Ann. chim.*, 83 (1812), 67—105. In a later study of the dissolution of lead in lead nitrate, explored below in section 9, Berzelius cited Chevreul's results in support of his own conclusions. For this, see below, p. 248, text accompanying n. 5.

[8] Berzelius's discussion of the basic lead nitrite resulting from this process is in *Versuch*, pp. 141—148; of the "überbasisch" nitrite, pp. 149—152.

[9] The "überbasisch" nitrite, like the corresponding nitrate, could not be made to conform to Berzelius's stoichiometric laws when ammonium was considered the radical of the acid. Only on the assumption of elementary nitrogen did its composition appear law-like. Berzelius held this fact not to support the elementary nature of nitrogen, but again the notion that, in basic salts saturated with base, the oxygen content of the nitrogen fell out of calculation (*Versuch*, p. 150).

was the radical of the acid.[1] When ammonium was taken to be the radical, however, Berzelius showed that the salt could be made to conform: the acid contained twice the oxygen of the base, and the latter contained the same amount of oxygen as the water of crystallization.[2]

Berzelius supported this evidence for the compound nature of nitrogen with a calculation of the quantity of lead necessary to convert a given amount of the neutral nitrate into the basic nitrite.[3] The table of the oxides of ammonium indicated that ratio of the oxygen of nitric acid, to that of nitrous acid, was as 5 to 4. Accordingly, one-fifth of the oxygen in the nitric acid of the nitrate must be used in oxidizing the metallic lead. The oxygen content of the acid was equal to six times that of the base, as Berzelius's studies of the neutral nitrates had shown. The oxygen available to oxidize the metallic lead, then, amounted to 6/5 the basic oxygen of the neutral nitrate. Berzelius then compared the quantity of lead oxide available—the sum of that from the neutral nitrate and that oxidized by dissolution of the metal—with the quantity of nitrous acid formed by reduction of the nitric acid. To form the basic nitrite, Berzelius's analysis had shown, the oxygen of the nitrous acid must be twice that of the base, lead oxide. The calculated amount of nitrous acid formed by reduction, however, exceeded twice the oxygen in the lead oxide. The dissolution therefore must have been further complicated, Berzelius concluded, by the decomposition of the excess nitrous acid, which would have manifested itself as an evolution of either nitrogen gas or nitric oxide (*Salpetergas*). Though Proust, Berzelius indicated, and subsequently Gehlen had denied the evolution of gas in the dissolution of lead nitrate, Berzelius repeated the experiment to see if evolution of gas could be detected. Keeping the temperature below the boiling point, he found that a gaseous product was evolved. He collected it and showed it to be nitric oxide. This having been the evolved product, the other decomposition product, oxygen, must have been used to oxidize a definite additional quantity of lead. Berzelius then compared the total quantity of lead that his calculation suggested could dissolve with the amount he had shown to dissolve in an experiment designed to determine this quantity.[4] The two figures agreed closely, leading Berzelius to what was evidently a gratifying conclusion:

[1] *Ibid.*, pp. 142—143. As noted earlier, what Berzelius saw as water of crystallization was often what now would be taken as hydroxide (corresponding to Berzelius's idea of basic water).

[2] *Ibid.*, pp. 143—144.

[3] This paragraph is a reconstruction of material developed in *ibid.*, pp. 144—147.

[4] For this experiment, see *ibid.*, pp. 146—147. Berzelius discovered that still more lead could dissolve than was indicated by calculation, resulting in the formation of an "überbasisch" salt. Consequently, the quantity of lead that dissolved to form the basic nitrate could not be precisely determined. He repeated the experiment several times, each time using slightly more lead than in the preceding experiment, and checked for the appearance of the "überbasisch" salt. By this technique, he determined, within an interval of 0.1 g., how much lead was used to form the basic nitrite without contributing to the formation of the "überbasisch" nitrite. It was this figure that agreed well with the calculated quantity.

Vielleicht gibt es unter den indirecten Beweisen für die wahre Zusammensetzung der Salpetersäure, und mithin auch des Ammoniaks und des Stickstoffs, keinen bündigeren als diesen. Denn die hier angeführten Beobachtungen, wenn sie auch nicht auf die grösste Genauigkeit Anspruch haben, können unmöglich so fehlerhaft sein, dass die Bildung des basischen salpetrigsauren Bleioxyduls sich mit einer andern Idee über die Zusammensetzung und die Sättigungs-Capacität dieser beiden Säuren [nitric and nitrous acids] vertragen sollte.[5]

Impressive as this evidence was for the compound nature of nitrogen, it later proved deficient in two respects: First, Berzelius's arguments rested on Gay-Lussac's erroneous analysis of nitric acid. Second, though Berzelius believed himself to have prepared a simple basic nitrite, corresponding to the modern formula $(PbO)_2(N_2O_3).2H_2O$, in fact, he had obtained the tetrahydroxydinitritodinitrate, $(PbO)_4(N_2O_3)(N_2O_5).4H_2O$, equivalent, stoichiometrically, to an equimolar mixture of the basic nitrite and the basic nitrate: $(PbO)_2(N_2O_5).2H_2O + (PbO)_2(N_2O_3).2H_2O$. Because the lead oxide content of this salt differs only minimally from that of the basic nitrite, Berzelius was likely to recognize this error from closer observation of the events of the experiment rather than from the analytical data. When he later returned to the salts of the nitrogen oxyacids, Berzelius corrected the analysis of nitric acid and attained an immensely improved understanding of the processes leading to the formation of the basic salts. This progress was an important step in the reconstruction of Lavoisier's correlation; it is described below, in section 9.

8. Double salts: toward hierarchical dualism

In undertaking his investigation of double salts, Berzelius was again breaking new ground, for these compounds occupied a still less conspicuous place in the prior history of chemistry than did basic salts.[6] Moreover, Berzelius altered the prevailing conception of double salts, and pressed, accordingly, for a change in their denomination. The current conception of these salts was Bergman's, who labelled them "triple salts," because they were composed of three saline components, most often one acid and two bases.[7] Berzelius's central conception of

[5] *Ibid.*, pp. 147—148.

[6] Bergman's recognition of complex salts, denominated "double," "triple," and so on, according to the number of their constituents, is noted above, p. 50, n. 7 and accompanying text. Scheele's recognition that microcosmic salt was "ein dreifaches Mittelsalz," is pointed out in Partington, *History*, III, p. 216. Fourcroy observed, during his extensive studies of salts in the 1790's, the formation of salts of two bases and one acid, like the sulfate of magnesia and ammonia, and, following Bergman's example, he called them "sels triples" or "trisules." For this, see Partington, *History*, III, pp. 75, 542. Berzelius's work on double salts is recounted in *Versuch*, pp. 196—205.

[7] The change in denomination appeared in the 1811 "Nomenclature," where Berzelius wrote in French of the "trisules" (266), to which he gave the Latin name "salia duplicia" (267). The change to the term "double salt" was suggested again in the *Versuch* (p. 196), where the older term "triple salt" was attributed not to Bergman, but to the antiphlogistic nomenclature.

these salts was that they were direct combinations of two neutral salts, an idea reflecting, Berzelius acknowledged, the usual method of preparing them by crystallization of a mixture of two solutions of each of two neutral salts.[8] Berzelius held that the designation "double salt" was more appropriate to this view.[9] By extension of the quantitative definition of neutrality, a neutral double salt was therefore one composed of two neutral salts.[1] By this departure from Bergman's precedent, Berzelius was taking a step away from the view of complex salts as compounds of multiple oxide components and toward his mature hierarchical dualism, in which double salts were third order compounds, consisting of a pair of second order constituents. However, the break was completed only in Berzelius's additions to his *Mineralogy,* as is shown in the next chapter. In the *Versuch,* the dualistic model of double salts as compounds of two simple salts coexisted with the multiple oxide model.

To develop his ideas about double salts, Berzelius explored several alternative conceptions in addition to his central idea that double salts were compounds of two simple salts. He considered double salts as formed from acid salts by addition or by displacement with another base, and as compounds of acids with binary bases. These alternative conceptions are considered here in turn. The formation of double salts by addition of base to acid salts appeared in Berzelius's introductory remarks on the subject of double salts consisting of two bases and one acid:

> Das Prinzip der Bildung dieser Doppelsalze ist nicht schwer zu finden, da wir wissen, dass die meisten sauren Salze, wenn man sie mit einer zweiten Basis sättigt, Doppelsalze darstellen, und es seitdem dargethan worden ist, dass jene sauren Salze die doppelte Menge von Säure der neutralen Salze enthalten. Die neue hinzukommende Basis muss folglich die *nämliche Menge Sauerstoff,* als die vorher im Salze gewesene Basis enthalten. So z.B. muss das Natron, das Eisenoxydul oder das Antimoniumoxydul, womit man in den pharmaceutischen Zubereitungen das säuerliche weinsteinsaure Kali [the acid tartrate of potash] neutralisirt (dessen Zusammensetzung aus der S. 178 angeführten Analyse bekannt ist), eben so viel Sauerstoff enthalten, als das zuvor im Salze vorhandene Kali in sich schloss, weil sie eine *gleiche* Menge Weinsteinsäure als dieses Kali neutralisiren.[2]

Unlike simultaneous crystallization of two neutral salts, this technique of the saturation of an acid salt with base had the advantage of permitting prediction of the combining proportions of the constituents.

The example of the acid tartrate of potash also displays the approach to the formation of double salts from acid salts not by addition, but by displacement. The analysis of this acid salt, the only such salt in the *Versuch,* appears in the

Though Berzelius did not claim credit for devising the new term, he cited no precedent. None is mentioned by Partington, either, who attributed the innovation to Berzelius, with a citation of the *Mineralogy* (*History,* III, p. 542).

[8] *Versuch,* p. 196.

[9] *Ibid.*

[1] As noted shortly below, in the discussion of Berzelius's study of alum.

[2] *Versuch,* pp. 196—197.

section on water of crystallization.[3] Berzelius's immediately preceding analysis of the neutral tartrate revealed that it contained no water of crystallization, but the analysis of the acid salt indicated the presence of water. Berzelius burned a sample of the acid salt and determined the quantity of potash remaining by preparing the muriate of potash from treatment of the ashes with muriatic acid. Then, following Wollaston's example, Berzelius found that the ashes of 10 g. of the acid salt were sufficient to saturate the same quantity of the unburned salt, a result permitting the conclusion, identical with that in the passage last quoted, that the acid salt contained twice the acid of the neutral salt. Calculation on this basis of the composition of the acid salt revealed that there was insufficient potash and tartaric acid to account for the full mass of the salt. The missing portion was assumed to be water. The oxygen contained in that portion of water was just equal, Berzelius showed, to that of the potash. Unlike water of crystallization, this water could not be driven off by heating, but disappeared from the analysis only by displacement from the acid salt with another base. Berzelius concluded, therefore, that the acid salt was a form of double salt in which the second base was water. Berzelius did not address directly in the section on double salts the full import of this conclusion. However, his citation in the passage last quoted of these experiments on the tartrates seems to have implied that all acid salts were species of double salts containing water in the role of base and that preparation of double salts by saturation was actually an example of displacement.[4]

Finally, the view of double salts as compounds of acids with binary bases was developed immediately after the account of the method of calculation of proportions of double salts by saturation of acid salts. The last quoted passage went on to argue of salts with one acid and two bases:

> Dass in diesen Fällen nicht so sehr die Säure das Dasein der Doppelsalze bestimmt, sondern dass dabei die Verwandtschaften der Basen zu einander die wirksamere Rolle spielen, werden wir aus der Analyse des Alauns ersehen; dann dieses Salz lässt sich dadurch, dass man in dem sauren schwefelsauren Kali die freie Säure mit Thonerde sättigt, nicht ohne einen grossen Ueberschuss an schwefelsaurem Kali erhalten. Gewisse Basen geben ferner beinahe mit allen Säuren Doppelsalze, ungeachtet die meisten Säuren mit keiner von diesen Basen ein saures Salze darstellen können. So z.B. giebt das Ammoniak mit Magnesia, mit Mangan-Oxydul, mit Kupferoxyd, mit Zinkoxydul, u.s.w. immer Doppelsalze, und wir haben grosse Ursache zu glauben, dass diese Basen unter sich immer das nämliche Verhalten beobachten, d.h., in den Doppelsalzen immer *gleiche* Sauerstoffmengen enthalten.[5]

Though this passage is not distinguished by its lucidity, a reasonable interpre-

[3] The analysis of the acid tartrate appears in *ibid.*, pp. 178—179. It is immediately preceded by that of the neutral tartrate, p. 178.

[4] The conclusion was reinforced by the observation (*ibid.*, p. 179) that the water of the acid salt was present in just the amount that the excess acid of the acid salt would have bound in its crystallized state.

[5] *Ibid.*, p. 197.

tation of it does suggest itself and can be supported with another text. Berzelius here argued that the "cement" that bound two neutral salts in a double salt was the mutual affinity of the two bases. This suggestion opposed the natural supposition that the mutual affinity between the acid and the bases was the binding agent. That supposition, more appropriate to the earlier view of these substances as triple salts, was undermined by two factors. First, Berzelius showed, in accordance with his view of double salts as binary compounds of simple salts, that the amount of acid saturated by the two bases in the double salt was the same as the amount saturated by the same quantities of the two bases in their simple salts. Accordingly, the mutual affinities of the constituent acid and bases of a double salt were already satisfied before the simple salts combined; there was no "extra" affinity left between the acids and bases to hold the two neutral salts together. Berzelius had proven this with the example of the tartrates of potash and with his subsequent analysis of alum. The former case revealed that, in double salts obtained by saturation of an acid salt, the two bases contained the same amount of oxygen, and that the quantities of acid saturated by these amounts of base in their simple salts equalled the quantities of acid present in the double salt. Although alum differed from the other double salts that Berzelius studied in that one of its bases, alumina, contained three times the oxygen of the other, potash, rather than the same amount of oxygen, this difference derived precisely from the composition of the sulfate of alumina: the quantity of alumina present in a given weight of alum saturated three times the sulfuric acid in its simple sulfate as the potash saturated in its simple sulfate. Therefore, double salts were generally formed so that the saturation of acid was the same as in the two constituent simple salts. This was only to be expected, given that double salts were binary compounds of simple salts. However, it implied that the saturation of the mutual affinities of the acid and the bases in the double salt was no greater than in the individual component salts. Something else, therefore, must hold the two constituent salts together.

The second factor undermining the assumption that double salts were held together by the affinity between the acids and the bases was the implications of the procedure for the formation of double salt by bases that did not form acid salts with the acids in question, as Berzelius was suggesting in the last quoted passage. Perhaps the existence of acid salts of a particular base in which more acid was saturated than in the simple salts suggested to Berzelius that such a salt must have that "extra" affinity that could bind together a double salt. If no acid salt formed, then this would not have been the case. The affinity of these bases for the acid could not be the binding agent of double salts. What then was? Berzelius believed it to be the mutual affinity of the two bases.

This belief was bolstered by Berzelius's recognition of a case of double salts formed by the combination of an acid with a binary compound of two bases, a compound capable of existing independently. The example was provided by the black oxide of iron and its salts, which Berzelius discussed after completion of the *Versuch* in a letter to Gilbert of 26 March 1812.[6] Berzelius showed that according to Gay-Lussac's recent analysis, the oxygen of the black oxide was to that of the lowest oxide of iron, for a given weight of metal, as 1 1/4 to 1, a ratio not in conformity with the usual proportions.[7] He argued that this apparent anomaly could be eliminated by regarding the black oxide as a binary compound of the two usual oxides of iron and the salts of the black oxide as double salts of this binary base.[8] In this letter, he recalled his speculation in the *Versuch* that the black oxide might be a binary compound and his suggestion elsewhere of this composition for a Swedish iron ore.[9] He offered persuasive experimental evidence to support the idea that the salts of the black oxide were double, not simple.[1] He pointed out, for example, that the decomposition of its salts always proceeded by separation of the two usual oxides of iron and not by straightforward displacement of the black oxide. Also, he noted the formation of Prussian blue by the precipitation of salts of the higher oxide of iron with the double salt (potassium ferrocyanide) composed of prussic acid, potash, and the lower oxide of iron.[2] He asserted that the higher oxide would much more likely have taken the place of the potash, forming the second base of the double salt, than to have produced with the lower oxide a new oxidation stage of iron, Gay-Lussac's black oxide. He maintained that to assume the latter process would have been to destroy the analogy between this precipitation reaction and those reactions involving the precipitation of recognized double salts of dissimilar bases. That the mutual affinity of the two bases was sufficiently strong to allow them to exist as a binary compound apart from the acid appears to have been the only distinction, in Berzelius's mind, between a double salt of a binary base and any other double salt. The salts of the black oxide revealed more clearly than other double salts that the mutual affinity of the bases, and not the mutual affinity of the acids and the bases, was the effective binding agent.[3]

[6] One of the "Zwei Schreiben," cited above, p. 212, n. 1. See also Berzelius's letter of 27 April 1812 to Gaultier de Claubry (*Bref*, IV,vii, pp. 103—106).

[7] "Zwei Schreiben," 278.

[8] *Ibid.*, 278—280.

[9] *Ibid.*, 278. The speculation is developed in *Versuch*, pp. 43—47. The binary composition for the iron ore is in "Auszug," 227—232.

[1] "Zwei Schreiben," 279.

[2] In modern terms, the reaction is:
$4Fe^{+++} + 3\,[Fe^{II}(CN)_6]^{4-} = Fe^{III}_4[Fe^{II}(CN)_6]_3$.

[3] Following "Zwei Schreiben," the character of compounds formed from two oxides of the same radical became a recurrent theme in Berzelius's work. That these compounds offered a means to account for proportions of oxygen in certain nonintegral multiples explains his interest in them. In the present instance, Berzelius showed that the apparently anomalous black oxide of iron, which contained one and one-quarter times the oxygen of the lowest oxide, was a compound of two oxides that did conform to his statement of the law of multiple proportions (cf. above, p. 169, n. 2). However, the conforming compounds were just those that exhibited the sesquialteral multiple in the proportions of their oxygen. Nevertheless, Berzelius used this technique repeatedly to account for anomalous fractional multiples and claimed that it was precisely the oxides that exhibited the sesquialteral multiple that combined to form apparently anomalous oxides. For example, in "Zwei Schreiben" (280—281), he used this idea

These views of the formation of double salts reinforced Berzelius's essentially dualistic model for them as the compounds of two simple salts. Acid salts, illustrated by the acid tartrate of potash, could be assimilated to the model by viewing them as double salts in which one base was water. Saturation of the acid salt with base, in this view, proceeded by displacement of the basic water, producing a neutral double salt. The two constituent simple salts of a double salt could be viewed as held in combination by the mutual affinity of their bases. The neutrality of a neutral double salt was determined by the neutrality of its constituent simple salts. However, the model had one deficiency, in that it failed

to account for the anomaly of nitrous acid in the progression of the oxides of nitrogen as they had been determined by Gay-Lussac in his follow-up to the combining volumes paper (see above, p. 210, n. 5). (These analyses, as Partington points out [*History,* IV, p. 80] are erroneous.) Other examples were provided by the work of Berzelius's students and of Hisinger. Identifying some of them as participants in the research, he gave them all full credit in the "Rechtfertigung" (143): Rothoff, for the study of cobalt and nickel (reported in "Cause of proportions," 355—356); Lagerhjelm for that of bismuth compounds (in "Rechtfertigung," 70—71 and in his "Försök rörande de bestämda förhållanden, uti hvilka wismut-metallen ingår föreningar med svafel och syre," *K.S.V.A. Handlingar,* [2] 34 (1813), 219—224); Sefström, for the study of mercury compounds (cited above, p. 212, n. 6); Arfwedson, for that of the manganese compounds (in "Rechtfertigung," 76—78); and Hisinger, for the study of cerium compounds (in Hisinger's "Försök att bestämma syrehalten i cer-oxidul och cer-oxid," *K.S.V.A. Handlingar,* [2] 34 [1813], 216—218). In the "Rechtfertigung" (143) Hisinger was erroneously credited with a study of uranium compounds, but the error was corrected on the errata page, *Schweigger's Journal,* 6 (1812), 143. The citation is correct in the Swedish version, in *K.S.V.A. Handlingar,* [2] 34 (1813), 186. These examples were discussed in Berzelius's notes to the French edition of his *Mineralogy* (1819), cited in full below, p. 294, n. 9). There, in his note 22 (pp. 265—268), Berzelius showed that a naturally occurring iron ore was a compound of two oxides of iron, and that by so considering it, an anomalous progression of oxidation could be avoided. He also added (pp. 267—268) the remark that "En étudiant plusieurs combinaisons de radicaux combustibles avec l'oxigène, qui ont paru s'écarter des proportions générales, je crois avoir reconnu qu'elles ont été composées de deux oxides du même radical, et que c'est sur-tout les oxides dont l'oxigène est dans le rapport de 1 : 1 1/2 qui peuvent se combiner de cette manière. Je considère comme de telles combinaisons, l'acide nitreux de M. Gay-Lussac, l'oxide vert de cobalt, l'oxide rougeâtre de manganèse, un oxide verdâtre d'urane, que l'on obtient en chauffant l'oxide d'uran pur, à une chaleur rouge, dans des vaisseaux ouverts, etc." In his note 27 (pp. 275—278), he remarked that Arfwedson had obtained in his analyses a compound of the two oxides of manganese, which contained oxygen in sesquialteral proportion. It was formed by displacement of water by one oxide from the hydrate of the other.

A frequent but not universal aspect of this pattern, Berzelius believed, was that the two oxides combined in such a proportion that each contained the same amount of oxygen. In "Zwei Schreiben," he noted that the black oxide of iron, Gay-Lussac's nitrous acid, and binary oxides of cobalt and nickel were compounds of two oxides united so that each contained the same quantity of oxygen. (For Berzelius on Gay-Lussac's nitrous acid, see also below, p. 248, n. 6). He even asserted (281) that this proportion was always observed in compounds formed from oxides of comparable electrochemical character, that is, of members of the same class in the classification of oxides.

The idea that certain oxides could be considered binary compounds of two oxides of the same radical was not novel. Several earlier examples may be found in Partington's *History.* Dalton (*History,* III, p. 784) considered nitrous acid as a binary compound of two lower oxides. Fourcroy (p. 543) considered fuming sulfuric acid a compound of sulfuric and sulfurous acids. Proust (p. 646) considered some oxides of metals to be compounds of two other oxides.

to discriminate stoichiometrically between double salts and mere mixtures of simple salts. There was no principle, corresponding to the rule of oxides, that dictated the proportions subsisting between the two constituent simple salts, which Berzelius later described as second order constituents. As shown in the next chapter, Berzelius's dualistic model amounted to a call for the construction of such rules for higher orders. In their absence, Berzelius had to devise another technique, one based on the multiple oxide model. The latter therefore co-existed in Berzelius's chemistry with the emergent dualistic approach.

The technique was to examine the proportions subsisting among the quantities of oxygen carried by each of the first order constituents (oxides or sulfides) in a double salt and then to devise a rule that governed these proportions. Berzelius examined three hydrated double salts in the *Versuch,* the sulfate of ammonia and magnesia, the sulfate of ammonia and copper, and alum. Alum was the example that interested Berzelius most, and it became his classic example of a complex compound.[4] As both a laboratory chemical and a naturally occurring mineral, alum later provided an important link between Berzelius's inorganic chemistry and his mineralogy. In his investigation of alum, Berzelius began by portraying it as agreeing with his binary model. He then returned to the multiple oxide model, showing that alum, like the other double salts he had investigated, conformed to the law of unitary oxygen, which he had first suggested provisionally for simple salts with water of crystallization.[5]

Berzelius's analysis of alum revealed that, like the two double sulfates mentioned in the preceding paragraph, alum was composed of four oxides (sulfuric acid, two bases, and water of crystallization). With reference to his view of double salts as binary compounds of simple salts, he partitioned the acid between the two bases and showed that the results agreed with the composition of two simple sulfates. Similarly, he applied his ideas about the neutrality of simple salts to alum, arguing that since the constituent simple salts were quantitatively neutral, the double salt they formed must also be neutral. Accordingly, Berzelius argued, the older view that the acidic reactivity of alum showed it to be an acid double salt had to be abandoned. The acidity resulted from that of the constituent sulfate of alumina, which was qualitatively astringent but quantitatively neutral.

The binary or dualistic model of double salts thus accounted for the principal characteristics of alum, but it offered no limits on the proportions subsisting between the constituent simple sulfates. Consequently, Berzelius next portrayed

[4] Alum was considered by contemporaries to be among the most complex inorganic compounds then known. For example, Thomas Thomson, in his long summary account of the atomic theory, "On the Daltonian theory of definite proportions in chemical combination," *Ann. Phil.,* 2 (1813), 32—53, 109—115, 167—171, 293—301, remarked (295n1) that alum was "the most complicated salt known." Berzelius's discussion of alum occupies pp. 198—201 in the *Versuch.*

[5] For this, see above, p. 223, text accompanied by nn. 6 and 7.

alum as a complex combination of multiple oxide components and claimed that such compounds were subject to the law of unitary oxygen. He showed that the quantities of oxygen in the potash, alumina, sulfuric acid, and water of alum to be in the proportions, respectively, of 1 : 3 : 12 : 24. These proportions differed from those shown by the other double salts Berzelius examined in the *Versuch,* in that one base contained three times the oxygen of the other, rather than an equal quantity.[6] However, in all the salts the oxygen contents of three of the components were integral multiples of that of the fourth. Accordingly, Berzelius invoked the law of unitary oxygen, which asserted that the oxygen contents of the various constituents were integral multiples of the oxygen content of the component possessing the least oxygen. Alum exemplified the rule:

> Der Alaun und die übrigen Doppelsalze geben uns interessante Beispiele von Verbindungen mehrerer oxydirten Körper. Das Kali enthält hier die geringste Menge Sauerstoff, und diese muss also für die Sauerstoffmengen der übrigen Bestandtheile der gemeinschaftliche Divisor sein. Setzen wir den Sauerstoffgehalt des Kali = 1, so ist der der Thonerde = 3, der der Schwefelsäure = 12 und der des Wassers = 24.[7]

In the concluding section of the *Versuch,* Berzelius listed simple salts with water of crystallization and double salts together under the category of compounds having three or more oxides, and he claimed that all members of this group were governed by the law of unitary oxygen. He presented the law there unaccompanied by his earlier scruples about its lack of significant experimental support, even though he had added but three double salts to the short list of conforming substances:

> Wenn sich drei oder mehrere oxydirte Körper miteinander vereinigen, so ist der Sauerstoff desjenigen Körpers, der davon die Sauerstoffgehalte der übrigen, welche also von dieser geringsten Menge Multipla nach ganzen Zahlen sind.
>
> Hierher gehören:
>
> a) Krystallwasser enthaltende Salze. In den Neutralsalzen enthält das Krystallwasser 1, 2, 3, 4, 5 etc., seltener nur 1/2, 1/3, 1/4 etc. Mal, so viel Sauerstoff als die Basis. In basischen Salzen, in welchen der Sauerstoff der Säure ein Submultiplum von dem der Basis ist, enthält das Krystallwasser den Sauerstoff in einem solchen Verhältnisse, dass er ein Multiplum sowohl von dem Sauerstoff der Basis als von dem der Säure ausmacht.
>
> b) Doppelsalze. Beispiele von solchen, worin der Sauerstoff der einen Basis dem der anderen gleich, oder davon ein Multiplum nach einer ganzen Zahl ist, geben, ersteres die Doppelsalze aus Ammoniak und Magnesia, letzteres der Alaun. Wenn ein solches Salz zugleich Krystallwasser enthält, so ist es eine Verbindung von vier oxydirten Körpern,

[6] Berzelius showed (*Versuch,* pp. 197—198) that sulfate of ammonia and magnesia and sulfate of ammonia and copper each consisted of four oxides (two bases, an acid, and water of crystallization) and that in both salts the oxygen contents of the constituents were, respectively, as 1 : 1 : 6 : 7.

[7] *Versuch,* p. 200. The text of both the *Ostwald's Klassiker* edition and of *Gilbert's Annalen* lists the oxygen of the alumina equal to eight rather than three times that of the potash. This was evidently a typographical or transcription error, the numeral 3, appearing in *Afhandlingar,* 5 (1818), 170, having been changed into the numeral 8.

von dem Wasser, der Säure und den zwei Basen. Der Sauerstoff desjenigen Bestandtheiles, welcher am wenigsten davon enthält, ist ein gemeinschftlicher Divisor nach einer ganzen Zahl der Sauerstoffgehalte der übrigen, wie wir im Alaun gesehen haben, wo der Sauerstoff des Kalis in der Thonerde dreimal, in der Schwefelsäure zwölfmal und in dem Wasser 24 Mal enthalten ist.[8]

Berzelius also believed the law of unitary oxygen was applicable to naturally occuring minerals. Though he offered this idea tentatively, acknowledging its lack of experimental foundation, he later granted the law the status of a universal imperative. Particularly after his adoption of an atomic theory, he demanded that in all compounds, however complex, the oxygen of one constituent had to serve as the unit for the oxygen contents of all the others. As shown in the next chapter, his insistence issued from his belief that this constraint would help him draw a sharp boundary between the organic and the inorganic. Moreover, he went on to developed other, quite specific rules governing the composition of complex inorganic compounds, including minerals. The special problems presented by minerals provoked the mature articulation of Berzelius's dualism, which called for rules governing higher order, rather than first order constituents.

9. Berzelius's reconstruction of Lavoisier's correlation of degree of oxidation and properties

a. Experimental foundations

The reconstruction of Lavoisier's correlation of degree of oxidation with properties was based on the connection that Berzelius expected to exist between the composition of acidic oxides and that of their salts. The reconstructed correlation thus involved both first and second orders. The rule of oxides was the connecting link between the two, and Berzelius's first examples documenting the rule were at once the source of this expectation and the paradigmatic case of the anticipated connection. The oxygen contents in the two oxyacids of sulfur (sulfuric and sulfurous acids) were as three to two, and the acids formed neutral salts so composed that the multiplier by which the basic oxygen of sulfates was contained in the sulfuric acid was to the multiplier by which the basic oxygen of sulfites was contained in the sulfurous acid also as three to two. Much of the research after the *Versuch* was inspired by the hope that the higher oxides of other radicals might have been acidic, the higher more strongly than the lower; that the proportion of oxygen in the higher might have been one and one-half times that in the lower; and that the salts of the higher oxides might have exhibited a corresponding coincidence of multiples in the proportions of oxygen with multipliers in the rule of oxides. These were the chief characteristics of the sulfur

[8] *Versuch*, pp. 207—208.

pattern, which Berzelius expected to find recapitulated in the chemistry of most other radicals.

Several circumstances contributed to the adoption of sulfur as an archetype. Berzelius's concern with lower oxides as a means to eliminate sesquialteral multiples, especially the development this concern acquired in the table of the oxides of ammonium, had attuned him to the occurence of proportions in multiples by three, six, or twelve. The case of the sulfur oxides, developed in the "Zugabe" and in the "Zusatz," exhibited the multiplier six in the proportion of oxygen of the highest oxide to that of the lowest. In the first *Fortsetzung,* he had achieved some success in bringing the arsenic oxides into conformity with the sulfur pattern. Last, in his discussion of suboxides in the "Rechtfertigung," Berzelius remarked that the multiple by six already seemed to him the fundamental one:

> Ich habe in meiner vorigen Abhandlung [the *Versuch*] angeführt, dass in der Oxydationsreihe alle Multiplicationen mit 11/2 vielleicht nur scheinbar sind und wirkliche Multiplicationen mit einer ganzen Zahl von einer unbekannten oder nicht untersuchten niedrigeren Oxydationsstufe ausmachen. Diese Zahlen können nun 3, 6, 12 u.s.w. seyn. Von der Analogie mit einigen, wiewohl sehr wenigen, Versuchen geleitet, habe ich vermuthet, dass in den mehrsten Fällen 6 der wirkliche Multiplicator sey. In dem Suboxyde dürfte das Blei [the radical under discussion in this part of the "Rechtfertigung"] wohl also 1/6 so viel Sauerstoff, als in der Mennige aufnehmen. Da es aber noch nicht erwiesen ist, dass nicht auch 5 der wahre Multiplicator seyn kann, so ist es noch zweifelhaft, ob in dem Suboxyde 100 Th. Blei 1,925 oder 3,85 Th. Sauerstoff aufnehmen.[9]

Thus, early on Berzelius had readied himself to see radicals as having two higher oxides, both acidic, the higher more so than the lower, with the oxygen of the higher oxide being a sesquialteral multiple of that of the lower. He had prepared himself to discover that the sesquialteral multiple was an appearance resulting from the existence of lower oxides and that the oxygen of the highest oxide was six times that of the lowest. Moreover, he expected to observe that the two acidic oxides contained the basic oxygen of their neutral salts by the multipliers two and three. This was the pattern that Berzelius hoped would replace Lavoisier's correlation.

In attempting this reconstruction, Berzelius set himself a formidable task. As of the first *Fortsetzung,* he could observe few analogies in the compounds of the acidic radicals he had exploited in documenting the rule of oxides. This apparent disorder gave rise to the statement already quoted (p. 198, accompanied by n. 9), in which Berzelius portrayed the old correlation of oxidation and properties as a shambles. Nevertheless, Berzelius was undeterred from attempting to reduce this diversity. He was able to bring many radicals into full or partial conformity with the sulfur pattern, and he discovered a second pattern, exhibited by nitrogen, arsenic, and phosphorus, which he juxtaposed with the first. The

9 "Rechtfertigung," 73.

reconstructed correlation was less tidy than Lavoisier's, but it was securely anchored to the stoichiometric and electrochemical characters of specific oxides and it linked the composition of oxides with the composition of salts.

b. The nitrogen oxyacids

Among the first steps toward a reconstructed correlation was a set of "Neue Untersuchungen" of the nitrogen oxyacids and their salts and of other compounds of the supposed ammonium radical, occasioned by the appearance in 1812 of new analyses of nitric acid.[1] Berzelius's original studies of these substances, in the second *Fortsetzung,* had been predicated on Gay-Lussac's erroneous analysis of nitric acid. In the "Neue Untersuchungen," Berzelius indicated that he found a differing analysis of the acid in Dalton's *New System,* which he had received in 1812. While on a visit to London in the same year he had obtained a new analysis from Davy that agreed with Dalton's and was published shortly thereafter in Davy's *Elements.*[2] Moreover, Berzelius stated that the results of his study of the thermal decomposition of ammonium nitrate, which produced only water and nitrous oxide, were consistent only with the analyses of the Englishmen.[3] The immediate consequence of these new results, Berzelius remarked, was that nitric acid, with nitrogen taken to be the radical, contained five times the oxygen of the base in neutral nitrates; or, if the oxide nature of ammonia was to be sustained, that the nitrogen of the acid contained one-sixth the total oxygen of the acid, rather than one-fifth as he had formerly believed. In effect, this change brought the nitrogen compounds into conformity with the sulfur pattern. However, Berzelius indicated that the numerical difference introduced into his earlier calculations was too small to enable him to

[1] Berzelius, "Neue Untersuchungen über die Natur des Stickstoffs, des Wasserstoffs, und des Ammoniaks; und einige Bermerkungen über die wahre Zusammensetzung der salpetrigen Säure und der Salpetersäure: Erster Nachtrag zu seinem Versuch, die bestimmten und einfachen Verhältnisse aufzufinden, nach welchen die Bestandtheile der unorganischen Natur mit einander verbunden sind," *Gilbert's Annalen,* 46 (1814), 131—175; Swedish, "Undersökning af quäfvets, vätets och ammoniakens natur; jemte några anmärkningar om salpetersyrans och salpetersyrlighetens rätta sammansättning," *Afhandlingar,* 5 (1818), 198—238; henceforth cited from the German text as "Neue Untersuchungen."

[2] Berzelius, "Neue Untersuchungen," 133, cited Dalton's *New system of chemical philosophy,* Pt. II (1810), p. 329, where Dalton reported that the smallest volume of nitrous gas (nitric oxide, NO) condensed by one volume of oxygen was 1.3. (Twice that volume of nitrous gas, when condensed by oxygen, would have given nitric acid, Dalton believed. For this, see Partington, *History,* III, pp. 790—792, 806.) Davy's new analysis (*Works,* IV, pp. 194—195), cited by Berzelius ("Neue Untersuchungen," 134—135), differed from his earlier one that had been sustained by Gay-Lussac's eudiometric studies.

[3] "Neue Untersuchungen," 134. The reaction is $NH_4NO_3 = N_2O + H_2O$. In the second *Fortsetzung (Versuch,* pp. 131—132) Berzelius indicated that nitrogen gas was among the products. The extent to which the reaction yields N_2 rather than N_2O depends on the reaction conditions. In the "Neue Untersuchungen," however, Berzelius decided that only nitrous oxide and water were formed.

judge the veracity of Dalton's results by their consequences for the stoichiometry of the salts of the nitrogen oxyacids.[4] It was symptomatic of Berzelius's low opinion of Dalton that he was moved to renew his study of the nitrogen compounds only with the appearance of Davy's corroboration of Dalton's results.[5] Berzelius pointed out that of his earlier analyses only those of the "überbasischen" nitrate and nitrite could not be made consistent with the new analysis of nitric acid.[6] Accordingly, he reinvestigated these first.[7]

In the *Versuch*, the "überbasischen" nitrate and nitrite conformed to the rule of oxides only if nitrogen, and not ammonium, was considered the radical of the acids. Now, Berzelius concluded that the "überbasisches" nitrate of the second *Fortsetzung* was a mixture of several basic salts, which he was now able to distinguish. He also showed that nitric acid formed a series of at least three basic salts, containing, respectively, two, three, and six times the quantity of base in the neutral salt. Granted that in the neutral salt the oxygen of the acid, including that of the nitrogen, equalled six times that of the base, the failure to find a basic salt with four times the base of the neutral salt could now be taken, Berzelius noted, as a consequence of the rule of oxides: in the missing salt, the acidic oxygen would not have been an integral multiple of that of the base.[8] This confirmed what was to be inferred from the composition of the highest basic salt: that only if nitrogen was an oxide would the rule of oxides have held for this salt (that is, would the ratio of acidic to basic oxygen have been 1 : 1 rather than 5 : 6).[9]

Berzelius regarded these experiments as proof of the compound nature of nitrogen.[1] The conformity of the nitrate salts with the version of the rule of oxides for basic salts demanded this conclusion. Berzelius indicated that calculation on this basis of the composition of nitrogen gave approximately 44.3% radical and 55.7% oxygen. However, Berzelius no longer believed the radical

4 "Neue Untersuchungen," 134.

5 *Ibid.*, 135: "Das Zeugniss Davy's für die Richtigkeit der Dalton'schen Angabe der Zusammensetzung der Salpetersäure, war von so grosser Wichtigkeit, dass eine Revision meiner ehemaligen Versuche nöthig wurde."

6 *Ibid.*

7 The discussion of the basic nitrates occupies "Neue Untersuchungen," 135—145.

8 *Ibid.*, 142. It would have been a multiple by 1 1/2, but Berzelius did not say so.

9 *Ibid.*, 138. Again, Berzelius himself did not state these numbers, but only indicated whether the rule held.

Berzelius also provided ("Neue Untersuchungen," 145—148) an experiment sustaining Dalton's figure for the composition of nitric acid and undermining Gay-Lussac's. He used a solution of nitrate of barytes to oxidize metallic copper. Believing the reaction to proceed by reduction of the nitric acid with evolution of nitric oxide, Berzelius showed that the nitric acid of the salt could oxidize more copper than could the oxygen in nitric acid according to Gay-Lussac's analysis. Dalton's figure, however, offered sufficient oxygen to account for the dissolution of all the metallic copper.

1 This paragraph describes arguments developed in "Neue Untersuchungen," 148—151. The section opens with this comment, of which all but the first clause was italicized: "Ich wage nun zu sagen, dass die Zusammensetzung des Stickstoffs nicht mehr eine blosse Hypothese ist, sondern dass sie, wenn man nur die Richtigkeit der Lehre von den chemischen Proportionen zugiebt, als eine völlig erwiesene chemische Wahrheit angesehen werden kann."

to be ammonium, the metal manifest in amalgamation with ammonia, but a still unknown substance, which he named "nitricum." Its lowest oxide was nitrogen, not hydrogen as he had believed in the *Versuch:*

> Der Stickstoff selbst muss in dem Geiste meiner Nomenclatur *Suboxydum nitricum* gennant werden, da er die erste, mit anderen oxydirten Körpern nicht vereinbare Oxydationsstufe des Nitricum ausmacht, und also alle Charaktere der Suboxyde an sich trägt.[2]

Like many suboxides that Berzelius believed he had discovered, nitrogen was electrochemically indifferent, that is, unreactive; and like them it was the lowest oxide, for, after the "Rechtfertigung," Berzelius no longer believed in the existence of oxides containing less than one-sixth the oxygen of the highest oxide.[3]

Berzelius developed his evidence that nitricum and not ammonium was the radical of nitrogen by examining once again the composition of ammonia.[4] Recalling the evolution in his beliefs about ammonia, Berzelius indicated that his new studies of nitrogen and its compounds had persuaded him only of the oxide nature of nitrogen. In the first *Fortsetzung* his commitment to the oxygen content of hydrogen was already wavering, and now he gave it up, particularly since its supposed oxygen content was smaller than the limits of experimental error. He went on to develop a series of calculations of combining volumes, made in accordance with the rule of oxides, to demonstrate that the oxygen content of ammonia lay entirely within the nitrogen (within the limits of experimental error).[5] This was an easy step for Berzelius to take. The oxide nature of nitrogen was clearly important to Berzelius, because it guaranteed the conformity of the composition of ammonia with the composition of other bases, and it brought the pattern of oxidation, reactivity, and salt formation of the nitrogen compounds into conformity with the sulfur pattern; but the oxide nature of hydrogen served no clear theoretical purpose.

Turning to nitrous acid and its salts, Berzelius first reaffirmed the value for its composition that he had originally employed in the *Versuch.* This value agreed neither with the one published by Gay-Lussac in his paper on the oxides of nitrogen nor with the value given by Davy in his *Elements.* Berzelius adduced some analyses that he had recounted in the *Versuch* of neutral and basic lead nitrites and that were consistent, he argued, only with his value for the composition of nitrous acid.[6] He also reevaluated another experiment reported in the *Versuch* concerning the thermal decomposition of lead nitrite. A residue of lead oxide containing 70% of the original mass remained after heating. The 30% that was lost had to be partitioned between nitrous acid and water of crystallization. "Ohne allen Zweifel ist in diesen Salze wenigstens doppelt so viele Säure

2 "Neue Untersuchungen," 151.
3 Cf. above, p. 240, n. 9 and accompanying text.
4 "Neue Untersuchungen," 165—169.
5 *Ibid.*, 167.
6 The experiments are in *Versuch*, pp. 141—149.

gegen die Basis als in dem basischen salpetrigsauren Bleioxydul enthalten," Berzelius remarked in the *Versuch,* the certainty apparently deriving from the failure of any other multiple to fit the data.[7] Recalling his experiment,[8] Berzelius indicated its consequence that in neutral nitrites the oxygen of the acid must have been an *even* multiple of that of the base. That this multiplier must have been two also appears to have resulted from the failure of any other to fit the analysis of the neutral nitrite by thermal decomposition. To this evidence, he added evidence derived from the phenomena of the thermal decomposition of nitrate of ammonia, which he had also studied in the *Versuch.*[9] The salt decomposed by evolution of pure nitrogen, although the decomposing mass remained neutral. Berzelius now claimed, correctly, that this effect was consistent only with his value for the composition of nitrous acid.[1]

By adopting this new value for the composition of nitric acid and by retaining his original value for nitrous acid, Berzelius could bring the nitrogen oxyacids into conformity with the sulfur pattern. In this connection, he remarked that his new results for nitric acid

> beweisen entscheidend die Richtigkeit der von mir angegebenen Zusammensetzung der salpetrigen Säure. Und daraus folgt, erstens, dass die Oxydationsreihe des Nitricum 1, 2, 3, 4 und 6 seyn, und zweitens, dass die Salpetersäure 1 1/2 mal so viel Sauerstoff als die salpetrige Säure enthalten muss, ganz so wie dieses mit den Säuren des Schwefels, des Arseniks, des Molybdäns u.s.f. der Fall ist.[2]

To sustain this result, Berzelius coupled his new investigation of nitric acid with additional evidence for his earlier conclusions about the composition of nitrous acid. This evidence derived primarily from a second study of the phenomena presented by the dissolution of metallic lead in a solution of lead nitrate. In his original study of this phenomenon in the *Versuch,* Berzelius had documented the existence of nitrite salts, including the anomalous "überbasisches" nitrite, which, like the "überbasisches" nitrate, conformed to Berzelius's stoichiometric laws only on the assumption that nitrogen was the radical of the acid. The "überbasisches" nitrite had made its appearance in the experiment when Berzelius attempted to confirm experimentally his calculation of the quantity of metallic lead that must be dissolved to convert all of the neutral nitrate into the basic nitrite. Berzelius now recognized that this prediction of the composition of the basic nitrite was undermined by his reliance on Gay-Lussac's erroneous analysis of the composition of nitric acid.[3] His new study of these phenomena allowed him to reconstruct the argument, to eliminate the anomaly, to establish his values for the composition of nitrous acid while undermining those of Davy and Gay-Lussac,

[7] The experiment is in *ibid.*, pp. 148—149. The passage quoted in the text is on p. 149.
[8] "Neue Untersuchungen," 153.
[9] *Versuch,* pp. 152—153.

[1] "Neue Untersuchungen," 153.
[2] *Ibid.*, 153—154.
[3] For this flaw as a motive for the new study, see *ibid.*, p. 154.

and to bring the oxides of nitrogen, or rather of "nitricum," into conformity with the sulfur pattern.[4]

This study marks perhaps the earliest appearance of Berzelius's symbolic notation and suggests its utility to him in clarifying the process by which basic salts were produced from neutral salts. Davy had argued in his *Elements* that nitrous acid was composed of one volume of nitrogen and four of oxygen, and nitric acid was composed of one volume of nitrogen and five of oxygen. Using his symbolic notation, Berzelius formulated the compounds differently:

> Es bedeutet A [from azote] Stickstoff, N Nitricum, O Sauerstoff, und [P] Blei; so ist nach Davy's Bestimmungen A + 5O die Salpetersäure, und A + 4O die salpetrige Säure. Nach meinen Bestimmungen dagegen sind N + 6O = Salpetersäure, und N + 4O = salpetrige Säure. Um die salzartigen Verbindungen kurz zu bezeichnen, setze ich in ihren Zeichen statt A + 5O u.s.f. A^5O, und bezeichne also z.B. das salpetersaure Blei oxyd mit A^5O + PO.[5]

If Davy's figures had been correct, Berzelius indicated, the first stage in the dissolution of lead in lead nitrate would have been simply an oxidation of the lead by reduction of the nitric acid of the salt to nitrous acid. The dissolution of an amount of lead equal to that in the neutral nitrate would have doubled the amount of base and reduced all the nitric acid to nitrous acid. The sole product, obtained immediately, would then have been the basic nitrite ([A^5O + PO] + P = [A^4O + 2PO]). Dissolution of still more lead ultimately would have produced the "überbasisches" salt, the final product of the reaction, which Berzelius recognized would have contained still more base than the salt he considered the "überbasisches" in the *Versuch*.[6]

[4] This study occasioned Berzelius's abandonment of the oxide nature of hydrogen and his corresponding shift to "nitricum" in place of "ammonium" as the radical of nitrogen. In this connection, see "Neue Untersuchungen," 150, 167—175 and cf. Berzelius's remarks to Hisinger of 22 May 1813 (*Bref*, V,viii, p. 35).

[5] "Neue Untersuchungen," 160.

[6] *Ibid.*, 156: "Wenn nun das salpetersaure Bleioxyd mehr Blei auflöst und in ein salpetrigsaures Salz verwandelt wird, so muss dieses durch einen Zusatz von *einem* P geschehen; dieses nimmt aber *ein* O von der Säure, und das neugebildete Salz is nun A^4O + 2PO. Es ist klar, dass durch diese Zerlegung kein Salpetergas [nitric oxide, NO] gebildet werden kann, und dass, wenn das Bleisalz noch einmal so viel Blei, als es voraus enthielt, aufgenommen hat, die ganze Menge von Salpetersäure in salpetrige Säure verwandelt seyn muss. Jeder Zusatz von mehr sich auflösendem Blei muss daher das in diesem Zeitpuncte hervorgebrachte Salz zerlegen, und ein anderes mit mehr Basis übersättigtes hervorbringen, bis zuletzt A^4O + 4PO entstanden ist. Es ist klar, dass, wenn dieses geschieht, Davy's Angabe richtig ist; dass aber, wenn etwas andres vorgehet sie unmöglich die wahre seyn kann." In this paper, Berzelius demonstrated that the reaction did not proceed according to this description.

However, the evolution of nitric oxide was a problem in its own right. It was inconsistent not only with Davy's figures, Berzelius indicated (156), but with any value for the composition of nitrous acid. Berzelius had shown in the *Versuch* (pp. 145—146), he recalled, that solutions of nitrite salts decomposed on heating to form basic nitrates and that nitric oxide was evolved in the process. However, in the "Neue Untersuchungen," he did not note that in the *Versuch* he explicitly required the evolution of nitric oxide, even though it had been denied by Proust and Gehlen, to complete his prediction of the composition of the basic nitrite.

However, Berzelius showed that during the initial stages of the recation, the new portions of lead oxide, as they were formed, combined with the nitric acid remaining undecomposed to produce the neutral nitrite and the first of the basic nitrates.[7] As the first portions of lead dissolved, *no* basic nitrite was produced. As the reaction continued to the end of its first stage, that is, up to the point at which as much lead had been dissolved as was originally present in the neutral nitrate, the new portions of lead combined with the already formed neutral nitrite producing the basic nitrite. The additional lead was now equally divided between the basic nitrite and the basic nitrate, which Berzelius formulated as $(N^6O + 2PO)$ and $(N^4O + 2PO)$, respectively; half the nitric acid at this point had been reduced to nitrous acid; and all the neutral nitrate, into basic nitrate and basic nitrite.[8]

Moreover, Berzelius noted that by Davy's formula dissolution of still more lead would have had to begin producing a second, higher basic nitrite; but Berzelius had shown in the *Versuch*,[9] he now recalled, that more lead than this, indeed, up to one-quarter more, could be dissolved without the appearance of a second nitrite. The implication was clear, Berzelius asserted: the reduction of

(For this, see above, p. 230, text accompanying nn. 3 and 4). Accordingly, it was important to know the source of the gaseous evolution that had been observed: could the dissolution of metallic lead take place without it, he asked ("Neue Untersuchungen," 157); could it be that the evolution of the gas resulted merely from decomposition of a part of the newly formed basic nitrite?

He repeated the dissolution experiment (157—158), collecting the gaseous products and observing the temperature. He discovered that gas was evolved only at temperatures above 80 degrees, where decomposition of the already formed nitrite occurred. Accordingly, the evolution of nitric oxide was not an essential aspect of the process. Moreover, he had shown that undecomposed nitric acid saturated the excess base formed by dissolution of the metal and produced basic nitrates (158). In the *Versuch*, Berzelius had not observed the production of basic nitrates. As noted shortly, this phenomenon proved to be a major component of his analysis of the events of the dissolution.

[7] This paragraph reconstructs the arguments in "Neue Untersuchungen," 158—159.

[8] Equations for this first stage of the reaction, though not given by Berzelius, may be written, using modern formulae, as follows:

1. $1/2(PbO.N_2O_5) + Pb =$ (neutral nitrate) $1/2(PbO.N_2O_3) + PbO$ (neutral nitrite)
2. $1/2(PbO.N_2O_5) + 1/2PbO =$ (neutral nitrate) $1/2[(PbO)_2.N_2O_5]$ (basic nitrate)
3. $1/2(PbO.N_2O_3) + 1/2PbO =$ (neutral nitrite) $1/2[(PbO)_2.N_2O_3]$ (basic nitrite)

Overall, the first stage may be written thus:

$PbO.N_2O_5 + Pb =$ (neutral nitrate) $1/2[(PbO)_2.N_2O_5)] + 1/2[(PbO)_2.N_2O_3]$ (basic nitrate) (basic nitrite)

As expressed, the products are stoichiometrically equivalent to the tetrahydroxydinitritodinitrate. Berzelius's understanding of the stoichiometry of the reaction was therefore excellent indeed, and it marked a significant improvement over his investigation in the *Versuch*.

[9] It was at this point in the reaction, that is, just when one and one-quarter times as much lead had dissolved as was originally present in the salt, that Berzelius in the *Versuch* had observed the formation of a second basic nitrite, the "überbasisch" salt.

nitric acid to form nitrous acid produced more oxygen than Davy's formula for the latter would have allowed. Accordingly, it must be rejected. Using his own formulae, Berzelius could describe a second stage in the reaction that

> dauert, bis 2/3 [actually, it would be 5/8] der Salpetersäure in salpetrige Säure verwandelt sind. Das neutral salpetersaure Blei hat nun 78 Procent [of own weight in] Blei aufgelöst, d.i., 1 1/4 Mal so viel, als es vorhin enthielt. Die heisse Auflösung ist vollkommen klar, setzt aber während des Abkühlens ein Gemisch von N^4O + 2PO mit N^6O + 3PO (d.i. mit dem zweiten basischen salpet[er]sauren Bleioxyde ab).[1]

Berzelius did not analyze the third and final stage of the reaction in detail. The third stage actually had two parts, though in the "Neue Untersuchungen" Berzelius did not make this clear. The first part involved the reduction of the remaining nitric acid to nitrous acid, with production of the "überbasisches" nitrite previously observed in the *Versuch*.[2] The second part was concluded this Berzelius did make explicit—when all the nitrous acid was fully saturated with lead oxide, that is, when a still higher basic salt, "im Maximo überbasisches," was produced. Conversion of all the original lead nitrate into this salt was achieved, Berzelius recognized, by dissolution of a total of two and one-fifth times more lead than was originally present in the neutral nitrate. He understood, moreover, that decomposition of nitrous acid and evolution of nitric oxide were essential only for the production of this highest basic salt in the final stage of the reaction.[3] By this reasoning, Berzelius calculated that the maximum amount of metallic lead that could be made to dissolve in a solution of 100 parts of neutral

[1] "Neue Untersuchungen," 153. Equations for the second stage, with coefficients reflecting the comsumption of one mole of the original lead nitrate, may be written:

1. $1/8[(PbO)_2.N_2O_5] + 1/4\ Pb =$ (basic nitrate) $1/8[(PbO)_2.N_2O_3] + 1/4\ PbO$ (basic nitrite)
2. $1/4[(PbO)_2.N_2O_5] + 1/4\ PbO =$ (basic nitrate) $1/8[(PbO)_3.N_2O_5]$ (higher basic nitrate)

The overall reaction for the second stage is:

$$3/8[(PbO)_2.N_2O_5] + 1/4\ Pb = 1/8[(PbO)_2.N_2O_3] + 1/4[(PbO)_3.N_2O_5].$$

This leaves $1/8[(PbO)_2.N_2O_5]$ still in solution, though Berzelius did not point this out. The overall reactions for the first and second stages may be written:

$PbO.N_2O_5 + (1\,1/4)Pb =$ (neutral nitrate) $1/8[(PbO)_2.N_2O_5] + 5/8[(PbO)_2.N_2O_3]$ (basic nitrate) (basic nitrite) $1/4[(PbO)_3.N_2O_5]$ (higher basic nitrite)

[2] In the "Neue Untersuchungen," Berzelius did not point out explicitly the production of the "überbasisch" salt except to comment on it in a note concerning Chevreul's studies (cited above, p. 229 in n. 7) of the dissolution of metallic lead in lead nitrate. There (163, in the note beginning on 161) Berzelius offered what he believed to be an improved analysis of this salt. In fact, the analysis in the *Versuch* was better. The salt in question corresponds to the modern formula $(PbO)_3.N_2O_5$. Conversion of all the original lead nitrate into this salt is accomplished by dissolution of two times more lead than was present in the neutral nitrate:

$$PbO.N_2O_5 + 2Pb = (PbO)_3N_2O_3.$$

[3] "Neue Untersuchungen," 161. The final reaction may be written:

$$(PbO)_3.N_2O_3 + 1/5\ Pb = 4/5[(PbO)_4.N_2O_3] + 2NO$$

lead nitrate was 137.5 parts.[4] Because, Berzelius noted, the metallic lead became encrusted with its own salts as it dissolved, and because, near the end of the reaction, the concentration of the decomposing "überbasisches" nitrite was diminished, it was difficult to bring the reaction to this final stoichiometric condition. Nevertheless, he had managed to dissolve as much as 127 parts, and Chevreul, 135.[5] Berzelius believed that his study of the dissolution of metallic lead in lead nitrate had fully vindicated his formula for the composition of nitrous acid.[6]

The results of this study placed Berzelius in what he doubtless experienced as a particularly gratifying position. Given the composition of nitrogen, the oxides of the nitric radical, of which nitrogen was the lowest and nitric acid the highest, contained oxygen in the proportions 1:2:3:4:6. The progression agreed with the law of multiple proportions and contained no fractional multiples. Second, all nitrate and nitrite salts, neutral and basic, agreed with the rule of oxides or its variant for basic salts. Third, the progression of oxides, the pattern of salt formation, and the variation in acidity of the oxides all agreed with the sulfur pattern. The lowest oxide of nitricum, corresponding to the suspected suboxide of arsenic and the suboxide of sulfur found in the lower muriate, was nitrogen itself, and, like those of arsenic and sulfur, contained one-sixth the oxygen of the highest oxide, the acid in *-ic*. Moreover, the proportion for all three radicals, of the acid in *-ous* to that in *-ic* appeared to be 1:1 1/2. Finally, the multipliers in the rule of oxides exhibited by the two sulfur oxyacids and the two nitricum oxyacids in their neutral salts (though not by the acids of arsenic) were, respectively, in the proportion of 1:1 1/2. This signal success suggested that the apparent diversity of the patterns of oxidation and salt formation could indeed be reduced to the paradigmatic case of sulfur.

c. The sulfur pattern in the "Cause of proportions"

In the "Cause of proportions," where Berzelius was attempting to translate his stoichiometric laws into formulae, his confidence in the generality of the sulfur

[4] In modern figures, the amount of lead that need be dissolved to achieve this final stage is 137.61 parts. In fact, still more lead can be made to dissolve, with production of various complex products. These reactions were slowly elucidated during the nineteenth and twentieth centuries. That Berzelius could not have recognized these complexities, however, in no way undermines his argument and does not reflect upon the conclusions that he drew from it.

[5] "Neue Untersuchungen," 160—161.

[6] *Ibid.*, 161—164: "Die grosse Menge Blei, die aufgelöst werden muss, um das überbasische salpetrigsaure Blei im Maximo zu bilden, ist der letzte und zugleich der unwiderleglichste Beweis für die Richtigkeit meiner, und für die Unrichtigkeit der Davy'schen sowohl als der Gay-Lussac'schen Bestimmungen der Zusammensetzungen der salpetrigen Säure." He went on to suggest that the error of these two researchers resulted from their observation of a *compound* of two of the nitrogen oxides conceived as formed by displacement of basic water (164—165). In this connection, cf. above, p. 235, n. 3.

pattern reached its zenith.[7] There, he tried to make the compounds of as many radicals as possible conform to it, in part if not *in toto.* His work exhibits the conditioning of his results by his expectations and the use he made of the correlation of composition and properties to establish formulae. Though this effort confronted Berzelius with at least one anomalous radical, antimony, his enthusiasm for the sulfur pattern did not wane until a later study, in which he failed to assimilate the compounds of phosphorus to it.

Typifying his approach in the "Cause of proportions" was a new study of the arsenic compounds. The study superseded the investigation in the first *Fortsetzung* of the *Versuch* that had helped to establish the rule of oxides.[8] Berzelius recalled that his original study had revealed certain anomalies. In particular, though the two acidic oxides seemed to contain oxygen in the ratio of 1 1/2 : 1, as did the acids of sulfur, the multipliers in the rule of oxides were reversed: by Berzelius's orginal analyses, the acid in *-ic* contained twice the oxygen of the base in its neutral salts, and the acid in *-ous* contained three times the oxygen of the base. Accordingly, if arsenic acid contained three volumes of oxygen, then its salts violated the rule of oxides. Berzelius therefore decided to examine the arsenic compounds more closely.[9]

This work reveals both how Berzelius could determine the composition of oxides from that of their salts using the rule of oxides, and how he deployed the rule to compel the conformity of the oxides to the sulfur pattern.[1] Berzelius first precipitated the neutral arsenate of lead (actually, the pyroarsenate, modern formula $(PbO)_2.As_2O_5 = Pb_2As_2O_7$),[2] finding that whether the acid contained

[7] Berzelius's success in bringing the nitrogen compounds into agreement with the sulfur pattern appears to have been the immediate inducement to develop in the "Cause of proportions" still more cases in which the pattern was observed. In April, 1813, Berzelius reported to Marcet (*Bref,* I,iii, p. 38) that he had just completed the nitrogen studies that were to be reported in the "Neue Untersuchungen," and in June he informed Marcet (p. 60) that he had completed a manuscript of the "Cause of proportions." The paper included a brief notice of his new figures for the composition of nitrogen, and a translation of his conclusions into formulae for the nitricum oxides ("Cause of proportions," 61): "*Nitricum,* radicle [*sic*] of *azote* (N).—I have proved in a former paper that nitric acid is composed of 11.71 of nitric radicle and 88.29 of oxygen; and that nitric acid is N + 6O. Hence the volume of nitricum must weigh 79.451 [in comparison to that of oxygen, which equals 100]." He then described the progression of oxidation (*ibid.*): "I have endeavoured to prove that azote, or the sub-oxide of nitricum, is composed of N + O; nitrous oxide, of N + 2O; nitric oxide, or nitrous gas, of N + 3O; nitrous acid, of N + 4O; and nitric acid, of N + 6O. The oxides of nitricum are the first series of oxides of which we know [as isolable chemicals] all the degrees."

Henceforth, all of Berzelius's arguments about oxidation and salt formation were developed in terms of his formulary.

[8] The new results are reported and discussed in "Cause of proportions," 93—101.

[9] For the motives underlying the new study, see *ibid.,* 93. For the anomalies revealed in the *Versuch* about the arsenic compounds, see above, p. 198, passage accompanied by n. 9, and below, p. 250, n. 5.

[1] *Ibid.,* 93—95.

[2] Berzelius frequently obtained the pyro-compounds because he usually heated precipitates to redness, both to dry them and to drive out their volatile impurities.

twice or three times the oxygen of the base, this composition did not agree with the results of earlier analysts, whose work had guided him in the first *Fortsetzung*. The analysis of the neutral arsenate of barytes (again, the pyroarsenate $(BaO)_2 . As_2O_5$) indicated that it contained the same quantity of basic oxygen as was in the arsenate of lead. Earlier analysts' work was sufficient to show, however, that arsenic acid contained more than twice that amount of oxygen.[3] Therefore, he concluded it must contain three times that amount. Because the rule of oxides demanded an integral multiple, the actual multiplier, two and one-half, never crossed Berzelius's mind. A rough—and erroneous—direct analysis of arsenious acid indicated, Berzelius asserted, that this value for arsenic acid was correct if the *-ic* : *-ous* ratio in the proportions of oxygen was 1 1/2 : 1.[4] Berzelius, then, had forced his results to conform with two expectations: the existence of an integral multiplier for the rule of oxides and the appearance of the sesquialteral multiple in the *-ous* : *-ic* ratio.

With these results Berzelius could bring the remaining arsenic compounds to appear, if not identical to the sulfur pattern, then verv much like it. He confirmed his analysis in the first *Fortsetzung* of the arsenite of lead, but concluded that the multiplier in the rule of oxides was four, not three as he had first thought.[5] Like the multiplier of the corresponding sulfites, the multiplier in the rule of oxides was even, though its value was double that of the sulfites. In the basic arsenites, exemplified by the subarsenite of lead, the quantity of base was doubled, and accordingly the multiplier halved.[6] Assembling the results for the composition of the two oxyacids of arsenic, Berzelius weighed the evidence for possible formulae:

It is evident that the number of volumes of oxygen contained in these acids must be either 2 : 3 or 4 : 6. The composition of the neutral arsenites [with multiplier four] speaks decidedly in favor of the last numbers; but that of the neutral arseniates [with the multiplier three] seems favorable to the first: and so much the more, because, except the suboxide of arsenic, we know of no other oxides of that metal than the two acids: but, on the other hand, the composition of the subarseniates [Berzelius having presented here analyses of those of barytes and lead, and found them to contain 1 1/2 times the base of the neutral salts (that is, the pyroarsenates)] do not agree with the supposition that arsenic acid is

[3] Berzelius did not cite the earlier analysts in the "Cause of proportions" but remarked only that "we know" (*ibid.*, 94) that arsenic acid contained more than twice the amount of oxygen in the base. The fact was known from the analytical work of Klaproth and Rose that Berzelius used in the first *Fortsetzung* (*Versuch*, pp. 116—117).

[4] "Cause of proportions," 94.

[5] *Ibid.*, 96. In the *Versuch* (p. 115), Berzelius remarked, "Meine erste Vermuthung, dass sie [arsenious acid] wie die schweflige Säure zweimal so viel Sauerstoff als die sie sättigende Basis enthalten dürfte, bestätigte sich aber nicht." He went on to state, however, "Wenn aber die arsenige Säure dreimal so viel Sauerstoff in sich enthält als das von ihr gesättigte Bleioxydul so ist ihre Zusammensetzung [in agreement with the analyses of Bucholz, Rose, Thenard, and Proust]" (pp. 115—116). It was this anticipated agreement and the determination (*Versuch*, p. 117) that arsenic acid contains twice the oxygen of its base that had provided Berzelius with an early departure from the sulfur pattern that he was now trying to eliminate.

[6] "Cause of proportions," 96—97.

As + 3O; for in that case (since the oxide of lead is P + 2O) the subarseniate of lead would be PO^2 + 1 1/3 AsO^3. Now the fraction 1/3 is nowhere else to be found in chemical compounds: but if arsenic acid be As + 6O, the subarseniate in question will be AsO^6 + 1 1/2 PO[^{2}]. These considerations alone would induce me to admit six volumes of oxygen in arsenic acid; but there are other proofs, more convincing, of which I shall now give an account.[7]

Those other proofs were evidence pointing to the existence of a hitherto unknown salifiable oxide of arsenic. In the first *Fortsetzung,* Berzelius had tried and failed to obtain an oxide of arsenic capable of serving as base, but he was induced to try again, he indicated, by a new observation suggesting its existence.[8] This was Laugier's analysis of realgar (As_4S_4), now confirmed by Berzelius, which indicated a composition corresponding to a lower, salifiable oxide. Berzelius described a series of unsuccessful attempts to isolate this supposed oxide. Though he failed to obtain it pure, he believed he had found a compound in which, like the lower oxides of sulfur in the muriates of sulfur, it existed as an ingredient by the intermediation of other bodies. In particular, he discovered a triple muriate, having arsenical and mercurial bases. In his conclusions, Berzelius painted a picture of the general conformity of arsenic compounds to the sulfur pattern:

All my attempts to convert this triple muriate into pure muriate of arsenic have been ineffectual. It follows from these experiments that arsenic, among other properties which it has in common with sulfur, has this also, of producing a salifiable oxide with muriatic acid, but which, like the [lower] oxides of sulphur, has no existence in a separate state, but the instant it is separated from muriatic acid it is decomposed into metallic arsenic and arsenious acid. Though I have not yet made analytical experiments on this muriate, there can be no doubt that it is proportional to the sulphuret at a minimum [realgar], that is to say, that it must contain half as much oxygen as the acid.

If we add the existence of this oxide to the composition of realgar, it appears to follow clearly that arsenic acid contains 6 volumes of oxygen. Hence a volume of arsenic will weigh 839.9 or at a maximum, 852.2.

The known degrees of oxidation of arsenic are, 1. The suboxide or black powder which forms upon metallic arsenic. In my former experiments [in the first *Fortsetzung*], I found that in it 100 parts of arsenic were united to 8.5 of oxygen. This is certainly either too much or too little, because this suboxide must be either 2As + O or AS + O. I shall make experiments on it hereafter. 2. The salifiable oxide As + 3O [found in the triple muriate]. 3. Arsenious acid, As + 4O. 4. Arsenic acid, As + 6O. Is there an oxide As + 2O [corresponding to the oxide of sulfur in the dichloride]?[9]

Other examples in the "Cause of proportions" in which Berzelius gave the

[7] *Ibid.*, 97; for the analyses of the subarseniates of barytes and lead, see 95.

[8] It was the suboxide of arsenic that, in the first *Fortsetzung* (*Versuch*, p. 120), Berzelius hoped was salifiable. This antedated his conclusion, ventured in the "Nomenclature" (261), that most suboxides were not salifiable. Berzelius's new discussion of the salifiable oxide of arsenic and its possible existence occupies "Cause of proportions," 97—100.

[9] *Ibid.*, 100. The value, 8.5, was one-sixth the amount of oxygen that, in the first *Fortsetzung,* Berzelius believed was contained in arsenic acid. He now (incorrectly) was revising upward the oxygen content of arsenic acid (97). Accordingly, the value for the suboxide also had to be revised.

sulfur pattern a normative function include the oxides of chromium, molybdenum, tungsten, and antimony. The case of antimony is particularly instructive, for it exemplified Berzelius's use of the paradigm of sulfur to correct earlier analytical results. The antimony compounds, however, do not follow the sulfur pattern; Berzelius had difficulty in attempting to make them conform. He obtained at least one analytical result that seemed anomalous but later revealed itself, in studies of other radicals, to belong to a second pattern of oxidation and salt formation.

Berzelius had first studied the antimony compounds in the "Rechtfertigung."[1] There he found what he believed to be four distinct oxides, a suboxide, an oxide, a white oxide, and a yellow oxide.[2] He was unable to produce the suboxide in sufficient quantity to analyze. The oxide appears to have been the trioxide, Sb_2O_3, and the yellow oxide seems to have been the pentoxide, Sb_2O_5. The white may have been the compound of the two preceding oxides, having a composition expressed as Sb_2O_4. An accumulation of what can now be recognized as three errors, both analytical and conceptual, led Berzelius to a conclusion that, in the light of the sulfur pattern, he sought to modify in the "Cause of proportions."

Dissatisfied, in the "Rechtfertigung," with the poor agreement among the results of several direct analyses of the oxide, he tried to use as a standard an analysis of the sulfide (Sb_2S_3) that corresponded to the oxide.[3] However, his analysis of the sulfide gave a slightly lower sulfur content than is correct and implied a slightly low value for the oxygen content of the oxide. Coupled with this inaccuracy was Berzelius's expectation, based on several direct but imprecise analyses, that the white oxide contained one and one-half times the oxygen of the oxide (the correct ratio is 4 : 3).[4] Berzelius obtained yellow oxides in several ways. A direct oxidation of the metal with nitric oxide produced 129—131 parts of yellow oxide for 100 parts of metal (for Sb_2O_5, the modern figure is 132.8). Another yellow oxide obtained by heating the metal with mercuric oxide was then decomposed by heat, leaving the white oxide and giving off one-third as much oxygen, Berzelius believed, as was in the residual white oxide. The yellow oxide obtained by mercuric oxide therefore contained more oxygen than the yellow oxide obtained by nitric acid. This difference suggested to Berzelius the erroneous conclusion that the yellow oxide obtained by nitric acid was a compound formed from the yellow oxide obtained by mercuric oxide with the white oxide and composed so that each of the constituent oxides contained the same amount of oxygen. These errors, combined with the assumption that the suboxide of antimony, by analogy with arsenic, contained one-sixth the oxygen of the highest oxide, permitted Berzelius to conclude that the four oxides of

[1] "Rechtfertigung," 144—176.

[2] For the suboxide, see *ibid.*, 145—146; the oxide, 146—148; the white oxide, 149—152; the yellow oxide, 152—157.

[3] *Ibid.*, 148—149, 151—152.

[4] *Ibid.*, 149.

antimony most likely exhibited the following progression in the proportion of oxygen for a given weight of metal: 1, —, 4, 6, 8.[5]

The two highest oxides, Berzelius demonstrated, were antimonic and antimonious acids, and he prepared many of their salts.[6] He believed himself to have demonstrated that in the antimonates and in the hydrate of the acid the acid contained six times the oxygen of the base; and, in the antimonites, four times. The pattern of salt formation appeared largely to conform with that of sulfur, but the progression in oxidation did not: the *-ous* : *-ic* ratio in the proportion of oxygen was 3:4, not 2:3.

In the "Cause of proportions," Berzelius explicitly labelled this ratio anomalous, and there he attempted to correct his earlier results in a way that would complete the analogy:

> I have already, in a preceding dissertation [the "Rechtfertigung"], described my experiments on [antimony]. I have pointed out the difficulties which prevented me from obtaining an exact result. It is only in consequence of experiments repeated according to the ideas which I have explained in this paper, that I consider myself as having gained more decisive results. I had found that 100 parts of this metal combine with [3]7.3 of sulphur, and that this sulphuret [the sulfide corresponding to the oxide and used by Berzelius as a standard] dissolves in concentrated murtiatic acid, and forms muriate of antimony (murias stibiosum) and sulphureted hydrogen, without any excess of either sulphur or hydrogen. Hence it follows that the oxide of antimony (oxidum stibiosum) ought to be composed of 100 metal and 18.6 oxygen. The next oxide, according to the notions which I entertained at that time, contained 1 1/2 as much oxygen, that is to say, 27.9 [parts to 100 parts of metal]. But in none of my experiments on the composition of antimonious acid, could I find that the metal absorbed that quantity of oxygen. This I accounted for by an imperfect oxidation... But when I consider the ratio of the oxygen in the acids in *ous,* to that in the acids in *ic,* which is generally as 2 : 3; but never as 3 : 4, I think that I endeavoured in these experiments to obtain a result which could not take place.[7]

In his efforts to sustain the generality of the sulfur pattern, however, Berzelius was still attempting to obtain a "result which could not take place."[8] Maintaining his original figure for the oxide of antimony, he argued now that the white oxide contained only as much oxygen as was indicated by direct experiment (production of the yellow oxide by oxidation of the metal with nitric acid, followed by heating to drive off enough oxygen to get the white oxide), rather than using the standard of the sulfide. Because this figure was also slightly low, the values for the next two higher oxides remained in the correct ratio of 3 : 4. If these values, Berzelius indicated, corresponded to the formulae Sb + 3O, and Sb + 4O (the latter agreeing, he pointed out, with the existence of salts of antimonious acid exhibiting a multiplier of four), antimonic acid, "by analogy," ought to

[5] *Ibid.*, 154—155 (p. 154 is mispaginated as 541).

[6] For salts of antimonic acid, see *ibid.*, 157—166; of antimonious acid, 167—173.

[7] "Cause of proportions," 248.

[8] The discussion of the antimony compounds occupies *ibid.*, 248—250.

have been Sb + 6O. However, this conclusion could not be confirmed by determination of the saturation capacity of the acid, Berzelius noted, because the difference in the amount of base neutralized lay within the limits of experimental error for antimony compounds. Instead, Berzelius tried to determine its composition by thermal decomposition of the hydrate. However, these experiments were at worst inconclusive, and at best they indicated a composition for antimonic acid of Sb + 5O, something he held to be "contrary to all experience."[9] Instead, Berzelius concluded about antimonic acid,

> I think analogy so strongly in our favor, that we cannot avoid adopting Sb + 6O, as its true composition. Especially as at present we do not know of a single example of a radicle combined with 5 volumes of oxygen.[1]

d. The phosphorus oxyacids

An example of a radical combined with five volumes of oxygen presented itself to Berzelius when he studied the oxides of phosphorus and their salts. He had previously studied few of these substances himself, relying instead on the analyses of others.[2] His work on the phosphorus compounds forced him to recognize the existence of a class of radicals exhibiting another progression in their oxides and producing salts not in conformity with the rule of oxides but agreeing with a revised form of it. Berzelius therefore wrote the *Proportions chimiques* mindful of three categories of radicals: those conforming with the sulfur pattern; those agreeing with the phosphorus pattern; those agreeing with neither, but producing salts in conformity with the rule of oxides in its original form. The first and last categories, containing radicals with oxides conforming in their salts with the rule of oxides, were presented together as a single group and juxtaposed with those of the second.[3]

The study of the phosphorus compounds was published in 1816 as a second "Nachtrag" to the *Versuch* (the first having been the study of the nitrogen compounds).[4] Recalling the experiments and conclusions of the first *Fortsetzung*

[9] *Ibid.*, 249—250.

[1] *Ibid.*, 250. Cf. the addition of 1815 to Berzelius's *Mineralogy* (cited below, p. 294, n. 9), *Schweigger's Journal*, 15 (1815), 294, where Berzelius remarked that an ore composed of oxidized iron, simply because its analysis suggested a multiple of 5 in the proportion of oxygen, must be a compound of two oxides. Such a "Zwischengrad [ist] ein für alle Körper bis jetzt unbekanntes Multiplum."

[2] As he made clear in the first *Fortsetzung* of the *Versuch*, pp. 111—113.

[3] See above, p. 200, passage accompanied by n. 1.

[4] Berzelius, "Untersuchungen über die Zusammensetzung der Phosphorsäure, der phosphorigen Säure und ihrer Salze: Zweiter Nachtrag zu seinem Versuche, die bestimmten und einfachen Verhältnisse aufzufinden, nach welchen die Bestandtheile der unorganischen Natur mit einander verbunden sind," *Gilbert's Annalen*, 53 (1816), 393—446; 54 (1816), 31—55. The phosphorus study appeared in Swedish as part of a long paper in the *Afhandlingar:* Berzelius, "Försök att närmare bestämma åtskilliga oorganiska kroppars sammansättning, till vinnande af en närmare utveckling af läran om de kemiska proportionerna," *Afhandlingar*, 5 (1818), 379—520. The study of the phosphorus compounds covers 401—444. This entire paper was translated from Swedish into German and published in *Schweigger's Journal:* "Versuche die Zusammensetzung verschiedener unorganischer Körper

about the oxyacids of phosphorus, Berzelius then indicated several circumstances that had induced him to study them further:

> Bei meinen ersten Versuchen über die chemischen Proportionen [that is, in the first *Fortsetzung*] suchte ich die Zusammensetzung der Phosphorsäure nach Analysen des phosphorsauren Bleyes und Baryts zu berechnen, und zog dabei den Schluss, dass die Phosphorsäure 55 Proc. Sauerstoff enthalten müsse, dass also 100 Th. Phosphor sich mit 122,2 Th. Sauerstoff verbinden... Und da ich gefunden zu haben glaubte, dass 100 Th. Phosphorsäure von einer Basis so viel sättigten, als an Sauerstoff 55/2 = 27,5 Theile enthalten, so schloss ich, das diese Säure, den Ansichten der Corpuscular-Theorie gemäss, aus 1 Atom oder Raumtheil Phosphor gegen 2 Atome Sauerstoff bestehen müsse. Nachdem ich indess in diesem Theil der Chemie mehr Erfahrung erlangt und besser über ihn nachgedacht hatte, war es mir wahrscheinlich geworden, dass die Phosphorsäure eine ähnliche Zusammensetzung mit andern Säuren haben, deren Basis sich mit dem Sauerstoff in zwei verschiedenen Verhältnissen verbindet, und dass also der Sauerstoff der phosphorigen Säure zu dem der Phosphorsäure in dem Verhältniss von 2 : 3 stehe; daraus aber würde folgen, dass die Phosphorsäure mehr als 2 Atome oder Raumtheile Sauerstoff enthalte. Diese Bemerkungen machten mich an der Richtigkeit meiner ersten Versuch zweifelhaft, und da ich für die zweite Ausgabe meine Elemente der unorganischen Chemie eine tabellarische Darstellung der chemischen Proportionen auszuarbeiten hatte, so entschloss ich mich, die Untersuchung der Zusammensetzung sowohl der Phosphorsäure und der phosphorigen Säure, als auch der Salze dieser Säuren, von neuem vorzunehmen.[5]

Berzelius's original results for the phosphorus compounds were in conflict with the sulfur pattern, which he had extended so successfully to a variety of radicals in the "Cause of proportions." Hoping to eliminate the conflict, Berzelius undertook this new research.

He began it with a set of astonishingly exact analyses of salts of phosphoric acid, achieved by overcoming, as he put it, "... Schwierigkeiten, welche ich nicht erwartet hatte, und sie kosten mir viel Zeit und Arbeit..."[6] Only the phosphoric acid compounds of lime, which present a higfhly complex chemistry, left Berzelius stumped: "Diese anomale Verbindungen bestätigen sich [in their peculiar

näher zu bestimmen, zur genaueren Entwickelung der Lehre von den chemischen Proportionen," *Schweigger's Journal,* 23 (1818), 98—122, 129—202, 279—299; phosphorus compounds, 117—122, 129—160. The investigation of the phosphorus compounds henceforth is cited according to the 1816 text in *Gilbert's Annalen* as "P-Untersuchungen."

Berzelius performed the experiments recounted in "P-Untersuchungen" during the spring of 1816. On 15 April, he reported to Gahn (*Bref,* V,ix, pp. 135—136), "Jag smörjer ännu med phosphorsyran..." and gave a hasty report of preliminary results. On 9 May, be reported to Marcet (*Bref,* I,iii, pp. 137—138) that he had been long at work on phosphorus compounds, and he summarized the major outcomes of his work (pp. 138—140). In a letter to Gahn of 10 June (*Bref,* V,ix, pp. 138—140), he reported that he had completed his study of the phosphorus compounds.

[5] "P-Untersuchungen," 394—395. The composition of phosphoric acid was first given in corpuscular terms in the "Cause of proportions," not in the first *Fortsetzung*.

[6] "P-Untersuchungen," 395—396. Cf. Berzelius's letter to Marcet of 9 May 1816, where he observed of his studies of phosphorus (*Bref,* I,iii, pp. 137—138): "...je me vis ... engagé dans un labyrinthe qui me fit oublier mes amis, et toute mon âme se remplissait d'acide phosphorique, dont cependant je commence à présente à me rendre quitte, ayant vu le jour poindre après une nuit fatigante."

nature] von allen Seiten," leaving Berzelius "... ganz ausser Stande, eine Erklärung von diesen Erscheinungen zu geben."[7] He did not allow these problems to detain him, however, for he had more than enough positive results on which to base his conclusions. The phosphate salts that he analyzed are summarized in this table, in which both ordinary modern formulae, and modified modern formulae designed to exhibit oxide components, are displayed.

Character, acc. to B.	*base*	*oxide-formula*	*unmodified formula*
1. neutral phosphate of	barytes	$(BaO)_2 . P_2O_5$	$Ba_2P_2O_7$
2. acid phosphate of	barytes	$BaO . P_2O_5 . H_2O$	$BaH_2P_2O_7$
3. intermediate[ly acid] phosphate of	barytes	$(BaO)_3 . (P_2O_5)_2$	$Ba_3P_4O_{13}$
4. neutral phosphate of	lead	$(PbO)_2 . P_2O_5$	$Pb_2P_2O_7$
5. intermediate[ly acid] phosphate of	lead	$(PbO)_3 . (P_2O_5)_2$	$Pb_3P_4O_{13}$
6. basic phosphate of	lead	$(PbO)_3 . P_2O_5$	$Pb_3(PO_4)_2$
7. basic phosphate of	silver	$(AgO)_3 . P_2O_5$	$Ag_3(PO_3)_2$
8. neutral phosphate of	soda	$(NaO)_2 . P_2O_5$	$Na_4P_2O_7$
9. basic phosphate of	lime	$(CaO)_3 . P_2O_5$	$Ca_3(PO_4)_2$

Berzelius understood very clearly that most of these salts were anomalies for the rule of oxides:

> Wenn man die Resultate meiner Zerlegungen phosphorsaurer Salze, welche ich in dem vorigen ... mitgetheilt habe, mit der von mir aufgefundenen Regel, über die Verhältnisse, nach denen sich die oxydirten Körper verbinden müssen, vergleicht, so übersieht man leicht, dass entweder diese Regel hier nicht mehr Stich hält, oder die Zusammensetzung der Phosphorsäure nicht die seyn kann, welche man bisher angenommen hat. Denn es lassen sich die verschiedenen Sättigungs-Grade der Phosphorsäure allein unter der Annahme, dass die Säure in den neutralen Salzen 3 Mal, in den basischen 2 Mal, und in den beiden sauren Salzen 4 und 6 Mal so viel Sauerstoff als die mit ihr verbundene Basis enthalten müsse, mit den Resultaten dieser Analysen in Uebereinstimmung bringen. In diesem Fall müsse aber die Phosphorsäure nicht weniger als 67,5 Procent Sauerstoff enthalten, d.h. etwas mehr als 200 Th. Sauerstoff gegen 100 Th. des Radikals, welches sich sehr von den durch Erfahrung gegebenen Resultaten [of analyses of phosphoric acid] entfernt, und beinahe das Doppelte von der Menge ist, die Rose gefunden hat.[8]

Accordingly, Berzelius was moved to settle the still undecided question of the oxygen content of phosphoric acid. All techniques previously employed in the analysis of this substance having given a variety of results, Berzelius chose another: he made use of the ability of phosphorus to reduce certain metals from their acidic or alkaline dissolutions.[9] At least one carefully described experiment, the reduction of a muriatic acid solution of gold, gave a reproducible and accurate result (55.95—56.18% oxygen):[1]

> Vergleicht man diese Zahlen mit den verschiedenen Stufen der Verbindung, welche die Phosphorsäure mit den Basen bildet, so findet sich, dass in den sauren phosphorsauren Salzen der Sauerstoff der Säure fast genau das 5-fache ist von dem der Basis; denn es ist

[7] "P-Untersuchungen," 420.
[8] *Ibid.*, 427.
[9] *Ibid.*, 428.
[1] For the experiment, see *ibid.*, 429—432.

11,24 × 5 = 56,2. Unter den übrigen Stufen findet sich keine, in welcher der Sauerstoff der Säure ein Multiplum nach einer ganzen Zahl von dem der Basis ist; im Gegentheil beträgt in den verschiedenen phosphorsauren Salzen der Sauerstoff der Basis 3/5, 2/5, (1 1/2)/5, 1/5 von dem Sauerstoff der Säure. Es ist daher der Frage, ob hier eine Ausnahme von der Regel statt findet, oder ob zu vermuthen ist, dass der Phosphor Sauerstoff enthalte, z.B. 1/5 von der Menge, welche er bedarf, um zu Phosphorsäure zu werden, in welchem Fall der Sauerstoff der Basis 1/2, 1/3, 1/4, 1/6 von dem der Säure seyn, und also der Phosphor unter der Regel der übrigen bis jetzt untersuchten oxydirten Körpern passen würde.[2]

Like the nitrogen compounds, those of phosphorus could be brought into conformity with the rule of oxides and with the sulfur pattern if their radical could be proven to be an oxide. Berzelius made the analogy with the nitrogen compounds explicit, pointing out, for the first time, what pattern they exhibited when nitrogen was taken to be elementary:

Die Phosphorsäure verhält sich in dieser Hinsicht wie die Salpetersäure, doch mit dem Unterschiede, dass, was bei der Salpetersäure das neutrale, bei der Phosphorsäure das saure Salz ist; auch giebt die Salpetersäure basische Salze mit grösserm Ueberschuss der Basis, als die Phosphorsäure. Der Sauerstoff der Basen in den verschiedenen salpetersauren Salzen steht zu dem Sauerstoff der Salpetersäure in den Verhältnissen 1/5, 2/5, 3/5, 6/5, und diese reduciren sich, wenn man zugiebt, das Stickstoff und Ammoniak oxydirte Körper sind, auf die Verhältnisse 1/6, 1/3, 1/2. Der Sauerstoff, der Schwefel, der Arsenik- und der Molybdän-Säure verhält sich zu dem Sauerstoff der schwefligen, der arsenigen und der molybdänigen Säuren, wie 3 : 2. Eben so verhält sich der Sauerstoff der Salpetersäure zu dem der salpetrigen Säure, wenn man die Zusammengesetztheit des Stickstoffs annimmt; sonst aber wie 5 : 3. Dagegen muss der Sauerstoff der Phosphorsäure zu dem der phosphorigen Säure, wenn der Phosphor selbst keinen Sauerstoff enthält, in dem Verhältniss von [5 : 3] stehen; dagegen wenn er (gerade so wie sich das vom Stickstoff vermuthen lässt) ein Atom Sauerstoff in Verbindung mit einem in isolirtem Zustande unbekannten Radikale enthält, indem Verhältniss von [3:2], wie der Sauerstoff in den beiden Stickstoffsäuren.[3]

Berzelius now asked whether phosphorous acid confirmed this departure from the sulfur pattern. Prior analyses by other chemists had given varying results for the *-ous* : *-ic* ratio in the proportion of oxygen. Berzelius therefore resolved to settle this question. He intended to use the composition of the lower muriate of phosphorus to determine the composition of phosphorous acid. As a check on his approach, he first confirmed his figure for phosphoric acid by analyzing the higher muriate.[4] The result agreed, and his study of the lower muriate confirmed

[2] *Ibid.*, 434.

[3] *Ibid.*, 434—435. The difference between the nitrates and phosphates that Berzelius noted at the beginning of this passage derives from the differing basicities of the acids. Stoichiometrically, the acid salt of the tribasic phosphoric acid, e.g., $BaO.P_2O_5.H_2O = BaH_2P_2O_7$, corresponds with the neutral salt of the monobasic nitric acid, e.g., $BaO.N_2O_5 = Ba(NO_3)_2$.

[4] For variability in prior results, see *ibid.*, 436. The higher muriate, with modern formula PCl_5, was considered a compound of two oxides, $1/2[P_2O_5.(MO_2)_5]$, where M represents the muriatic radical and the coefficient 1/2 makes the equivalent weight of the formula agree with PCl_5. Similarly, the lower muriate, PCl_3, was conceived as $1/2[P_2O_3.(MO_2)_3]$. The analysis of the higher muriate is in *ibid.*, 436—437.

the *-ous* : *-ic* ratio of 3 : 5.[5] Apparently hoping to salvage something of the sulfur pattern, Berzelius now asked whether phosphor*ous* acid conformed with it (since, like sulfuric acid, it contained three volumes of oxygen).[6] To this end he presented a series of analyses of the phosphites of lead and barytes. Oxidation of lead phosphite by nitric acid produced a lead phosphate containing the same amount of lead oxide as was in the neutral phosphate analyzed earlier in the paper (and listed in the table above). This was his old oxidation technique, based on the principle of the preservation of combining proportions. Like the corresponding sulfur compounds, Berzelius pointed out, the neutral salts of phosphorous acid could be oxidized to neutral salts of the acid in *-ic*. From this he deduced the composition of the phosphite and showed that the acidic oxygen was two-thirds the basic. The phosphite of barytes, similarly oxidized, gave the same results.[7] Accordingly, Berzelius concluded that phosphorous acid did not behave like other acids with three volumes of oxygen. No part of the sulfur pattern could be salvaged.

Hoping to eliminate the anomalies by demonstrating the oxide nature of phosphorus, Berzelius attempted an experimental test of this assumption.[8] He expected that phosphorus, if an oxide, would have lost its oxygen upon combining with a metal, and that accordingly more oxygen would have been required to produce phosphoric acid from this combination than from phosphorus itself. Upon oxidizing a combination of iron and phosphorus, however, Berzelius showed that this was not the case: phosphorus combined with iron without altering its composition.[9] The phosphorus compounds, therefore, had to be admitted as exceptions to the rule of oxides. What explanation could account for this departure?[1]

Previously, Berzelius recalled, only the basic nitrates and nitrites had presented exceptions to the rule. The assumption of the compound nature of nitrogen had eliminated these anomalies, and Berzelius here stated that if these salts were the only exceptions, the compound nature of nitrogen was definitely proved.[2] With the additional exceptions provoded by the phosphorus compounds, Berzelius now conceded that it was not proved, but highly probable.[3] His belief was quite shaken, however. There was less evidence for the oxide nature of

[5] The analysis of the lower muriate is in *ibid.*, 434—440. The results agreed very closely with the ratio of 3 : 5, leading Berzelius to remark (440): "Diese vollkommene Uebereinstimmung des Resultats mit der Berechnung übertrifft alles, was wir bei der Schwierigkeit dieser Versuche erwarten durften. Es erhellt daher, dass die Ursache der Abweichung des Phosphors von der Regel für die Verbindungen oxydirter Körper der analog seyn könne, welche bei der Salpetersäure statt findet."

[6] *Ibid.*, 440—441.

[7] For oxidation of lead phosphite, see *ibid.*, 441—444; of phosphite of barytes, 444—446. The phosphite of lead was $1/2[(PbO)_2.P_2O_5.H_2O] = PbHPO_3$.

[8] *Ibid.*, 34—40.

[9] *Ibid.*, 37.

[1] *Ibid.*, 38.

[2] *Ibid.*

[3] *Ibid.*, 39.

phosphorus than for that of nitrogen, as he went on to state, and if the phosphorus compounds could stand as exceptions, so too could the nitrogen compounds.[4] Moreover, disclaimimg any attempt to make plausible the idea of the oxide-nature of phosphorus merely to justify the rule of oxides, he pointed out that even if phosphorus were an oxide, then the two muriates would become exceptions to the rule.[5] Still, Berzelius noted that the exceptional cases were few, and the violations of the rule seemed correlated with the *-ous* : *-ic* ratio of 3:5:

Wenn es aber auf der einen Seite entschieden ist, dass es Verbindungen zwischen oxydirten Körpern giebt, in denen der Sauerstoff des einen nicht ein Multiplum von dem des andern ist, so ist es auf der andern Seite eben so wahr, dass diese Ausnahmen sehr selten sind. Irre ich mich nicht, so finden sich unter den bis jetzt gemachten Analysen keine andern Beispiele, wo die, die Oxyde betreffende Regel nicht Stich hält, als die welche sich auf den Phosphor und auf den Stickstoff beziehen. Da diese Regel aber doch, wenn ihr Allgenmeinheit zukäme, von grosser Wichtigkeit seyn wurde, so ist es gewiss sehr der Mühe werth, zu untersuchen, ob die Ausnahmen sich wirklich auf die Verbindungen der Oxyde dieser beiden Körper beschränken, oder ob es deren noch andere giebt. Wären wohl diese Ausnahmen eine nothwendige Folge der Progression ihrer Oxyde, nämlich des Sprunges derselben von 3 zu 5?[6]

Shortly before the *Proportions chimiques,* Berzelius indeed discovered another case of precisely the same sort. He showed in 1817 that the arsenic compounds, originally adduced in the first *Fortsetzung* as the earliest example of another radical besides sulfur that conformed to the sulfur pattern, in fact exhibited the *-ous* : *-ic* ratio of 3:5 and presented the same violations of the rule of oxides.[7] If established, the oxide nature of arsenic would reduce these anomalies,

[4] *Ibid.*, 39—40.

[5] *Ibid.*, 41. If phosphorus was the radical, then the muriates agreed with the rule: in both the higher muriate, $P_2O_5.(MO_2)_5$, and in the lower, $P_2O_3.(MO_2)_3$, the acidic oxygen is an integral multiple of the basic. This is not true if phosphorus is an oxide of an unknown radical. If the latter is represented by $P' = P_2{-}O$, the muriates become $P'O_6.(MO_2)_5$ and $P'O_4.(MO_2)_3$, respectively.

[6] *Ibid.*, 42.

[7] Berzelius's further studies of the arsenic compounds were occasioned, initially, by Thomson's criticisms of Berzelius's earlier work on them. For this, see Berzelius's letter to Marcet of 14 April 1817 (*Bref,* I,iii, pp. 149—152, esp. p. 151, and Söderbaum's n. 112, on p. 254). Berzelius attained his new results for arsenic by 15 May 1817, when he reported them to Gahn (*Bref,* V,ix, pp. 164—165). A preliminary report of the result, showing that the *-ous:-ic* ratio was 3:5, appeared in Berzelius's "Nouvelles recherches sur les proportions chimiques," *Ann. Chim.*, 5 (1817), 174—181, esp. 179—180. This paper was submitted by Berzelius in manuscript to the *Annales de Chimie* (*ibid.*, 174n). The full account of his new studies of arsenic and its compounds appeared in the long paper in the *Afhandlingar,* cited above, p. 254, n. 4. The discussion of arsenic compounds covers 455—475; and in the German translation, cited in the same note, 170—186. The conformity of arsenic to the phosphorus pattern is considered in the translation on 178—180.

The new figures for the composition of arsenic that Berzelius obtained were inconsistent with the composition of the supposed suboxide of arsenic (into which the metal degenerated on exposure to air). According to his former figures for arsenic acid, which permitted assimilation of arsenic to the sulfur pattern, arsenic acid contained six times the oxygen of the suboxide. According to his new figures, which

but Berzelius conceded that the oxide nature of arsenic was even less likely than that of phosphorus. Still, he refused to believe that the last word had been spoken in this connection.[8]

This was effectively the state of Berzelius's thinking as he exhibited it in the *Proportions chimiques.* Having articulated the rule of oxides, in a passage quoted above (p. 200, accompanied by n. 1), he signalled the exceptions presented by the radicals conforming to the phosphorus pattern. Later, when he exhibited the use of the rule of oxides in determination of formulae, he indicated that to the anomalous acids and their salts belonged anomalous formulae, but he refused to agree that the question was settled:

je ne dois pas omettre la conjecture qui a été faite, que lorsqu'un radical donne deux acides dans lesquels les rapports mutuels de l'oxigène sont comme 3 à 5, ces acides peuvent contenir deux atomes de radical sur 3 à 5 atomes d'oxigène, et qu'ainsi cette différence de composition peut causer l'anomalie que l'on observe dans ces acides, dont les rapports aux bases salifiables diffèrent de ceux que l'on trouve ordinairement dans les autres acides. Dans une matière comme celle-ci, où presque tout est conjecture, il ne convient pas de rejeter une idée que n'est pas reconnue décidément fausse; on doit la considérer comme possible, et la ranger auprès des autres. La décision de cette question dépend entièrement de celle où il s'agit de savoir si l'azote est un corps simple. S'il l'est, ses deux acides sont composés, suivant des preuves positives, de deux atomes de radical sur 3 et 5 atomes d'oxigène, et dans le cas contraire, d'un atome de radical sur 4 et 6 atomes d'oxigène. Dans ces derniers temps, on a beaucoup soutenu, en chimie, l'assertion que l'on doit considérer comme inexact ce qui ne peut pas être prouvé par des expériences directes; en conséquence, fermant les yeux à toutes les probabilités qui font croire que l'azote est un corps composé, on est décidé à le considérer comme un corps simple, jusqu'à ce qu'il puisse être décomposé. Je ne puis pas admettre la justesse de cette méthode; il serait même plus exact de considérer comme possible tout ce qui n'est pas démontré impossible; et d'ailleurs nos recherches nous conduisent souvent à des conclusions qui ne peuvent être prouvées qu'à la suite de nouveaux travaux. On ne peut pas contester au jugement de l'homme la faculté d'induire, de circonstances indirectes, l'existence de faits qui, pour le moment, ne peuvent pas être directement prouvés. Si, parce qu'on peut abuser de cette faculté, l'on voulait exclure son application de la chimie, où elle est plus nécessaire que dans beaucoup d'autres sciences, on tomberait sans doute dans un excès nuisible.[9]

Despite his disquiet with the phosphorus pattern, Berzelius had succeeded in reconstructing Lavoisier's correlation of oxidation and properties. He exhibited shared patterns in the progression of oxidation of radicals and pointed out the qualitative changes that accompanied the quantitative ones: many radicals formed generally nonsalifiable suboxides, middling oxides that were salifiable, and higher, acidic oxides. Accordingly, the names of the categories in the classifi-

compelled assimilation of arsenic to the phosphorus pattern, the oxygen of arsenic acid was not a multiple of that of the suboxide (180—181). Therefore, Berzelius attempted a new analysis of the suboxide (180—183) and reported that "zu meinem grossen Verwunderung" he could not produce it (181).

[8] In the German text of the arsenic study, 178—179.

[9] *Ibid.*, 120—121.

cation of oxides presented in the "Nomenclature" could retain their utility as designations of stages exhibiting shared qualitative and quantitative patterns in their oxidation, even as the radicals retained their primacy as the seats of specific properties. Moreover, these patterns were tied with patterns of salt formation that were applicable in extraordinarily wide domains. By invoking these patterns, Berzelius had shown he could specify both quantitatively and qualitatively virtually all saline compounds encountered in the laboratory, no matter how complex, and many that did not display typically saline characters. Berzelius's successes in using the theory of salts to specify artificial products inspired him to deploy his methods in studying natural products: organic chemicals and minerals. These deployments are the subject of the next chapter.

CHAPTER VI

Full Articulation of the System: Deployments in Mineralogy and Organic Chemistry

1. Introduction

a. The timing and character of the deployments

By the middle of 1813, Berzelius had accomplished the bulk of the research on which the main lines of his chemistry rested, and he had yet to encounter phenomena that later compelled him to qualify his beliefs. Two of the principal post-*Versuch* projects were complete, the "Neue Untersuchungen" on nitrogen compounds and the "Cause of proportions."[1] The "Neue Untersuchungen" suggested the generality of the sulfur pattern of oxidation and salt formation and the "Cause of proportions" documented the widespread occurrence of the sulfur pattern. The study of the phosphorus compounds, revealing another pattern of oxidation and salt formation, one inconsistent with the rule of oxides, was still three years in the future.[2] In mid-1813, Berzelius doubtless believed himself fully in control of an elaborate chemistry, based on the electrochemical generalization of the ideas of acidity and basicity and on a body of analytical material supporting precisely articulated laws. Confident of the value and durability of his achievements, Berzelius was ready to test them in new deployments. He turned to mineralogy and organic chemistry.

[1] Berzelius, "Neue Untersuchungen über die Natur des Stickstoffs, des Wasserstoffs, und des Ammoniaks; und einige Bermerkungen über die wahre Zusammensetzung der salpetrigen Säure und der Salpetersäure: Erster Nachtrag zu seinem Versuche, die bestimmten und einfachen Verhältnisse aufzufinden, nach welchen die Bestandtheile der unorganischen Natur mit einander verbunden sind," *Gilbert's Annalen,* 46 (1814), 131—175; henceforth cited as "Neue Untersuchungen." For completion of the studies reported in the "Neue Untersuchungen" in April, 1813, see above, p. 249, n. 7. For Swedish version, see above, p. 241, n. 1.

Berzelius, "Essay on the cause of chemical proportions, and on some circumstances relating to them; together with a short and easy method of expressing them," *Ann. Phil.,* 2 (1813), 443—454; 3 (1814), 51—62, 93—106, 244—257, 353—364; henceforth cited as "Cause of proportions." For completion of the manuscript version of the "Cause of proportions" in June, 1813, and for its delayed publication, see above, p. 213, n. 7.

[2] Berzelius, "Untersuchungen über die Zusammensetzung der Phosphorsäure, der phosphorigen Säure und ihrer Salze: Zweiter Nachtrag zu seinem Versuche, die bestimmten und einfachen Verhältnisse aufzufinden, nach welchen die Bestandtheile der unorganischen Natur mit einander verbunden sind," *Gilbert's Annalen,* 53 (1816), 393—446; 54 (1816), 31—55; henceforth cited as "P-Untersuchungen." For completion of the studies reported in the "P-Untersuchungen" in the spring of 1816, as well as for Swedish version and an additional translation, see above, p. 254, n. 4.

Both subjects had been among his earliest interests. Some of his first researches were studies of minerals, often conducted with Hisinger, who, with Gahn, had introduced Berzelius to mineralogy. With Hisinger in 1802, Berzelius discovered cerium in a rare mineral, and henceforth he published several papers annually devoted to mineralogical topics.[3] Also during the years prior to the *Versuch,* Berzelius devoted much attention to chemical investigation of organic matter, especially animal products. A book recounting his researches in animal chemistry appeared in Stockholm in 1806—1808,[4] and a collection of numerous analyses of animal substances that he published in England while visiting there made his work widely known.[5] His lecture on the state of animal chemistry, given upon his relinquishing the presidency of the Swedish Academy of Sciences in 1810, also became widely known through translations of the published Swedish version.[6] Though persistent, these interests in mineralogy and organic chemistry were eclipsed by Berzelius's involvement in electrochemistry and by his growing preoccupation with the analytical project recounted in the *Versuch* and subsequent papers. However, when his electrochemical and analytical work reached a temporary culmination in the "Cause of proportions," Berzelius returned to his earlier concerns, hoping to apply to them his hard won chemical achieve-

[3] For the discovery of cerium, see J. Erik Jorpes, *Jac. Berzelius: his life and work* (Berkeley: 1970), pp. 26—29; henceforth cited as Jorpes, *Life and work.* For an indication of the development of Berzelius's research interests, suggesting his persistent concern with mineralogy, see the bibliography of his works, organized by year of publication, in Arne Holmberg, *Bibliografi över J.J. Berzelius* (I, *Tryckta arbeten av och om Berzelius,* Stockholm and Uppsala: 1933; I, suppl. i, Stockholm and Uppsala: 1936; I, suppl. ii, Stockholm: 1953; I, suppl. iii, Stockholm: 1967; II, *Manuskript,* Stockholm: 1936; II, suppl., Stockholm: 1953), I, I, suppls. i—iii.

[4] Berzelius, *Föreläsningar i djurkemien,* 2 vols. (Stockholm: 1806—1808); henceforth cited as *Föreläsningar.*

[5] Berzelius, "General views of the composition of animal fluids," *Medico-chirurgical transactions,* 3 (1812), 198—276; republished as "On the composition of animal fluids," *Ann. Phil.,* 2 (1813), 19—26, 195—208, 377—387, 415—425; henceforth cited from the text in the *Annals of Philosophy* as "Animal fluids." In the opening of the paper (19), Berzelius remarked, "Most of these observations [on animal chemistry] have been published in the Swedish language;* but as they have not been translated into any other language, and as they have appeared to those who have seen them, to contain some new views, I am induced to offer them to the [Medico-Chirurgical] Society, in the hope that they will be received with indulgence." The note signalled by the asterisk reads (19n): In my Förelasningar [*sic*] i Djurkemien, 2 vol. Stockh. 1808. And also in Afhandlingar i Fysik, Kemi och Mineralogi, 3 vol. Stockh. 1810." The latter was probably a reference to Berzelius, "Försök till en allmän öfverblick af varmblodiga djur-vätskors sammansättning," *Afhandlingar,* 3 (1810), 1—105.

[6] Berzelius, *Oefversigt af djur-kemiens framsteg och närvarande tillstånd; tal, hållet för Kungl. Vetenskaps-Akademien vid Praesidii nedläggande den 13 Augusti 1810* (Stockholm: 1812); English, *A view of the progress and present state of animal chemistry,* tr. Gustavus Brunmark (London: 1813; reissued, 1818). Partington (*History,* IV, pp. 146—147) erroneously identifies the translations of the lecture as translations of the *Föreläsningar.* Though similar to the latter, the lecture exhibited a different tone and emphasis. In this connection, see Sten Lindroth, "Berzelius och hans tid," *Vetenskaps-Societeten i Lund, Årsbok,* 1964 (1964), 15—40, esp. 24—25; henceforth cited as Lindroth, "Berzelius."

ments. During the winter of 1813—1814, Berzelius devoted his attention exclusively to mineralogy and organic chemistry.[7]

Gaining access to the specific level, the leading theme of the deployments of Berzelius's theory of salts in mineralogy and organic chemistry, is the focus of the present chapter. As noted in the Introduction to Part II, establishing the specificity of a substance required demonstrating two things: that the substance could be viewed as a salt and that as such it was composed in accordance with one of Berzelius's stoichiometric laws for salts. A substance could be assimilated to the category of salts if some of its oxide (or sulfide) constituents could be characterized qualitatively as electronegative in relation to other constituents that could be characterized as electropositive, that is, if some constituents played the role of acid while others played the role of base. Any substance formed from

[7] The shift in Berzelius's interests from mid-1813 can be documented in his correspondence. Once his inorganic chemistry had attained the state represented by the "Cause of proportions," he turned his attention to quantitative organic analysis, and he did not actively engage mineralogy until the organic work was largely under control. Thus, in a letter to Hisinger of 22 May 1813 (*Bref,* V,viii, p. 35) he announced the impending conclusion of his inorganic studies: "Jag håller ännu på med bestämda proportioner, men lider nu til et complett slut med den oorganiska naturen." By October, he had become thoroughly immersed in organic analysis, for he remarked in a letter to Gahn of 6 December 1813 (*Bref,* V,ix, p. 93) that he had been occupied for two months with the analysis of the vegetable acids, and he told Marcet in a letter of 7 January 1814 (*Bref,* I,iii, pp. 87—90, esp. p. 88) that he had been pursuing organic analysis for nearly four months. Though in the letter to Marcet and in another to Gahn written almost a week later (13 January 1814, *Bref,* V,ix, pp. 95—97, esp. pp. 95—96) he claimed that he was in a position to deduce the stoichiometric laws governing organic nature, he reported to Marcet some three months later (letter of 12 April 1814, *Bref,* I,iii, pp. 93—97, esp. p. 94) that he was still working on organic proportions. Finally, in a letter to Marcet of 16 June 1814 (*Bref,* I,iii, pp. 100—104, esp. p. 103), Berzelius stated that he had sent a manuscript of his paper on organic proportions to Thomas Thomson for publication in the *Annals of Philosophy*. (Berzelius's undated covering letter to Thomson is in *Bref,* III,i, pp. 14—17). The paper appeared as Berzelius, "Experiments to determine the definite proportions in which the elements of organic nature are combined," *Ann. Phil.*, 4 (1814), 323—331, 401—409; 5 (1815), 93—101, 174—184, 260—275; Swedish, "Försök öfver de bestämda förhållanden, hvari elementen för den organiska naturen äro förenade," *Afhandlingar,* 5 (1818), 520—646; henceforth cited from the English verssion as "Organic proportions."

Late in the winter, Berzelius turned to mineralogy. The immediate inducement was the visit of William MacMichael (1784—1839), a young English physician who was recommended to Berzelius by Marcet (*Bref,* I,iii, pp. 55, 68). MacMichael's interest in mineralogy (*Bref,* I,iii, pp. 68, 76; V,ix, p. 79) led him to purchase Ekeberg's mineralogical collection (*Bref,* III,i, p. 95; V,ix, p. 97). He gave Berzelius the duplicates of those minerals that he had obtained elsewhere (*Bref,* III,i, pp. 94—95; V,ix, pp. 97—98), extracting a promise from his mentor to exploit them in a scientific study of mineralogy. Berzelius developed his mineralogical ideas rapidly, saying to Gahn in a letter of 28 February 1814 (*Bref,* V,ix, p. 97), "Nu studerar jag ifrigt mineralogien och har under mina studier trott mig finna nyckeln til et vetenskapligt system, som är ganska simpelt och som på en gång förliker mineralogien med electrochemiska theorien och med läran om bestämda proportioner." This work culminated in a little book on mineralogy, of which the Swedish version was published by 12 April 1814, when Berzelius told Marcet (*Bref,* III,i, p. 95) that his friends "me l'ont fait imprimer." The *Mineralogy* is cited fully below (p. 293, n. 7 and p. 294, n. 9).

acidic and basic constituents was a salt and therefore was covered by Berzelius's theory of salts. The specificity of a substance composed of acids and bases was established by showing that the oxygen (or sulfur) contents of its constituents were related to one another as dictated by one of the laws of Berzelius's saline theory. The qualitative characterization was *necessary* to establish species, for it permitted the application of the saline theory, but it was not sufficient, for the complete determination of the specificity of a salt required stoichiometry. If the composition of a substance did not conform to stoichiometric law, then the substance was a mixture, not a species.

The mineralogical deployment of Berzelius's chemistry, explored in detail in this chapter, reveals these differing functions of qualitative characterization and quantitative law particularly clearly. Berzelius's recognition of the acidic role of silica, a largely unreactive substance, was a pillar of the mineralogical deployment in that it permitted assimilation of the numerous siliceous minerals to the category of salts by portraying them as complex silicates. Having demonstrated that most minerals were salts, Berzelius could determine whether the composition of minerals conformed with his stoichiometric laws by comparing the oxygen of the silica with that of the other constituents. Minerals composed in conformity with stoichiometric laws were specific compounds and could be established as distinct mineral species. Those that did not conform had to be considered mixtures. The identification of silica as electronegative, as an acid, allowed the application of Berzelius's stoichiometry as a fundamental criterion of specificity.

The central role that Berzelius accorded stoichiometry in gaining access to the specific level rendered his science antithetical to natural history. The conflict between Berzelian science with natural history is evident in the discussion portrayed below in which Berzelius argued with the Wernerian critics of his mineralogy about the role of chemistry in discrimination of species. Allying himself with Haüy, Berzelius claimed that, in mineralogy, stoichiometry defined species unequivocally without reference to the doctrine of transitions. His case was strong, but it was not immune to his Wernerian opponents, who responded to him much as Berthollet had responded to Haüy. Though Haüy had conceded some ground to his critics, having recourse to morphology in the discrimination of species, Berzelius attempted to meet his critics with new articulations of his chemistry, including his mature expression of the doctrine of dualism.

Dualism, however, was insufficient to insure the victory of stoichiometry over natural history. Berzelius's success in mineralogy was significant, but not total. He established the foundations for the assimilation of minerals to inorganic chemicals by showing that minerals could be treated in terms of his saline theory. Further progress in approaching the specific level depended upon the recognition that chemistry could not dominate mineralogy totally without recourse to crystal morphology. In particular, Mitscherlich's discovery of isomorphism in 1819 cleared the path to the specific level in both mineralogy and inorganic chemistry. Elaborating the discovery under Berzelius's direct tutelage and later under the aegis of the Swede's ideas, Mitscherlich and Rose—another of Berzelius's pupils—learned how to define species by invoking both chemical and crystallographic criteria.[8] This work showed that neither a purely chemical definition of species nor a fully chemical classification of inorganic matter was to prove possible.[9] Though controversy about this conclusion continued well into the nineteenth century, ultimately it became clear that the means to attain the specific level were at hand in a combination of Berzelian stoichiometry and morphological analysis.[1]

Berzelius hoped to invoke the same principles in organic chemistry that he used in his mineralogy, but his successes in organic chemistry, if no less important, were far more limited. As shown in this chapter, Berzelius was responsible for two innovations in organic chemistry. First, he argued that hitherto the focus of organic chemistry had been misplaced. Preoccupied with the categories of animal and vegetable substances that figured in traditional extractive analyses and taxonomies of organic products, chemists had failed to realize that there was a specific level in organic chemistry. Organic species, Berzelius claimed, were fundamental in the same way as inorganic species were fundamental and were defined in the same way as inorganic species were defined: as individual chemical compounds, composed in definite proportions. In organic chemistry, however, access to the specific level was far more difficult to attain than in mineralogy and inorganic chemistry.

Berzelius's second innovation was to identify the obstacles in the path to the specific level in organic chemistry and to account for them. In particular, he pointed out that first order organic compounds (that is, oxides of supposed compound radicals) were subject to fewer stoichiometric restrictions than in-

[8]This work was accomplished by accumulation of cases of isomorphic substitution, by delimitation of the possibilities of isomorphism and by analysis of cases of polymorphism. Mitscherlich's discovery thus inaugurated a tradition of "chemical crystallography," as Paul Groth points out in his *Entwicklungsgeschichte der mineralogischen Wissenschaften* (1906; rpt., 1968), pp. 126—132, esp. p. 129. For the role of Berzelius and his students in starting this tradition, see Berzelius, "Ueber die Veränderungen in dem chemischen Mineral-Systeme, eine nothwendige Folge der Eigenschaft isomorpher Körper, sich einander in unbestimmten Verhältnissen, zu ersezzen," *Leonard's Taschenbuch für die gesammte Mineralogie*, 1825,ii (1825), 169—182, 193—218. Also in connection with the elaboration of the principle of isomorphic substitution, see Ida Freund, *The study of chemical composition: an account of its method and historical development* (1904; rpt., New York: 1968), pp. 409ff.; and Franz von Kobell, *Geschichte der Mineralogie von 1650—1860*, Geschichte der Wissenschaften in Deutschland. Neuere Zeit, Vol. II (Munich: 1864), pp. 317—331. Groth's book is henceforth cited as Groth, *Mineralogische Wissenschaften;* and Kobell's, as Kobell, *Geschichte der Mineralogie.*

[9]For these points, see Groth, *Mineralogische Wissenschaften*, pp. 154—155.

[1]For the impact of Berzelius's chemistry on mineralogy, see Freund, *Study*, pp. 409—411; and Kobell, *Geschichte der Mineralogie, loc. cit.* This subject invites further study.

organic compounds. In the absence of elaborate law-like restraints on composition, numerous species were possible, and proximate ones that formed generic mixtures were insufficiently differentiated qualitatively to permit ready experimental resolution of the genus into its constituent species. That is, transitions occurred between proximate species, even though species were composed in definite proportions. In contrast, even complex minerals formed genera of proximate species that were distinct and separated by jumps in composition. Analysis, under the control of stoichiometric principles that delimited the possible patterns of combination, permitted discrimination of mineral species. However, in organic chemistry, where a precondition of elemental analysis was resolution of the genus, the experimental techniques were not at hand to accomplish this. Deploying precedents in Lavoisier's chemistry, Berzelius traced the fecundity characteristic of organic matter to the operation in the life functions of unknown causes that the chemist was unable to invoke or imitate. Organic chemistry was largely outside the reach of chemistry because it depended upon agents that, unlike those of mineralogy, were not comparable to the forces at the disposal of the laboratory chemist.

Berzelius's attitude toward his limited successes in organic chemistry reinforces the interpretation of his chemistry as a quantitative science of stoichiometrically defined specific substances, a science antithetical to natural history. Unable to achieve in organic chemistry the same level of stoichiometric control that he had attained in mineralogy, Berzelius claimed, as shown below, that organic chemistry would have to remain within the purview of natural history. Where stoichiometry could not prevail, natural history remained dominant. Though he accepted this defeat, believing that there would be no early progress in the resolution of generic mixtures of organic of compounds, Berzelius was anxious to minimize his losses. His concern derived from his perception that in the intractability of organic chemistry lay a potential threat to his otherwise quite successful quantitative chemistry of specific substances. He responded to this perception by articulating a sharp distinction between the organic and the inorganic, placing inorganic chemistry and mineralogy in the latter category and organic chemistry, along with the other two traditional branches of natural history, under the category of the organic. Because it traced the special problems of organic chemistry to physiological causes that defied imitation or conceptual analysis, Berzelius's distinction between the organic and the inorganic entailed the discrimination of natural history from science.

Though this pessimism was ultimately proven unjustified, Berzelius was not soon shown wrong. He persisted in his ideas about the boundary line between the organic and the inorganic through most of his career,[2] and his views in this

[2] The perseverence of Berzelius's vitalist position through successive editions of his textbook is examined by Bent Søren Jørgensen in his "Berzelius und die Lebenskraft," *Centaurus,* 10 (1964), 258—281; henceforth cited as Jørgensen, "Lebenskraft." This insightful study is usefully

connection became the standard doctrine of chemists in the first half of the nineteenth century.[3] The doctrine remained in view even as chemists increasingly proved able to solve the problem that had loomed so large to Berzelius, the isolation of specific compounds from generic mixtures, and built on this achievement the so-called older radical theory of the 1830's. The first advance in organic chemistry that exceeded Berzelius's expectations, the older radical theory was constructed fully in the spirit of Berzelian science. By making organic radicals, stable groups of elementary atoms, perform the role played by the elements in inorganic chemistry, the theory permitted discussion and investigation of species of organic matter by transferring to the radicals the still inaccessible core of organic chemistry. Even after the decline of the older radical theory, as chemists hoped to penetrate that core, practitioners in organic chemistry realized that they could not dispense with radicals. Much of the subsequent history of organic chemistry centered on debate about whether atoms or groups of atoms should have been accorded primacy. In this debate, the Berzelian legacy, of which Kolbe became the standard bearer, remained a powerful force.

b. The status of stoichiometry in the deployments

Berzelius's need to draw this sharp boundary line derived from the manner in which he invoked stoichiometry to specify compounds. Because he was establishing the specificity of compounds at the same time as he was constructing the stoichiometric doctrines that permitted these specifications, his laws depended upon his data just as much as his data depended upon his laws. Berzelius made every effort to place his laws and his analyses in mutually supportive positions. For example, he had used the law of the constancy of basic oxygen, developed in the first half of the *Versuch,* as a check on the direct analyses of the alkalies presented in the second half; he had confirmed the rule of oxides with analyses of a series of acids and their salts and then attempted both to sustain the rule and prove the compound nature of nitrogen by imposing it upon his analyses of nitrate salts. The variants of the rule of oxides for more complex compounds, including basic salts, occupied a similar position in Berzelius's chemistry, as did the rule of unitary oxygen for water of crystallization and double salts, despite its provisional nature.

This mutual dependence could be utilized in two ways, that is, by adopting either the laws as the arbiter of the analyses or the analyses as the arbiter of the laws. By adopting the laws as standards, Berzelius used them as a control upon

read in conjunction with this chapter. In this connection, cf. also Lindroth, "Berzelius," pp. 37—40, and Hugo Theorell, "Berzelius och livskraften," *Nordisk Medicin,* 60 (1958), 1537—1540.

[3] In this connection, see E. Benton, "Vitalism in nineteenth-century scientific thought: a typology and reassessment," *Stud. HPS,* 5 (1974), 17—48; henceforth cited as Benton, "Vitalism."

his analytical work. From his earliest researches in chemical proportions, Berzelius had pointed out the immense advantages accruing to the analytical art from the recent innovations in stoichiometry. Only those analytical results that agreed with stoichiometric principles could be accepted as accurate. In the first instance, the responsibility for nonconforming results must be laid not to the laws but to the analyst, to his skill and his techniques. For example, with this attitude, Berzelius attempted in the "Cause of proportions" to bring his results into conformity with the sulfur pattern. Though ultimately the analyses reflected upon the laws, forcing Berzelius to recognize other patterns, he permitted this to happen only after strenuously attempting to make the analyses conform to the laws.

There was another aspect to this controlling function of the laws. Though the reliability of the analytical data rested on their conformity with laws, the laws themselves effectively delimited the range of chemical phenomena and provided essential techniques for the conceptual and experimental management of the diverse content of chemistry. That is, the laws limited the number of possible species. The importance of this limitation increased with the complexity of the combinations. The smaller the differences in the percentage composition of similar complex combinations, the more difficult was the discrimination of particular compounds, species, from one another. By limiting possibilities, the laws gave the analyst a clearer choice among possibilities for the composition of definite compounds.

The mutual dependence of laws and analytical data could function in the other direction. The analytical results could determine the domain in which given laws were applicable or even whether given laws were viable. Consequently, Berzelius was frequently in the position of having to revise his laws to accommodate analyses that would not conform, no matter how carefully controlled. The emergence of the phosphorus pattern is a case in point, as is Berzelius's use of still more patterns in constructing the atomic weight tables of 1826.[4] It was in the light of the impact of the analytical results on the status of the laws that Berzelius felt the need to draw his sharp boundary between the organic and the inorganic. Though he found certain stoichiometric regularities in organic chemistry, they were insufficient to limit species in the same way as his laws of inorganic chemistry limited inorganic species. However, Berzelius believed that without a clear boundary marking the domain in which his inorganic laws applied, there was no reason to suppose that the lack of elaborate stoichiometric limitations on organic chemistry should not have reflected upon the viability of the inorganic laws.

[4]Cf. above, p. 196, n. 7.

2. Restrictions on inorganic composition: the unitary laws and the threat to stoichiometry

Berzelius enunciated his concerns about the need for restrictions on inorganic species in his comments on the Daltonian atomic theory, especially on the Daltonian rules of simplicity. In discussing Dalton's ideas, Berzelius asserted his analyst's conviction that the possible patterns of inorganic combinations were much more tightly limited than Dalton's rules suggested and that these limitations must be a central part of any theory of chemical combining proportions. Berzelius had two principal limiting laws in mind, which are called here the "law of the unitary atom" and the "law of unitary oxygen," the latter having been examined in part in the last chapter. The law of the unitary atom applied chiefly, but not exclusively, to first order compounds. It stated that in any compound atom, at least one of the elementary constituents must be represented by no more than a single atom. The law of unitary oxygen applied to higher order compounds. A special case of it was the rule of oxides, applicable to second order compounds.

The period from the *Versuch* to the *Proportions chimiques* witnessed changes in the status of the law of unitary oxygen and in its relation to the law of the unitary atom. Berzelius nowhere offered a concise statement that both laws, as he had originally formulated them, must simultaneously govern all inorganic combinations. He made a close approach to that sort of statement in the *Mineralogy*, but its impact was minimized. By translating the law of unitary oxygen into corpuscular terms, he nearly identified it in the *Mineralogy* with the law of the unitary atom. In an 1815 addition to the *Mineralogy*, moreover, Berzelius attained his mature expression of his dualistic doctrine, which later overshadowed the law of unitary oxygen, because it suggested another way to perform the functions originally intended for the law. In the 1815 addition, while discussing the role of natural-historical principles in mineralogy, Berzelius cited only the law of the unitary atom as the source of limitation on inorganic combinations, failing to mention the law of unitary oxygen. In the *Proportions chimiques*, the oxygen law, though noted, occupied a decidedly minor position.[5] The emergence of Berzelius's commitments to these overarching contraints before the *Mineralogy* is documented in the remainder of this section. The subsequent evolution of his beliefs about restrictions on combinations is pursued in other portions of this chapter.

[5] For the near identification of the two laws in the 1814 *Mineralogy*, see below, p. 308, n. 6 and accompanying text. For the oxygen law in the 1815 *Mineralogy* (cited fully below, p. 294, n. 9), see *Schweigger's Journal*, 15 (1815), 308–311. For the oxygen law in the *Proportions chimiques*, see Berzelius, *Essai sur la théorie des proportions chimiques et sur l'influence chimique de l'électricité* (Paris: 1819), p. 39; cited fully above, in the Introduction to Part II, p. 145, n. 1; henceforth cited as *Proportions chimiques*.

Most prominent among the earlier works in which Berzelius's convictions crystallized were the "Cause of proportions" and his "Address," written in reply to Dalton's "Remarks" on the "Cause of proportions."[6] Berzelius's ideas about restrictions on combinations were developed further in an article on definite proportions that he wrote, sometime late in 1816, for Brewster's *Edinburgh Encyclopedia;*[7] in several of his analytical papers; and in the *Mineralogy,* where the law of unitary oxygen was eclipsed by the law of the unitary atom. The discovery of the phosphorus pattern of oxidation and salt formation and the recognition of some exceptions to the law of the unitary atom compelled some moderation, visible in the *Proportions chimiques,* in Berzelius's insistence on the narrowest of restrictions. However, only after the *Proportions chimiques* did he acknowledge the occurence of additional patterns of combining proportions. Moreover, it was his commitment to the tightest constraints and the

[6]The first installment of the "Cause of proportions," which contained the bulk of the theoretical material, elicited a response from Dalton, read before the Manchester "Lit and Phil" on 24 December 1813 and subsequently published in the *Annals of Philosophy,* after the third installment of the "Cause of proportions" had appeared, as "Remarks on the essay of Dr. Berzelius on the cause of chemical proportions," *Ann. Phil.,* 3 (1814), 174—180; henceforth cited as Dalton, "Remarks." Berzelius replied to this with "An address to those chemists who wish to examine the laws of chemical proportions, and the theory of chemistry in general," *Ann. Phil.,* 5 (1815), 122—131; henceforth cited as "Address." A French—language draft of the "Address," in the form of a letter to Thomson, was published in Berzelius's correspondence (*Bref,* III,i, pp. 26—40). The draft and the published version differ. Citations are drawn from the published version, except where the draft is used to provide clarification. Aspects of this exchange between Berzelius and Dalton are examined by Anders Lundgren, in his *Berzelius och den kemiska atomteorin,* Acta Universitatis Upsaliensis. Studier i idé- och lärdomshistoria, Vol. II, utgivna av institutionen för idé- och lärdomsahistoria vid Uppsala Universitet (Uppsala: 1979), pp. 145—147; by Colin A. Russell, in his "Berzelius and the development of the atomic theory," in D.S.L. Cardwell, ed., *John Dalton and the progress of science: papers presented to a conference of historians of science held in Manchester, September 19—24, 1966, to mark the bicentenary of Dalton's birth* (Manchester: 1968), pp. 259—273; and by Partington, *History,* IV, pp. 161—162. Lundgren's book is henceforth cited as Lundgren, *Berzelius och atomteorin.*

[7]Solicited by David Brewster, editor of the encyclopedia, in a letter to Berzelius of 15 June 1816 (*Bref.* V,ix, pp. 145—146, esp. p. 145), under the title, "Proportions, determinate," in the *Edinburgh Encyclopedia, conducted by David Brewster,* XVII (1830), pp. 175—190; henceforth cited as "Prop. det." The published version of the covering letter that Berzelius sent to Brewster along with the article (*Bref,* III,ii, pp. 6—8), like many of the letters in the published correspondence, was an undated draft retained by its author, the original having been lost. Accordingly, the article must be dated by other evidence. In a letter to Gahn of 16 October 1816 (*Brief,* V,ix, pp. 145—146, esp. p. 145), Berzelius remarked thet he was then at work on an article on definite proportions that he had promised in the summer for the "Encyclopedia Britannica"—evidently a reference to Brewster's request—and said that upon completing it he would begin work on the fifth volume of the *Afhandlingar.* On 13 January 1817 (*Bref,* V,ix, pp. 158—159, esp. p. 159), he reported to Gahn that he had made a start on the *Afhandlingar.* Accordingly, the article was likely written during the final months of 1816. It can have been written no later than April, 1817, for by then Berzelius had found that the arsenic compounds follow the phosphorus pattern (letter to Marcet of 14 April 1817, *Bref,* I,iii, pp. 149—152, esp. p. 151), but the article states that only nitrogen and phosphorus presented anomalies to the rule of oxides.

underlying concerns motivating that commitment that guided Berzelius's pioneering work in both mineralogy and organic chemistry.

In the "Cause of proportions," Berzelius denied that compounds of the form 2A + 3B, as well as other, more complex arrangements could exist, even though they were permitted by the Daltonian rules of simplicity:

> It is contrary to sound logic to represent a single compound atom of the first order as composed of two or more atoms of A combined with 2 or more atoms of B; as for example, 2A + 2B, 2A + 3B, 7A + 7B, &c.: for in such a case there is no obstacle, either mechanical or chemical, to prevent such an atom from being divided, by means purely mechanical, into 2 or more atoms of more simple composition.[8]

In his "Remarks" on Berzelius's paper, Dalton tried to overcome Berzelius's objections by speculating how the complex arrangements that Berzelius rejected might have been held together.[9] In his reply to Dalton's paper, Berzelius explicitly declined to debate the plausibility of different speculative ontologies and accordingly offered no theoretical reasons to sustain his opinion.[1] He affirmed only that his accumulated experience permitted the elevation of his belief to the level of principle:

> My idea, that in every inorganic combination one of the elements enters as unity, is founded on the circumstance that in all inorganic bodies which I have analyzed, and I have analyzed a great number, I have found it to be so.[2]

That neither Dalton, in his "Remarks," nor Thomson, in correspondence with Berzelius, considered his objection unreasonable, suggests that it was consistent with contemporary notions of common sense.[3] Clues to Berzelius's thinking in this connection may be found in an elaboration of the point appearing in the article in Brewster's *Edinburgh Encyclopedia.* There, Berzelius remarked that

> On surveying the analyses hitherto made of inorganic substances, Berzelius thinks he has found that nearly all of them are compounded in such a way, that one of the elements occurs only in a single molecule; and from this fact he has drawn the conclusion, that such substances are characterized in their composition by the circumstance of one element, in comparison with the rest, always entering by a single molecule. Some exceptions to this rule were at first considered by him as proceeding from our inadequate acquaintance with the number of molecules in such bodies; but after a more comprehensive examination, he is of [the] opinion, that there exist combinations of two molecules of one element, with three of another; which, however, he inclines to believe do not occur between two elements alone, but require the presence of a third [intermediating] body to effect their union. Thus the subsulphate of oxide of copper [a basic cupric sulfate] is composed of $3CuO^2 + 2SO^3$,

8 "Cause of proportions," 447.

9 Dalton, "Remarks," 176. Dalton's paper reveals much about his otherwise scarcely articulated ontological commitments.

1 "Address," 123—124.

2 *Ibid.*, 124.

3 Dalton, "Remarks," 176. Thomson's letter is dated 18 August 1815, and is in *Bref,* III,i, pp. 40—42.

whilst no combination of 3Cu + 2S, is known to exist, or can exist, as Berzelius thinks; because nothing more than a force purely mechanical seems requisite to divide it into one molecule of S + Cu and one of S + 2Cu. *A priori,* it cannot be decided whether, in the compass of inorganic nature, molecules do not combine in still more complicated numbers; such as 3A + 4B, 3A + 5B, &c. &c.; but so far as actual experience allows us to decide, it is reasomable to suppose that no such combinations occur. On attempting to discover the cause why the molecules, in all inorganic substances, combine only according to proportions so limited, we arrive at certain boundaries, over which it is not permitted for the present to extend the empire of science.[4]

Invoking here his conception of intermediation, Berzelius implied that a compound such as 3Cu + 2S was a lower sulfide, which, like many lower oxides, was incapable of an independent existence. Accordingly, second order compounds (basic salts) exhibiting the 3 : 2 proportion of the radicals were possible because of the presence of the intermediating oxygen, but a compound of the unoxidized radicals was impossible. Moreover, because basic salts violated the principle of the preservation of combining proportions, there was no reason to infer the existence of a sulfide having the formula 3Cu + 2S from the existence of a basic salt containing the radicals in the same proportion.

For first order compounds, then, or primarily for these, Berzelius had a "law of the unitary atom:" that one ingredient must enter as a unit, a single atom. As shown below, Berzelius adhered to this rule tenaciously, especially in constructing the *Mineralogy* and in discussing the division between organic and inorganic chemistry. In the *Proportions chimiques,* he relented slightly in this connection, acknowledging the possibility of first order compounds in the proportion of 2A + 3B, but suggesting that such cases were merely apparent. He did acknowledge that some second order compounds containing the radicals in the proportion 2A + 3B existed, but claimed that they few. Moreover, he flatly denied that still more complex arrangements could occur (such as two atoms of one kind with three, four, five, and so on atoms of another).[5] Up to the *Proportions chimiques,* then, the only permissible pattern of first order combinations was A + B, A + 2B, A + 3B. . .[6]

As the preceding discussion suggests, Berzelius's arguments against the existence of complex first order compounds were marked by an easy passage from

[4] "Prop. det.," pp. 184—185.

[5] These points are presented in the *Proportions chimiques,* pp. 30—32. Berzelius's concession that some compounds of the form 2A + 3B may exist and his denial that more complex arrangements could exist are both precedented in the "P-Untersuchungen," pp. 51—53.

[6] Berzelius also pointed out two other shortcomings of Dalton. First, Berzelius claimed that Dalton's assumption of a preference for one-to-one combination among atoms was unsupportable. Second, Berzelius noted that Dalton's rule failed to point out a preference that *was* observed: almost invariably, it was the electronegative element that combined in multiple proportions with the electropositive one, the latter serving as the unitary atom. These points were presented prominently in the introduction to the *Proportions chimiques* (pp. xii—xv). An earlier statement that Dalton's suppositions were inadequate appeared in "Prop. det." pp. 180—181.

first to higher order composition. Connecting the levels in Berzelius's thinking was the principle of the preservation of combining proportions and the analysis of its violations.[7] Indeed, Berzelius seemed incapable of confining his arguments to their principal focus, first order compounds, finding it necessary to note exceptions, like the basic salts, among higher order compounds. The corresponding passages of the *Proportions chimiques* were labelled as applying "principally" to elementary atoms, but some higher order compounds were explicitly included in the discussion, and only known violations among higher order compounds were explicitly excluded.[8] The law of the unitary atom, then, stating that at least one constituent of an inorganic compound must be represented by no more than a single atom, applied to all inorganic combination, of both first and higher orders (apart from the few noted exceptions). This was not the sole sense in which Berzelius, in arguing against Dalton's approach, conceived higher order composition as subject to a unitary rule. There was also a need for a unit of oxygen.

Though Berzelius argued this point at first in connection with the rule of oxides, his discussion tacitly came to presuppose the more general law of unitary oxygen. This argument was developed largely in the same papers as were the arguments applying chiefly to first order compounds. In his seventh rule of simplicity, Dalton had merely asserted the applicability to combinations of compound atoms of the first six rules articulated for elementary atoms. However, this assumption permitted the existence of compounds not in conformity with the atomic theory, as Berzelius remarked in the "Cause of proportions:"

> I think I have proved that when two oxides combine they always unite in such proportions that each contains either an equal quantity of oxygen or the one contains a quantity

[7] This same focus on the proportions of the radicals was one rationale for Berzelius's much maligned "dot notation," suppressed, for the sake of typographical convenience, in the passage quoted above (p. 273 accompanied by n. 4) from "Prop. det." The notation indicated the number of atoms of oxygen combined with one atom of a radical by placing that number of dots above the symbol for the radical; (the notation was later extended to indicate sulfur atoms with commas, selenium atoms with dashes, and tellurium atoms with crosses). In dot formulae of higher order compounds, the most prominent information was the proportion of the radicals. For example, the subsulfate of copper oxide mentioned in the quotation was formulated as $3\dot{Cu} + 2\dddot{S}$.

The dot notation was originally introduced to shorten formulae, especially unwieldy mineralogical ones. In the 1815 addition to the *Mineralogy*, cited below, p. 294, in n. 9. Berzelius remarked (*Schweigger's Journal*, 15 [1815], 289): "Zum Ausdruck mannichfaltig zusammengesetzter Körper passt es besser, das Zeichen des Oxygens O auszuschliessen und mit Punkten über dem Radical anzudeuten, dass es oxydirt sey, und wie viel Volumina Sauerstoff es enthält." However, Berzelius indicated (289—290) that uncertainty about the composition of many oxides of which minerals were composed prohibited the widespread use of this notation in mineralogy, for which Berzelius had developed a separate system of notation (for which, see below, p. 305, n. 1). In chemistry, Berzelius made greater use of dot formulae, remarking in the *Proportions chimiques* (p. 112) that this sort of formula showed "... tout-à-la-fois le rapport entre l'acide et la base, entre les radicaux et entre l'oxigène des corps oxidés."

[8] *Proportions chimiques*, pp. 30—34.

which is a multiple by a whole number of the oxygen of the other. This law, though in itself conformable to the corpuscular theory, admits, on the one side, of combinations inconsistent with that theory; and, on the other hand, it excludes combinations perfectly conformable with that theory. I shall explain this by an example. Let O be oxygen, A and B two combustible bodies. The law which we are considering admits of a combination of A + 3O with 1 1/2 BO because $1\,1/2 \times 2 = 3$; and we shall see immediately that such combinations exist, though, according to the corpuscular theory, they appear absurd [because containing half-atoms]. On the other hand, the law does not admit the combination of A + 3O with B + 2O, though such a combination be conformable with the theory of atoms. The black oxide of copper is composed, according to our present knowledge, of 1 atom of metal and two atoms of oxygen, and sulphuric acid of 1 atom of sulphur and 3 atoms of oxygen. We know that there is a subsulphate of copper in which the acid and the oxide contain each equal quantities of oxygen. Of course, this subsulphate must contain for every atom of sulphuric acid an atom and a half of oxide of copper [$1\,1/2\ CuO^2 + SO^3$].[9]

If the second difficulty was a consequence of the law of the unitary atom, the first was a consequence of the rule of oxides. Dalton replied rather testily to this question, saying

I may observe that it is not the peculiar business of the atomic theory to explain why A + 3O do not unite with B + 2O, any more than to show why all the metallic oxides do not mutually combine with each other; for it may be said there is nothing apparent in the atomic theory to prevent such combinations.[1]

It was enough for Dalton that any combinations of oxides with one another that did exist were conformable with the seventh rule of simplicity. Which ones could or could not occur, and for what reasons, did not need to concern the atomic theory, which, as Dalton saw it, offered no limitations. Though Dalton did not deny the rule, he evidently did not think it imperative to uphold it. Berzelius replied in no uncertain terms, indicating the importance he attributed to the rule of oxides as a constraint upon inorganic combining proportions:

I have considered it as a great difficulty attached to the atomic theory that it does not explain the laws concerning the combinations of oxides with each other, namely, why the oxygen in one is always a multiple by a whole number of that in the other. Mr. Dalton removes this difficulty by saying, "It is not the peculiar business of the atomic theory to explain it any more than to show why all the metallic oxides do not mutually combine with each other." But we must recollect that the principal circumstance in favor of the atomic theory is, that it gives a mechanical and very satisfactory cause why elementary atoms unite only in proportions which are multiples of each other. We have observed likewise that the compound atoms, that is to say, most of those which contain oxygen, combine likewise in a *multiplex ratio,* provided we attend only to the oxygen which they contain. It is clear that this must be owing to a cause similar to that which occasions similar

[9] "Cause of proportions," 448; cf. above, p. 214 n. 9. Berzelius's failure to mention the possibility of eliminating the half-atom by doubling the coefficients of the two oxides was a consequence of his adherence to the law of the unitary atom. This is the same salt that he cited in "Prop. det." as a case of the occurence of the pattern 2A + 3B in higher order compounds by virtue of the intermediation of the oxygen.

[1] Dalton, "Remarks," 177—178.

proportions between elementary atoms; but the atomic theory throws no light on the subject. This must be admitted as something still wanting in the theory; for when a theory only explains one half of the phenomena which result from the principle which regulates these phenomena, it is surely incomplete. I differ, then, from Mr. Dalton, and must continue to consider the atomic theory as imperfect, and as clogged with difficulties, till it give us satisfactory explanations of all the phenomena relative to the chemical proportions.[2]

The rule of oxides, then, was a law of multiple proportions applied to the oxygen of the compound atoms, not to the elementary atoms. Seen in this light, the rule appears to have been a special case of the law of unitary oxygen, a status that Berzelius was increasingly inclined to accord the rule. As he pursued his discussion with Dalton, he tacitly moved from second order compounds governed by the rule of oxides to compounds of multiple oxide components governed by the law of unitary oxygen. Such compounds were typical of mineralogy, and, Berzelius suspected, of organic chemistry. It was the threat to stoichiometry posed by these fields that led Berzelius to insist on the unitary laws.

Berzelius commonly juxtaposed somewhat muted expression of his fears for stoichiometry with the sort of arguments against Dalton already recounted here. For example, the passage quoted above (p. 272, accompanied by n. 2), continued:

> Besides, if this [that one element of every inorganic combination is present as a unit] were not the case, it is evident that all traces of chemical proportions would disappear in combinations which consist of various oxides, *comme p. ex. des sels à plusieurs bases,* and could only be perceived in the most simple combinations.[3]

Later in the same paper, pursuant to his complaint that the Daltonian theory failed to consider limitations upon higher order composition, Berzelius assembled in a lengthy passage his reasons for urging further investigation of these limitations. The passage exhibits Berzelius's shift from speaking of the rule of oxides, as he had in the "Cause of proportions," to the more general rule of unitary oxygen:

> Chemical proportions depend upon two cardinal points 1. The proportions in which the elementary atoms unite. 2. The proportions in which the compound atoms combine. [This is Berzelius's discrimination between first and higher order composition.] The first of these is a necessary consequence of the atomic theory; and as the number of combinations which it includes is very limited, almost the whole of them have been examined, without finding a single exception to the law [of multiple proportions]. The case is very different with the second point. It includes an almost infinite number of compounds, varying in their elements as to the number of compound atoms which they contain. I have examined a great number of these substances, and as they are for the most part

[2] "Address," 125—126.

[3] *Ibid.*, 124. The phrase in French did not appear in the "Address" as published, but only in the draft (*Bref,* III,i, p. 28). Cf. the continuation of the passage from the "Cause of proportions" (above, p. 272, accompanied by n. 8): "Besides such a composition would almost totally destroy chemical proportions."

composed of oxides, I paid particular attention to the oxygen which they contain. The laws which I considered myself as entitled to establish from these experiments are well known.

The first cardinal point being established we must endeavor to establish or refute the second. If the accurancy of my experiments be admitted. I think their number is such that the second point also [that is, the restrictive laws governing higher order composition] may be considered as verified...

This law is of much more importance than may be supposed at first. It deserves therefore the most careful and impartial examination. Were it not for this law, no combination composed of several oxides could be calculated, nor any analysis verified in a decisive manner for the theory [of chemical proportions]. For it is clear, that if there be combinations expressible, for example, by $2AO^3 + 4BO^3 + 7CO + 5HO$; or if nature were to allow us to change the numbers in the formulas in any manner whatever, in such a case, all ideas of determinate proportions would disappear, in proportion as these combinations became more complicated. It is likewise by means of the law concerning the combination of oxides with each other, that we can hope to penetrate into the secrets of organic composition. When that law is demonstrated, the theory of chemical proportions may be considered as established.[4]

The law of unitary oxygen was thus more than a means to discriminate species. By serving as a limit or constraint on the proliferation of species, the law rendered the discrimination of species possible.[5] As Berzelius indicated in this passage, he perceived organic chemistry as the source of this threat to the power of stoichiometry. His concern was not merely that stoichiometry would prove powerless in

[4] "Address," 124. Passages similar to the one quoted here may be found in several of Berzelius's works. As Lundgren points out (*Berzelius och atomteorin*, p. 135, text and n. 4), the argument was precedented (without reference to either the corpuscular theory or the law of combining volumes) in Berzelius's "Om de bestämde proportioner hvari den oorganiska naturens beståndsdelar finnas förenade; summariskt resultat af de försök som häröfver blifvit anställde," *K.S.V.A. Handlingar*, [2] 32 (1811), 169—197, esp. 188n; (cf. German translation by Berzelius, cited fully above, p. 172, n. 6, in *Schweigger's Journal*, 2 (1811), 318n). The argument was made in terms of the law of combining volumes in "Cause of proportions," 450—451. In corpuscular terms, the argument was offered in juxtaposition with the need for the unitary law in the passage from "P-Untersuchungen," quoted below, p. 278, in n. 6; in the *Mineralogy*, in the passage quoted below, p. 319, accompanied by n. 4; and in "Prop. det.," pp. 184—185. Finally, the same points were developed in the *Proportions chimiques*, pp. 28—29.

[5] As the passage quoted in the next note illustrates, Berzelius frequently referred to his unitary laws as limiting laws. Moreover, he regarded the whole of his inorganic stoichiometry as a system of limitation on species. For example, after reviewing the main lines of his inorganic stoichiometry in the *Proportions chimiques*, Berzelius remarked (p. 40): "Nous venons de parcourir les lois découvertes jusqu'ici, suivant lesquelles les combinaisons des atomes tant simples que composés sont limitées dans la nature inorganique, et c'est dans la connaissance de ces lois que consiste la théorie des proportions chimiques." Thus, limitation of species and discrimination of species were nearly concident tasks.

As for organic chemistry, by the time of the *Proportions chimiques*, Berzelius had concluded that there were virtually no limiting laws (pp. 40—41). However, his fear that the diversity of organic combinations was a threat to chemical proportions was alleviated by the security that his inorganic laws acquired from the sharp boundary he established between the organic and the inorganic and from his demonstration that the rule of oxides applied to second order organic compounds; for the latter point, see below, p. 289, n. 9 and accompanying text; but see also the next note.

organic chemistry, but also that its potential failure to discriminate organic species threatened its viability as a discriminator of inorganic species as well.[6] In the absence of a sharp distinction between the organic and the inorganic, the domain in which Berzelius's inorganic laws applied was also not sharply delimited. His confidence in their applicability to inorganic cases was shaken by the existence of an ill-defined region where they did not hold. If they failed to limit species in one place, what was to prevent their failure in another? Accordingly, as Berzelius pursued his deployments in organic chemistry and mineralogy, he sought to render organic chemistry tractable by studying the stoichiometry of organic matter; he strove to keep mineralogy within the pale of inorganic chemistry by imposing upon it the tightest constraints; and he attempted to draw a sharp boundary between the organic and the inorganic, hoping to safeguard his inorganic stoichiometry.

3. Berzelius's organic chemistry

a. Berzelius's early organic analyses: the pinnacle of qualitative techniques

Berzelius's early work in the qualitative analysis of organic matter represents the highest level of skill attained in traditional organic analysis prior to the advent of stoichiometry.[7] During the earliest years of the nineteenth century,

[6] Even after Berzelius's "Organic proportions" had exhibited some stoichiometric regularities in organic chemistry, his fear about the implications of unlimited speciation of organic matter persisted. Thus, he observed in the "P-Untersuchungen" (pp. 50—52): "Giebt man zu, dass die Körper aus untheilbaren Moleculs zusammengesetzt sind, so folgt hieraus nicht, dass es in der Chemie nur solche bestimmte Proportionen geben müsse, wie wir deren in der unorganischen Natur aufgefunden haben; denn wenn die Molecüle die Eigenschaft hätten, sich in jeder beliebigen Zahl zu verbinden, so würden wir uns vergebens bemühen, die Spuren davon aufzusuchen. Wir finden aber, dass bei den organischen Substanzen der Verhältnisse, in denen sich ihre elementaren Molecüle verbinden können, und mithin dieser Verbindungen selbst, beinahe unzählige sind, während dieses in der unorganischen Natur nicht statt findet; woraus folgt, dass in dieser letztern, der grossen Anzahl von Körpern ungeachtet, die für jetzt als elementare betrachtet werden, und aus denen sie besteht, die Anzahl der Verbindungen doch sehr beschränkt ist. Sucht man die Ursache hiervon auf, so bietet sich sogleich der Umstand dar, dass die Elementar-Molecüle sich nicht in allen Verhältnissen verbinden lassen, d.h. dass eine unbestimmten Anzahl Molecüle eines gewissen Elements sich nicht mit einer unbestimmten Anzahl Molecüle eines oder andern Elemente vereinigen können. Es ist daher sehr natürlich, zu suchen, welches die Zahlen sind, in denen die Molecüle der verschiedenen Elemente (oder die verschiedenen chemischen Aequivalente) sich vereinigten finden. Sind sie 1A mit 1, 2, 3, 4, 8B? und welches ist die grösste Zahl von B, die sich mit einem einzigen A verbinden kann? Oder sind sie 2A + 3B, 3A + 4B, 5A + 6B, 9A + 10B . . . 99A + 100B etc.? Es ist klar, dass in der Reihe, in welche nur ein einziges Molecül von dem einen der Elemente eingeht, die Verbindungen die am meisten beschränkten sind, und zwar auf dieselbe Art beschränkt, wie wir es in der unorganischen Natur finden."

[7] Several of Berzelius's most prominent publications on qualitative organic analysis were cited early in this chapter (above, pp. 263—264, nn. 5—7). Other works are cited shortly. For Berzelius's qualitative organic analyses, see Jorpes, *Life and work*, pp. 34—35; and Jørgensen, "Lebenskraft," 265—268.

Berzelius conducted many analytical studies of animal and vegetable matter, using primarily extractive techniques. These studies were fully characteristic of comtemporary practice insofar as they consisted of separations of animal and vegetable constituents from samples of tissues and humors. Each stage in the analyses gave rise to substances that were generic mixtures, as Berzelius later pointed out, though the generic substances were often accompanied by clearly discrete compounds, especially mineral substances. From the generic substances, still more materials, those later recognized as generic as well as others considered distinct compounds, could be obtained in subsequent steps. This work differed from Berzelius's inorganic analyses, for in general there was no question of determining the proportions of the elements in an organic substance. The analyses were intended to determine only the qualitative characteristics and the relative, if imprecise, quantities of the components obtained by analysis. Moreover, as Berzelius later pointed out, a quantitative analysis had no significance unless performed upon a specific compound. This was a major lesson that Berzelius taught practitioners: a precondition of organic quantitative analysis was gaining access to the specific level.

Even before his stoichiometric achievements allowed him to formulate this claim, Berzelius manifested a full awareness of the possibilities for further resolution of generic products and a concern for differences among generally similar analytical products. In these respects, Berzelius's analyses were uncharacteristic of contemporary practice. His skills were manifest in the collection of analyses of animal fluids that Berzelius published in English in 1812 and in his study of icelandic moss, first published in 1808.[8] In the studies of animal fluids appeared Berzelius's demonstration that the coloring matter of blood did not derive from an iron salt, but that the iron was contained in the coloring matter in a fashion not understood.[9] Among the proofs was a series of chemical tests with aqueous inorganic reagents that would have detected the iron oxide as a saline constituents.[1] A similar procedure was also employed in the paper on icelandic

[8] The paper on icelandic moss is "Versuche über die Mischung des isländischen Mooses und seine Anwendung als Nährungsmittel," *Schweigger's Journal,* 7 (1813), 317—352; Swedish, "Undersökning af island-mossans sammansättning och dess användande såsom näringsmedel," *K.S.V.A. Economiska Annaler,* 5 (November, 1808), 53—94; henceforth cited from the German text as "Moos."

[9] In the *Föreläsningar* (I, pp. 130—132), Berzelius had presented an analysis of the coloring matter of blood that agreed with Fourcroy's claim (in his *Système de connaissances chimiques,* 11 vols. (Paris: An 9—10 [1801—1802]), IX, pp. 152—153), that iron was present in the form of a basic phosphate. However, Berzelius reported to Gahn in February, 1808 (*Bref,* V,ix, p. 25) that he had disproved this claim. He developed his argument in detail in the second volume of the *Föreläsningar* (pp. xxii—xxxix, lvi—lvii), where he showed that the observed phosphate was an artifact of the analysis and that the iron was present in some unknown form. This work constituted the point of departure for Berzelius's idea of intermediation. For this, cf. above, p. 176, n. 8 and accompanying text. In "Animal fluids," the study of iron in the coloring matter of blood occupies 24—26, 195—200.

[1] "Animal fluids," 197—198.

moss to exhibit its content of inorganic compounds as well as of its content of simple vegetable acids and their salts.[2] Supported by combustion analyses,[3] these tests enabled Berzelius to maximize the number of products obtained. In both cases, the method was generally to separate substances by precipitation with minimal alteration in the composition of the materials under analysis.

The paper on icelandic moss, which had attracted Berzelius's attention because of its medicinal virtues and its potential as a nutrient, reveals Berzelius's concern for differences among generally similar analytical products and his criticism of traditional characterizations for their failure to accommodate these peculiarities. One of his techniques was well-suited to support this criticism: performing analyses by alternate routes to provide an internal check on the results. For example, Berzelius chose two routes for the analysis of icelandic moss, each one giving an extract, that is, a substance belonging to the generally recognized category of extracts. That the two extracts displayed qualitative differences led Berzelius to criticize the traditional characterization of extracts as insufficiently discriminating:

> Die Chemiker haben die Eigenschaften des Extracts mit Bestimmtheit noch nicht angegeben; oder genauer, sie haben so viele ähnliche Körper unter den Namen Extractivstoff aufgestellt, dass man schwerlich einen bestimmten Begriff damit verbinden kann. Eine seiner gewöhnlichsten Eigenschaften ist, dass er sich beim Abdunsten an der Luft allmählig als ein schwer auflösliches braunes Pulver abgesondert. Eben diese Eigenschaft zeichnet auch den Stoff aus, den ich unter den Namen Extract aufgestellt habe. Aber die beiden Stoffe, die ich in den zwei Analysen auf gleiche Weise erhielt, waren von einander so wesentlich abweichend, dass man sie schwerlich in eine und dieselbe Reihe stellen kann.[4]

Berzelius went on to describe the very different properties of the two extracts, remarking,

> Da ich diese so unähnlichen Stoffe in beiden Analysen in gleicher Verhältnissmenge [in comparison with the total mass of the material analyzed] erhielt, so ist wohl kein Zweifel, dass beide von demselben Stoffe des Mooses herrühren, aber in beiden Fällen verändert durch den in der Analyse genommenen Gang.[5]

Extractive analysis therefore could not be assumed to be free of the difficulties presented by fire analysis. The products of analysis might have been preexistent as such in the material analyzed; but they also might have undergone some alteration during the analysis. Different extractive substances, even if obtained from the same vegetable, were sufficiently different to undermine the basis for the traditional category of extracts.

Even granting the utility of the generic categories, assignment of a substance to one of them must be made with care. Berzelius made this point by showing

[2] "Moos," 321—322.

[3] E.g., *ibid.*, 334—335.

[4] *Ibid.*, 335.

[5] *Ibid.*, 336.

that a gelatinous material obtained from the moss, previously obtained by Proust and Westring, was a form of starch.[6] Proust had reckoned it among the gums, and Westring had compared it to the gel of hartshorn. Berzelius showed first that its behavior was entirely distinct from that of animal gels and that its behavior with acids suggested a relation not with gums, but with starch.[7] Berzelius then investigated samples of the gelatinous material of sago and of ordinary starch and revealed a great similarity, though not absolute identity, in their behavior.[8]

> Diese Erfolge beweisen noch mehr, dass der gallertartige Stoff des Mooses eine besondere Abänderung des Stärkmehls ist, die ich zur Unterscheidung von den übrigen Abänderungen Moss-Stärkmehl (Lafstärkelse) nennen will.[9]

Qualitative categories could retain their utility, but they had to be employed with caution.

This quotation also reveals another sort of criticism that Berzelius levelled at qualitative organic analysis. As noted in Chapter III (p. 122, n. 1 and accompanying text), Fourcroy, while teaching chemists to classify the products of vegetable analysis according to chemical criteria rather than according to plant species of origin, lost sight of differences between similar products obtained from different plants. In contrast, Berzelius, while classifying vegetable matters chemically, retained a sense for specific differences among similar products of differing origins. Accordingly, he distinguished the starch in icelandic moss from other starches. Later, he could explain these differences in terms of one or two causes. Either similar products from different plants were generic mixtures, their specific properties dependent upon their particular mix of species; or they were distinct compounds, differing from one another as the result of slight differences in proportions.

In his use and elaboration of techniques for studying animal and vegetable matter, Berzelius displayed uncommon skill and critical caution, qualities that earned him the esteem of contemporaries. Later, in the light of stoichiometry, Berzelius translated his early concern with particular differences among qualitative categories into recognition of a specific level in organic chemistry, that of individual compounds in definite proportions, and he hoped to redirect the attention of contemporaries accordingly. The new goals that he set for himself and his fellow practitioners were to attain the specific level and to analyze specific compounds quantitatively. Berzelius elaborated his views of organic matter in the context provided by the prior novelties, described in Chapter III, that Lavoisier introduced into organic chemistry.[1]

[6] *Ibid.*, 336—342.
[7] *Ibid.*, 338—340.
[8] *Ibid.*, 341—342.
[9] *Ibid.*, 342.
[1] In the "Organic proportions" (330—331), Berzelius himself indentified Lavoisier as his sole conceptual predecessor, "who began to consider the composition of organic bodies in a proper point of view," and he dismissed "the experiments of the phlogistic chemists to decompose

b. Berzelius's return to Lavoisier's views of organic combination

Berzelius raised the question of the applicability to organic matter of the laws of inorganic combination as early as the "Spätere Zugabe" to the *Versuch*. Originally believing, like Lavoisier, that organic compounds such as the vegetable acids contained compound radicals, Berzelius tried to obtain the radicals by electrolysis. As already noted (Chapter IV, subsection 2.*g*.), the failure of this effort led him to suggest instead that organic compounds were binary unions of lower oxides of elementary radicals, each incapable of an independent existence, but each existing in combination by virtue of the intermediation of the other. However, the "Zugabe" had preceded the rule of oxides. With the establishment of the rule, Berzelius abandoned his binary theory of organic oxides and returned to Lavoisier's view that organic oxides were unions of compound radicals with oxygen.

The motive for this change appears to have been the expectation that these organic acids ought to conform to the rule of oxides no less than any other acids. Had Berzelius remained with the view of the "Zugabe," he would have had to consider salts of organic acids as "triple salts," in the language of Bergman, having three oxide components. Prior to his studies of water of crystallization, Berzelius did not have any law to cover such cases. Moreover, even if he did, the salts of the vegetable acids were evidently analogous with the simple salts of mineral acids, not with higher order compounds. Having offered the salts of the acids of carbon, phosphorus, arsenic, tungsten, and molybdenum, in the first *Fortsetzung* in documentation of the rule, Berzelius argued for enlarging the domain governed by the rule:

> Da alle Säuren, deren Zusammensetzung wir einigermaasen kennen, der angeführten Regel [of oxides] gehorchen, so können wir, glaube ich, mit Sicherheit die nämliche Berechnung anwenden, um die Zusammensetzung auch von den Säuren aufzufinden, welche wir durch directe Versuch nicht analysiren können. Es wird mir daher erlaubt sein, dieses Gesetz auf die Salzsäure und auf die Säuren mit zusammengesetztem Radikal auszu-

substances by naked distillation, in order to obtain their acid and empyreumatic spirits, their empyreumatic oils, their essential salts, gases, etc." as having "no tendency to decide the question under consideration." However, he also noted that Lavoisier's analytical technique was inadequate for determining the proportions of the elements, and later in the paper (401—403), Berzelius acknowledged his indebtedness to Gay-Lussac and Thenard for their analytical methods. They burned organic substances with hyperoxymuriate of potash (potassium chlorate), a substance that Berzelius adopted in place of the brown oxide of lead that he first tried. However, Berzelius objected to some features of their apparatus that, he believed, would introduce errors; to their determination of hydrogen by loss, "which in some cases may be owing to other unforeseen circumstances"; and to their failure to take into account what he considered to be basic water.

For experimental innovations in quantitative analysis of organic compounds in the early nineteenth century, see Partington, *History*, IV, pp. 233—239; and Ferenc Szabadváry, *History of analytical chemistry*, tr. Gyula Shvela, International series of monographs in analytical chemistry, Vol. XXVII (New York: 1966), pp. 288—290.

dehnen. In den letzten werden wir den Uebegang von der unorganischen zu der organischen Natur sehen, mit Modification der nämlichen Gesetze, welche die Natur in beiden befolgt.[2]

The substances with unknown simple or compound radicals to which Berzelius wished to extend the rule were just the sorts of compounds that Guyton and Lavoisier had included in the purview of their radical theory. Berzelius's calculation, using the rule of oxides, of the oxygen content of muriatic acid followed here immediately, and the view of vegetable acids as saline compounds of basic water with acidic oxides of compound radicals was presented in the third *Fortsetzung*. To uphold the universality of the rule and to exhibit a law common to both the organic and the inorganic spheres, Berzelius had returned to Lavoisier's view.

However, this approach also demanded abandonment of another analogy between organic and inorganic matter developed in the "Zugabe." If organic acids were oxides of compound organic radicals, they must have been composed differently from inorganic compounds. They were not binary unions of oxides of elements, but ternary and higher compounds of their elements. Berzelius understood that the radical theory of Guyton and Lavoisier entailed this distinction. Accordingly, he went on to offer in juxtaposition two different principles of composition, one for each of the two spheres. This pair of notions remained characteristic of Berzelius's science. At first, the inorganic principle was the rule of oxides and its corollary, the principle of the preservation of combining proportions:

In einer chemischen Verbindung von zwei oder mehreren oxydirten Körper (mag sie aus Säuren mit Säuren, oder aus Säuren mit Basen, oder aus Basen mit Basen bestehen) ist der Sauerstoff der in grösserer Menge gegenwärtigen Körper ein Multiplum nach einer ganzen Zahl (d.i. nach 1, 2, 3, 4, u.s.f.) von dem Sauerstoffe des in geringerer Menge gegenwärtigen Körpers; und in jeder chemischen Verbindung zwischen zwei brennbaren Körpern sind diese in solchem Verhältnisse vorhanden, dass, wenn die Verbindung oxygenirt wird, eine neue Zusammensetzung entsteht, welche nach dem eben angeführten Gesetze gebildet ist.[3]

Organic matter, Berzelius went on, must be conceived differently:

Um nicht unrichtig verstanden zu werden, stehe hier vorläufig aus den die organische Natur betreffenden Fortsetzungen meiner Versuche das *Prinzip für die Bildung der organischen Producte*. Es lautet: "In den organischen Producten sind zwei, drei oder mehrere brennbare Körper gemeinschaftlich vereinigt um eine Portion Sauerstoff, welche nur zur Oxygenation eines einzigen von ihnen hinreicht, und diese Zusammensetzung kann nicht in nährere Bestandtheile getrennt oder daraus zusammengesetzt werden."[4]

[2] *Versuch*, p. 121.
[3] *Ibid.*, pp. 125—126.
[4] *Ibid.*, p. 126. It may be asked how Berzelius could set the rule of oxides, as a description of inorganic composition, in opposition to another principle for organic matter, when he had just called for the application of the rule to salts of acids with compound radicals. As shown below, Berzelius himself resolved the contradiction in 1814 by stating that the distinctions between the

The unitary approach of Lavoisier is clearly manifest here, the speculations of the "Zugabe" about the binarism of organic compounds having been given up.

If the insufficiency of the oxygen in organic compounds to oxidize all the constituents to the degrees customarily observed in inorganic chemistry were no longer to be explained by a binarism of minimal oxidation stages, then some other pattern of combination must prevail. Berzelius went on to emphasize the point explicitly:

Diese Verbindungsart gehört der organischen Natur so ausschliessend, dass wenn dergleichen Verbindungen in der unorganischen Natur vorkommen, wir ihnen immer einen organischen Ursprung zuschreiben... Da die unorganische Natur eines Theils aus brennbaren Körpern ohne Sauerstoff, und andern Theils aus oxygenirten Körpern besteht, so hat in den letzteren ein jeder brennbarer Körper seine Portion Sauerstoff, die ihm folgt, wenn er von den andern abgesondert wird. Die organischen Producte sind aber, obgleich sie niemals ohne Sauerstoff existiren, doch alle verbrennlich, weil der Sauerstoff, wenn er gleich dem einen nicht mehr als dem anderen angehört, doch nur hinreicht einen einzigen von den brennbaren Körpern, aus denen sie zusammengesetzt sind, auf eine bestimmte Oxydationsstufe, die dazu noch sehr selten die höchsten ist, zu versetzen.[5]

The echoes of Lavoisier were stronger in the 1811 "Nomenclature." There, Berzelius pointed to the differences in combination distinguishing organic from inorganic matter and bolstered the distinction with Lavoisierian views of the ternary and higher composition of organic substances. Of inorganic bodies Berzelius observed,

Ils sont composés, a) des corps combustibles avec l'oxigène; b) des corps combustibles deux, trois ou plusieurs ensemble sans oxigène; c) des corps combustibles combinés avec des corps oxidés; d) des oxides combinés entre eux, de telle manière, que chacun peut être séparé des autres en conservant toutefois la portion d'oxigène qui lui appartient. Toutes ces substances se combinent en des proportions déterminées, d'après des lois invariables, que j'ai tâché d'exposer dans les différentes continuations [the *Fortsetzungen*] de mes expériences sur ces mêmes proportions déterminées.[6]

In organic products, on the other hand, the oxygen was shared by the combustibles:

organic and the inorganic lay only on the level of the first order and that the rule of oxides, a statement about second order compounds, applied to both the organic and the inorganic. That the contradiction was not immediately evident to him may be explained by recalling that the rule, having emerged from Berzelius's experience with inorganic matter, not only embodied an explicit statement about the proportions of oxides in second order compounds, but implicitly represented Berzelius's accumulated experience with *inorganic* oxides. In the first *Fortsetzung,* seeking to articulate the distinctions between the organic and the inorganic, he readily adduced the rule as the principle that expressed his understanding of inorganic chemistry. Later, after having recognized the contradiction, Berzelius offered a different statement of the peculiarities of inorganic composition. For this, see below, pp. 288—289, nn. 5—7 and accompanying text.

[5] *Versuch,* pp. 126—127.

[6] Berzelius, "Essai sur la nomenclature chimique," *J. de physique,* 73 (1811), 253—286, esp. 260; henceforth cited as "Nomenclature."

Ceux-ci sont composés de deux ou quelquefois de plusieurs corps combustibles combinés avec une quantité d'oxygène qui leur est commune et qui ordinairement ne suffiroit qu'à oxider un seul de ces corps combustibles. Leurs parties constituantes ne peuvent pas être séparées sans qu'ils se combinent de plusieurs manières nouvelles, en produisant des combinaisons binaires, et en partageant l'oxigène entre elles d'après des affinités très-compliquées.[7]

The notion of complex affinities and the primacy accorded decomposition into simple inorganic products, both characteristic of Lavoisier's treatment, are in evidence here. Also characteristic of Lavoisier was Berzelius's belief in the impossibility of organic synthesis:

Le principale différence entre ces deux classes consiste dans ce que la première ne contient que des combinaisons binaires, seules ou combinées de différentes manières plusieurs ensemble, tandis que la dernière contient toujours des combinaisons ternaires, quaternaires, etc. Les corps composés d'après le principe des productions organiques, ne peuvent être produits que très-rarement de substances purement inorganiques par le moyen de la Chimie. Leur production est réservée à la nature organique, et paroît être le but chimique de l'organisation, où l'intervention du système nerveux, qui préside aux procédés organochimiques, et qui détermine la nature et les qualités des produits, nous dérobe l'intérieur de ses opérations. La condition principale de la formation organique paroît être une modification électrochimique dans les élémens, qui diffère de celle qui leur appartient originairement dans la nature inorganique. Dès qu'ils sont éloignés du corps organique qui les a produits, ils tendent toujours à regagner cette modification électrochimique originaire, et c'est par l'effet de cette tendence qu'ils fermentent et qu'ils se décomposent en contact avec l'air et l'eau, et quand on les expose à une température élevée.[8]

Berzelius had thus begun the separation of the organic from the inorganic on which he later insisted in stoichiometric terms. However, as of the "Nomenclature," Berzelius had little to offer in elaboration of his views. Though he stated expressly that organic matter must be governed by the law of definite proportions, he admitted that the details of organic stoichiometry remained unknown.[9] The only certainty was the conviction, derived from his early, qualitative organic analyses, that the diversity of organic matter would present difficulties for chemistry. Accordingly, one belief that he did not share with Lavoisier was his predecessor's complacent expectation that the nomenclature of organic compounds might be made to rest on composition in a fashion analogous with inorganic nomenclature. The only recourse was natural history:

[7] *Ibid.*, 260.

[8] *Ibid.* For Berzelius's views on the role of the nervous system in the life processes, see Jørgensen, "Lebenskraft," 267, 269—271.

[9] The passage from the "Nomenclature" quoted on this page, accompanied by n. 7, continued (260) to say that "Les élémens qui constituent les corps organiques obéissent à la même loi générale qui préside à la formation des combinaisons non organiques; mais les variations innombrables qu'admet cette loi dans la nature organique, ne sont pas encore assez clairement déterminées."

J'ai parlé de la manière dont les produits organiques sont composés; comme ils ont presque tous les mêmes parties constitutives, ne différant que dans les proportions, ils ne peuvent pas être nommés d'après le même principe que les corps inorganiques. Dans mes analyses de quelques végétaux, par exemple, de la mousse d'islande et d'autres espèces de lichen qui produisent de l'amidon, du quinquina et de l'écorce intérieure du pinus sylvestris, j'ai tâché de prouver que les substances végétales, que nous considérons comme des parties constitutives générales des végétaux, varient dans leur nature, autant que les plantes d'où elles sont tirées. J'ai prouvé, par exemple, que ce que nous appelons *tannin,* quoiqu'ayant des propriétés chimiques générales, diffère d'une manière très-sensible d'après le plante qui l'a produit, et que le tannin de galles, du cachou de l'uva urse, de l'écorce de bouleau, etc., sont autant d'espèces différentes de tannin. Il en est de même des huiles volatiles, des huiles grasses, des résines, de l'amidon, etc.

Il faut donc ranger ces substances à la manière du système botanique, en formant des *genera* et des *species,* et en conservant à chaque genre le nom qu'il a eu jusqu'ici.[1]

Having been compelled by his own studies to acknowledge that the products of traditional vegetable analysis differed from species to species, Berzelius had to invoke natural-historical principles. The precedent provided by Fourcroy was inadequate, however, for as shown in Chapter III, it ignored specific differences among comparable "immediate materials" obtained from different species. Accordingly, Berzelius turned to his contemporary in botany, Göran Wahlenberg (1780—1851), who provided a (medically oriented) taxonomy of extracts for Berzelius's "Nomenclature" that made explicit the specific differences among such products as gums, sugars, and starches.[2]

Even after Berzelius had begun to develop techniques of quantitative organic analysis during the winter of 1813—1814, he used his analytical results to intensify rather than diminish the distinction between the organic and the inorganic. In inorganic chemistry, composition varied through a small number of definite proportions in integral multiples, with large changes in properties; but in organic chemistry, especially within generic categories, composition varied only subtly in proportions with small changes in properties. Inorganic chemistry exhibited variation by saltation; organic chemistry, variation by transition. Accordingly, organic chemistry remained within the purview of natural history.

c. Quantitative organic chemistry

The results of the earliest of Berzelius's quantitative organic analyses were reported in the third *Fortsetzung* (1812) of the *Versuch* in exemplification of the idea of basic water, as noted in Chapter V, subsection 6.*a.* Berzelius used the example of three organic acids to bolster his claim that many acids consisted ot two oxides, an oxidized organic radical, the acidic component, united with

[1] "Nomenclature," 270.

[2] *Ibid.* On Wahlenberg, see Gunnar Eriksson, "Wahlenberg, Göran (Georg)," *DSB,* XIV, pp. 116—17; and E. Wikström, "Biografi öfver Georg Wahlenberg," *K.S.V.A. Handlingar,* [3] 39 (1851), 431—505.

water as the basic component. Two subsequent developments induced Berzelius to pursue his quantitative organic analyses more intensively: his encounter with Dalton's rules of simplicity in 1812 and growth of his mineralogical interests in 1814. These concerns inspired the program described above at the conclusion of subsection 2: separating the organic and the inorganic, retaining mineralogy within the inorganic sphere, and finding the laws of proportions governing organic matter. The mineralogical part of this program is examined further below. Here, the objective is to exhibit the organic side.

Berzelius constructed his theory of organic combining proportions at the same time as he drew the boundary between organic and inorganic matter. The stoichiometric doctrine of the distinctions between the two spheres was developed in detail in the "Organic proportions," was briefly atticulated in the *Proportions chimiques,* and was fully elaborated in the next edition of Berzelius's textbook.[3] The outcome of this theoretical development was to realize the implications of Berzelius's return to Lavoisier's views of organic matter. In particular, Berzelius recognized that the chief area of difference between the organic and the inorganic lay in compounds of the first order, and he provided detailed elaboration of this point.

The importance of the first order to the distinction between realms was adumbrated in the "Cause of proportions," where Berzelius introduced the vocabulary of "orders." Appearing before Berzelius attained the mature expression of his dualism, he stopped at the second order:

> We may divide the atoms [of bodies] into two classes: 1. Elementary atoms: 2. Compound atoms. The compound atoms are of three different species, namely, 1. Atoms formed of two elementary substances united. We shall call them *compound atoms of the first order.* 2. Atoms composed of more than two elementary substances: and as these are only found in organic bodies, or bodies obtained by the destruction of such organic matter, we shall call them *organic atoms.* 3. Atoms formed by the union of two or more compound atoms, as, for example, the *salts.* We shall call them *compound atoms of the second order.*[4]

[3]In the *Proportions chimiques,* the discussion occupied pp. 40—47. It was chiefly from the fuller discussion of his ideas on organic composition in subsequent editions of his textbook that contemporaries learned Berzelius's views, and it was through successive editions of the textbook that Jørgensen ("Lebenskraft") traces the persistence of Berzelius's vitalistic beliefs. Accordingly, some citations below are drawn from the most widely known version of the next edition of the text after the *Proportions chimiques,* itself a version (as noted above, p. 145, n. 1) of the third volume of the first edition of the Berzelius's *Lärbok i Kemi:* Berzelius, *Lehrbuch der Chemie, nach des Verfassers schwedischer Bearbeitung der Blöde-Palmstedt'schen Auflage übersetzt von F. Wöhler,* 4 vols. in 8 (Dresden: 1825—1831); henceforth cited as *Lehrbuch.* This was a descendant of the second Swedish edition of the *Lärbok.* For bibliographic details, see Partington, *History,* IV, p. 145. The discussion of the general characteristics of organic matter, of the distinctions between organic and inorganic matter, and the means for analysis of organic matter appears in *Lehrbuch,* III,i, pp. 135—175.

[4]"Cause of proportions," 446; by contrast, in the *Proportions chimiques* (pp. 26—28), Berzelius cited examples of compounds as high as the fourth order, and he added some obser-

In the paper on organic proportions, Berzelius elaborated his pair of principles of combination, one inorganic and the other organic, in terms of first order atoms. There he combined two previous ideas, the law of the unitary atom and the statement in the "Nomenclature" that all inorganic combinations were binary.[5] Taken together, these ideas permitted a narrow definition of inorganic chemistry, one that could replace the rule of oxides, which, in the first *Fortsetzung,* had served that purpose:

> The two circumstances, that, first, the compounds of the first order are always binary, and that, secondly, in all these combinations at least one of the constituents enters only in the quantity of a single atom or volume, constitute the exclusive characters of inorganic nature, and form what I call the principle of inorganic formation.[6]

Lack of a chemical unit in chemical combination distinguished organic chemistry

vations on the limits of the hierarchy: "Nous les [compound atoms] diviserons en atomes du premier, du second, du troisième ordre, etc. Ceux du premier ordre sont composés d'atomes simples élémentaires; ils sont de deux espèces[,] organiques et inorganiques. Ceux-ci ne contiennent jamais que deux éléments; les autres en contiennent toujours au moins trois. Les atomes composés, du second ordre, naissant des atomes composés du premier ordre, les atomes du troisième, de ceux du second, etc. Par exemple, l'acide sulfurique, la potasse, l'alumine et l'eau, sont tous des atomes composés, du premier ordre, parce qu'ils ne contiennent que le radical et l'oxigène; le sulfate de potasse et le sulfate d'alumine sont des atomes composés, du second ordre; l'alun sec, qui est une combinaison de ces deux derniers sels, offre un exemple d'un atome du troisième ordre; et enfin, l'alun cristallisé, contenant plusieurs atomes d'eau, combinés avec un atome de sulfate double, peut être cité comme un exemple d'atomes composés, du quatrième ordre. On ne sait pas encore jusqu'à quel nombre les ordres peuvent s'élever. L'affinité entre les atomes composés, décroît d'une manière bien rapide, à mesure que le nombre des ordres augmente, et le degré d'affinité qui existe encore dans les atomes du troisième ordre, est le plus souvent trop faible pour pouvoir être aperçu dans les opérations promptes et troublées de nos laboratoires. Cette affinité ne se manifeste pour l'ordinaire que dans les combinaisons qui se sont formées pendant que le globe passait lentement et tranquillement à l'état solide, c'est-à-dire dans les minéraux. Pour bien connaître leur nature, il serait important de savoir jusqu'où peut aller la combinaison des atomes composés, et quel est le dernier ordre. Quant aux atomes composés organiques, on ignore également en combien d'ordres différents ils peuvent se combiner, soit entre eux, soit avec des atomes composés inorganiques."

Intervening between the statement in the "Cause of proportions" and that of the *Proportions chimiques* were the prototype of hierarchical dualism in the mineralogy, discussed below, in subsection 5.*b.* and reformulations in the "Organic proportions" of the opposing principles of organic and inorganic combination. The reformulations are exhibited immediately below in the text and notes of this subsection.

[5] In the passage quoted above, p. 285, accompanied by n. 8. The point was reiterated in "Organic proportions" (324) (but was now informed by the mature expression of Berzelius's dualism attained in his *Mineralogy*): "In inorganic nature compound bidoes of the first order* never contain more than two elements; so that we may say that inorganic nature never contains more than binary combinations, and bodies composed of binary combinations.

"All inorganic bodies in which we find more than two elements are evidently composed of binary combinations of these elements, which may be separated from each other without decomposition, and generally they may be united again so as to form the compound substance anew." The note signalled by the asterisk explained the vocabulary of orders, proceeding up to the third order (though it referred the reader to the "Cause of proportions," where Berzelius had proceeded only to the second order).

[6] "Organic proportions," 325.

from inorganic. Recalling his abandonment of his earlier binary, saline view of organic compounds and his return to Lavoisier's unitary conception, Berzelius observed that organic compounds (of the first order) must not be binary, but ternary and higher combinations of the elements; "but it is not necessary that any of these elements should be considered as unity."[7]

> The exclusive characters, then, of organic bodies are, that those of the first order contain more than two elements, none of which must of necessity be considered as unity. This is what I mean by the principle of organic formation.[8]

However, Berzelius claimed that the rule of oxides applied equally well to both realms. The difference between them was confined to the first order: inorganic combination was tightly limited by restrictive laws but organic chemicals needed only to conform to multiple proportions.[9]

[7] *Ibid.* This statement was italicized. It was part of the passage quoted on this page, in n. 9.

[8] *Ibid.* Cf. *Proportions chimiques,* p. 42: "On peut . . . admettre comme le principe fondamental de la formation organique, que les atomes composés, du premier ordre, contiennent au moins trois éléments (l'oxigène, l'hydrogène et le carbone), et que leurs atomes peuvent se combiner dans toutes les proportions, sans que l'un d'eux y joue nécessairement le rôle de l'unité; circonstance qui caractérise le plus grand nombre des substances inorganiques."

[9] "Organic proportions," 325: "The laws of proportions in organic nature may be comprehended under the two following general rules:—

"1. When three or more elementary bodies, of which oxygen is always one, combine so as to produce a ternary, quaternary, &c. oxide, a certain number of atoms or volumes of one of the elements combines with a certain number of atoms or volumes of the others; but it is not necessary that any one of these elements should be considered as unity.

"2. When these oxides combine with each other or with binary oxides, the oxygen in the one is always a multiple by a whole number of that in the other.

"This last rule is common both to organic and inorganic bodies . . ." The rule of oxides was the sole limiting law that Berzelius found for organic combinations. He placed great emphasis on the applicability of the rule of oxides to organic matter, apparently because it alleviated his fear that, in the absence of limiting laws, there would have been no determinate proportions. For example, Berzelius remarked on the first of the two rules quoted in this note ("Organic proportions," 325—326) that it "would seem at first sight to indicate that determinate proportions do not exist in organic combinations, for that rule admits of combinations in all proportions, but this by no mean excludes determinate proportions, because the ternary, &c. oxides are subjected to the same law of combination [the rule of oxides] as the binary; and this must be ascribed to the same cause in organic and inorganic bodies; namely to the existence of determinate proportions in both kingdoms of nature."

The applicability of the rule of oxides to organic matter served as the foundation of Berzelius's quantitative organic analyses. He recalled ("Organic proportions," 326) the sort of argument used in the "Cause of proportions" by which the oxygen of an acidic oxide could be determined by analyzing its neutral and basic salts and by determining the multipliers in the rule of oxides. He then went on to explain (326—327) that the oxygen of an organic oxide could be determined in the same way, by analyzing its combinations with a metallic base, such as lead oxide. By combustion analysis (for which, see Partington, *History,* IV, pp. 235—238), the relative proportions of the carbon and hydrogen could be determined. Comparison of these figures with the quantity of oxygen established the absolute quantities of the other two elements.

This application of the rule of oxides to organic matter was expressed in the *Proportions chimiques* (p. 45) as the third of three general observations about organic matter: "Lorsque des atomes organiques du premier ordre se combinent avec des atomes composés inorganiques du premier ordre, ils suivent (d'après notre ex-

Berzelius saw foundations for this distinction in a protovitalist modification of his electrochemical theory, already adumbrated in the 1811 "Nomenclature."

> There is still another essential difference between organic and inorganic nature. This essential difference consists in the electrochemical modification of the organic products, which does not appear to depend immediately on that which the elements have in [in]organic nature; that is to say, on the original electrical modifications of the elementary substances. We cannot even form conjectures respecting the manner in which nature modified the electrical properties of the elements in the economy of living bodies; nor of the means which she employs to combine the elements in other proportions, and in other numbers, than in inorganic bodies.[1]

périence actuelle) les mêmes lois que les atomes composés inorganiques entre eux; de manière que l'oxigène de l'atome organique est un multiple, ou moins souvent un sous-multiple par un nombre entier, de l'oxigène de l'atome composé inorganique. Nous ignorons s'il existe aussi, dans la nature organique, des exceptions analogues à celles des acides de l'azote, du phosphore et de l'arsenic. En tout cas, l'expérience a prouvé que les combinaisons des atomes organiques avec d'autres atomes composés, sont limitées [again, the rule of oxides "saved" organic proportions] par les même lois auxquelles obéissent les atomes composés inorganiques, et donnent par conséquent naissance aux mêmes phénomènes de proportions fixes que ceux de la nature inorganique; en sorte que ce n'est que dans la formation des atomes organiques du premier ordre, que les combinaisons sont possibles dans presque toutes les proportions."

[1] "Organic proportions," 328. The relevant passage in the "Nomenclature" is accompanied by n. 8, p. 285. Cf. *Proportions chimiques,* pp. 95—96: "J'ai dit que la structure des atomes organiques est tout-à-fait différente de celle des atomes inorganiques. Leurs rapports électrochimiques diffèrent aussi essentiellement. Dans la nature inorganique, les éléments ont une certaine polarisation spécifique [a term drawn from Berzelius's microscopic electrochemical ontology; it was the label for the specific electrochemical endowment of a radical] qui est invariable; par exemple, le soufre est toujours un corps fortement électro-négatif, et l'hydrogène est tellement sur la limite entre le négatif et le positif, que l'on ne peut déterminer de quel côté il doit être placé. Or, ces substances conservent invariablement leur nature électrique dans toutes les combinaisons inorganiques; mais, dans la nature organique, les éléments diffèrent tout-à-fait sous ce rapport. Là, tantôt les oxides ternaires du carbone et de l'hydrogène ont une polarité négative qui rivalise avec celle des oxides binaires les plus négatifs, comme, par exemple, les acides acétique, tartarique, oxalique; tantôt elle est si faible qu'elle est à peine sensible, comme dans le sucre et dans l'amidon; et dans ces deux cas, elle ne dépend ni du nombre relatif des particules, ni de la quantité d'oxigène."

This last point, that the properties of organic compounds could not be deduced from their composition, did more than illustrate the doctrine that organic matter owed its peculiarities to a peculiar electrochemical modification. It also suggested that the lack of relation between composition and properties was a character of organic compounds, one making a chemical science of organic matter that much more difficult, as noted below in subsection 5.*c*. Berzelius had developed this idea earlier, in the "Organic proportions," where he remarked (329) that because the life processes had modified the electrochemical properties of the elements, "Consequently it is impossible to determine in organic bodies from the elements and proportions of which they are composed whether they be acid or not. Benzoic acid, for example, which only contains the fifth part of its weight of oxygen, is an acid; while gum arabic and sugar, which contain nearly half their weight of oxygen, are not acids. I have not hitherto been able to perceive that either the number of volumes, or their relation to each other, determines anything respecting the electrochemical properties of the ternary oxides; for the saclactic acid, for example, which contains eight volumes of oxygen, is a weaker acid than the citric acid, which contains only a single volume, while the oxalic acid, which contains 18 volumes is stronger than either of the two just mentioned. I shall show hereafter that the ratio between the oxygen in the ternary oxides, from which Gay-Lussac and Thenard have endeavoured to derive the properties which characterize these substances,

He attributed the ability to produce these substances to the vital processes of plants and animals. Organic matter, though inanimate, was organic for Berzelius, because it was produced by the life processes. In its application to matter, the distinction between the organic and the inorganic deriving from life function assumed paramount importance.

This view had several implications. Artificial synthesis was likely impossible, a point Berzelius made in the opening paragraph of the paper.[2] Moreover, the complexity of organic compounds made them weak, and their dependence on the life processes for their peculiar electrochemical characteristics made them vulnerable:

> In consequence of the great number of elementary volumes which enters into the composition of these ternary and quaternary oxides, and from the peculiar and special modification of the electro-chemical properties of these elements, organic bodies in general constitute but feeble compounds, which often begin to undergo decomposition as soon as they escape from the influence of the organ in which they were produced. Almost all organic bodies are decomposed by the united influence of air, water and heat. Their elements resume their ordinary electrochemical modifications, and there finally results a number of binary or inorganic combinations; but this seldom takes place all at once. They generally pass through a series of ternary oxides, the composition of which is varied. Thus sugar forms first alcohol then vinegar, then a mucous matter, &c. before its organic composition is completely destroyed.[3]

determines in fact nothing whatever. It appears, then, that at the instant of formation of each ternary and quaternary oxide in organic nature, its elements receive in combining a new electro-chemical modification, on which their chemical properties chiefly depend. Though we must acknowledge that the proportion of the elements ought not to be without some influence." Cf. also "Cause of proportions," 129, and *Lehrbuch*, III,i, p. 142.

[2] Berzelius, "Organic proportions," 323.

[3] *Ibid.*, 329. Cf. the continuation of the passage from the *Proportions chimiques* quoted in the last note (p. 96): "La nature organique a sa manière particulière de produire des oxides de radicaux composés, et de donner à leurs principes constituants une polarité électrique tout-à-fait indépendante et différente de celle qui leur appartient orginairement dans la nature inorganique, et que la plupart ne conservent que sous l'influence organique, mais que tous quittent tôt ou tard pour reprendre celle qui leur est propre dans la nature inorganique. De là proviennent, dans tous les produits organiques, les phénomènes de destruction que nous appelons fermentation et corruption, au moyen desquels les éléments reprennent peu-à-peu leurs caractères électro-chimiques originaires, et la nature organique restitute sans cesse à l'autre ce qu'elle lui a emprunté."

In the "Organic proportions," 262, he gave this elaboration of the phenomena of organic decomposition: "There is still a circumstance relative to the combination of the ternary oxides with the saline bases, which deserves to be mentioned here. An excess of these bases, especially the stronger ones, decomposes a great part of the ternary oxides in different ways, usually producing a great quantity of carbonic acid. The ternary oxides least exposed to this decomposition are these: 1. whose atoms of oxygen are equal in number to those of some one of the other elements, or surpass them in number; and 2. when the ratio of the hydrogen to the oxygen is less than in water. Such is the case with most of the acids [here] analyzed..."

"This is the reason why, when the ternary oxides are gradually decomposed, either by the influence of air and water, or of acids, their atoms arrange themselves in such a manner as to form new products, in which, on one side, the oxygen is in excess, and, on the other, the atoms of the combustible elements greatly surpass those of oxygen. Thus sugar is converted into

Weak affinities, ready decomposition through stages into inorganic products, and the impossibility of synthesis—standard features of this conception of organic matter—had their basis in the electrochemical modification and complex composition of the constituents of organic bodies.

The same considerations allowed Berzelius to rationalize the formation of generic substances as well as the difficulty of separating their constituent specific substances. Having first analyzed the vegetable acids, all of them individual compounds, Berzelius later remarked that

> Before proceeding further it will be necessary to make some observations on the substances which have been analyzed. They are all composed of a small number of volumes, compared with those that are to follow; so that the addition of a single volume or atom of one of the elements would produce a great change in the composition. This is the reason why the greater number of vegetable acids do not constitute genera of analogous substances having the same general character, but differing in some points, so that they must be considered as different *species* [Berzelius's emphasis] of the same genus. There is only one acetic acid, one gallic acid, and, wherever these bodies occur in the vegetable kingdom they constitute absolutely the same substance.
>
> As the number of volumes increases in the ternary oxides, they acquire the property of producing a certain number of varieties, or of different species of the same genus, which appear to owe their origin to the adhesion of one or more [additional] volumes of one of the elements. This addition, however, in consequence of the great number of volumes of the element contained in the substance, only produced an insensible alteration in its composition when we consider 100 parts only.[4]

Organic matter displayed generic categories of extremely similar substances because a change in the composition of organic compounds produced a small change in their properties. Organic matter therefore exhibited transitions in properties with transitions in composition, and, like the objects of the naturalist, had categories of entities exhibiting minimal differences. The remaining analyses in the paper concerned just those substances that Berzelius then included among the generic, mucous acid, benzoic acid, tannin of nutgalls, common sugar, mild sugar, gum arabic, and potato starch. At this point, he attempted to study them only at the generic level, since studies of specific composition required experiments "too delicate to be executed with success before the generic differences

carbonic acid and alcohol by fermentation; and gallic acid, by the influence of an alkali, yields on the one side water and carbonic acid, and, on the other, an extractive matter abounding in carbon, to which it owes its dark color. Perhaps it would be useful to class the ternary oxides into: 1. Oxides not easily decomposed in consequence of their excess of oxygen; such as oxalic, citric, tartaric acids, &c. 2. Oxides easily decomposed in consequence of the equilibrium of their elements, which is easily destroyed by a small force; such as tannin, gallic acid, sugar of milk, common sugar, &c. 3. Oxides not easily decomposed in consequence of the excess of their combustible atoms; such as benzoic acid, resins, fat oils, &c."

4 "Organic proportions," 179; cf. 326. The same point was developed in the *Proportions chimiques*, pp. 43—44. Cf. also Berzelius's letter to Marcet of 23 September 1817, written between the "Organic proportions" and the *Proportions chimiques* and containing a similar discussion (*Bref*, I,iii, pp. 155—159).

are known."[5] Though Berzelius could not now reach the last term of organic analysis, except in a few cases, he had at least determined what the last term was—it was the level of particular, individual compounds, of which the usual generic substances were mixtures. As he would in mineralogy, so in organic chemistry, Berzelius defined a species as a single compound, eventually capable of separation, analysis, and characterization.

Exceptional cases of organic composition were the few substances, notably some of the vegetable acids, that had early been recognized as distinct compounds by the majority of practitioners. These compounds represented the extreme of simplicity in organic combination; as Berzelius remarked in the last quoted passage, a change of only a single volume in their composition produced a dramatic change in their properties. Accordingly, though organic matter in general exhibited smooth transitions in characters, among the simpler compounds, characters changed by saltation. Despite this similarity with the inorganic, Berzelius did not permit the simpler organic compounds to blur the boundary between the organic and the inorganic. While retaining the simpler substances within the realm of the organic, Berzelius regarded them as the point of transition between the two.[6] Their nearest neighbors on the inorganic side—doubtless to the astonishment of any naturalist—were the most complex of minerals.

4. Berzelius's mineralogy

a. Introduction

Berzelius's *Mineralogy* consists of two monographs. The first one, a small book of 1814, was another "Versuch:" *An attempt, by application of the electro-chemical theory and the theory of chemical proportions to establish a pure scientific system of mineralogy.*[7] For "scientific," one may read "chemical,"

[5] "Organic proportions," 179.

[6] E.g., *ibid.*, 330: "I have chosen the vegetable acids as the first substances to begin [organic analyses] with, because they constitute, as it were, the transition from inorganic to organic nature."

[7] The Swedish original was *Försök att genom avändandet af den electro-kemiska theorien och de kemiska proportionerna grundlägga ett rent vettenskapligt system för mineralogien* (Stockholm: 1814). Berzelius sent a copy of the Swedish edition to Thomas Thomson (*Bref,* III,i, pp. 14—17) with the request that Thomson have it translated into English, correct the translation, and have it published "en forme de brochure à part" (p. 16). This Thomson did, having engaged the services of one John Black to do the translation, which appeared as Berzelius, *An attempt to establish a pure scientific system of mineralogy by the application of the electro-chemical theory and the chemical proportions,* tr. John Black (London: 1814). Berzelius was dissatisfied with the quality of the translation, remarking, in the introduction to the French edition (p. iv), cited below, p. 294, n. 9, "La traduction anglaise . . . fut faite malheureusement par quelqu'un qui n'a sans doute aucune notion de minéralogie, car elle est pleine d'erreurs; et dans beaucoup d'endroits, le sens d'original est tout-à-fait défiguré. Elle parut cependent précédée d'une préface du docteur Thomson, qui assurait l'avoir confrontée à l'original, et qu'elle était parfaitement fidèle." A German translation from the Swedish appeared serially in *Schweigger's Jour-*

for Berzelius demanded that to become scientific, mineralogy must be reduced to a branch of chemistry, and he described nonchemical mineralogies as performing the same function in mineralogy as did an index in a book.[8] With the addition of a second, supplementary monograph in 1815, the *Mineralogy* looked less like a mere "attempt," and more like a finished product, one that may be recognized in retrospect as a draft of the *Proportions chimiques*. The additional monograph, entitled "An attempt at a purely chemical mineralogical system," was provoked by the critical review of the original monograph by Berzelius's contemporary, the geologist and mineralogist, J. F. L. Hausmann (1782—1859). Though the 1814 monograph merely proposed and exemplified the method for the construction of a chemical mineral system, the 1815 monograph presented a fully articulated system. Moreover, it offered direct responses to Hausmann's contentions, as well as further articulations of both chemical and mineralogical themes (doubtless compelled by Hausmann's criticism) and a critical analysis of antecedent mineral systems.[9]

nal: "Versuch, durch Anwendung der electrisch-chemischen Theorie und der chemischen Proportion-Lehre ein rein wissenschaftlishes System der Mineralogie zu begründen," *Schweigger's Journal,* 11 (1814), 193—233; 12, (1814), 17—62; 15 (1815), 277—300. The first two installments were translated by A.F. Gehlen; the last by W. Pfaff. The third installment consists of five appendices that were part of the Swedish original of 1814; they appeared in *Schweigger's Journal* as "Beilagen zu der Bd. XI. S. 193 und Bd. XII. S. 17 mitgetheilten Abhandlung von Berzelius: Versuch ..." The German text of the 1814 *Mineralogy,* in the three installments just cited, is used in this study. To distinguish it from the additional monograph that Berzelius published on mineralogy in 1815, cited on this page, in n. 9, the 1814 *Mineralogy* is called in the text the "original monograph," the "1814 monograph," or the "1814 *Mineralogy*" and is cited in the notes as "1814 monograph."

8 For mineralogy a branch of chemistry, see "1814 monograph," 194—198, e.g., (194): "Die Kenntniss der Verbindungen an sich, ihrer Mischungsweise und chemischen Eigenschaften, erlangen wir durch die Chemie, so dass die Mineralogie, aus einem wissenschaftlichen Gesichtspunkt betrachtet, nur als ein Theil der Chemie angesehen werden kann, oder als ein Anhang zu derselben." Cf. this statement (196—197): "... hat man oft bei Abwägung der wechselseitigen Vorzüge der Werner'schen und Haüy'schen Methode die Vertheidiger der letzten gefragt, soll denn der Mineralog immer einer chemischen Analyse bedürfen, um ein Mineral zu bestimmen? An dieser Frage kann man stets den Mineralsammler vom Mineralogen unterscheiden; jener sucht blos Namen für die Fossile, wo dieser das Bedürfniss hat, ihre Natur zu kennen."

For the index metaphor, see the 1814 monograph, 197—198; cf. also the additional monograph of 1815 (cited in the next note), *Schweigger's Journal,* 15 (1815), 338.

9 The Swedish version is "Försök till ett rent kemiskt mineralsystem," *Afhandlingar,* 4 (1815), 1—148. It was completed by April, 1815, as Berzelius reported to Gahn (*Bref,* V,ix, p. 127). W. Pfaff's German translation, appearing in *Schweigger's Journal,* is used in this study: "Versuch eines rein chemischen Mineralsystems," *Schweigger's Journal,* 15 (1815), 301—363, 419—451; 22 (1818), 274—302. The last installment, containing the notes to the systematic arrangement of the minerals as they appeared in the original Swedish (117—148), was published in *Schweigger's Journal* under the title, "Anmerkungen zu der Abhandlung: Versuch eines rein chemischen Mineralsystems." In this study, the German translation is called the "1815 monograph" or the "1815 *Mineralogy*" in the text and is cited as "1815 monograph" in the notes. Aside from the notes, the 1815 monograph contained four parts: the first (301—323) offered direct responses to Hausmann's criticisms; the second (323—336) presented further articulations of chemical and mineralogical concepts; the third (337—

The assimilation of mineralogy to inorganic chemistry rested on Berzelius's belief in its underlying ontological identity with inorganic chemistry. The demonstration of this point was among the chief objectives of the 1814 monograph. Despite what Berzelius there frequently acknowledged to be the preliminary nature of his "attempt"—the analytical techniques were not yet available to accomplish the full subordination of mineralogy to inorganic chemistry—Berzelius saturated his prose with professions of faith in the commensurability of the two subjects and with illustrations of the power of this assumption. Not only were his major theoretical innovations in chemistry transported into mineralogy, but also the sensibilities and experience of a master of preparative inorganic chemistry. Berzelius regurlarly invoked his knowledge of the behavior and properties of precipitates, of solubility effects, and of problems in separation and purification of materials to deal with conceptual issues of mineral genesis and practical and conceptual problems in mineral analysis. The requirement for dividing the two fields was only pragmatic.[1]

363) contained a critical discussion of antecedant mineral systems; and the fourth (419—451) contained the full mineral system itself.

Hausmann's criticism is "[Review of] *Försök att genom användandet af den electrokemiska theorien och de kemiska proportionerna grundlägga ett rent vettenskapligt system för mineralogien,* af J. Jacob Berzelius, M.D. Chemiae Prof. etc. 1814. 103 S. in Octav," *Göttingische gelehrte Anzeigen,* 1814,ii (1814), 1089—2001 (the last two pages are mispaginated as 2000 and 2001 instead of as 1100 and 1101); henceforth cited as Hausmann, "Review."

Two subsequent editions of the *Mineralogy,* one in French and one in German, contained both monographs in a single tome. The French edition, containing slight alterations in the text, new mineral analyses, and many more notes than appeared in the 1815 monograph, was prepared while Berzelius was visiting Berthollet at Arcueil in 1818—1819 (the same visit during which the *Proportions chimiques* was published). It appeared as Berzelius, *Nouveau système de minéralogie* (Paris: 1819); henceforth cited as *Nouveau système de minéralogie.* (For the nearly simultaneous preparation of the French version of the mineralogy and the *Proportions chimiques,* see *Bref,* I,iii, pp. 185—187, 197—199; and Berzelius, *Autobiographical notes* [Baltimore: 1934], pp. 101—102). The German edition of the mineralogy was a *Sonderabdruck* of the translation of both 1814 and 1815 monographs by Gmelin and Pfaff. It contained the notes, which appeared in it before they were published in *Schweigger's Journal,* and it appeared as *Neues System de Mineralogie von J.J. Berzelius. Aus dem Schwed. von Gmelin und Pfaff* (1815). It has not been seen by the present author.

[1] The content of this paragraph is documented in detail below. Here, a preliminary example suffices to convey the flavor of Berzelius's argument. See, for example, his letter to Thomson requesting an English translation of the 1814 *Mineralogy* (*Bref,* III,i, pp. 14—17, esp. pp. 15—16): "Le but de ce mémoire est de prouver que les minéraux, étant des combinaisons chimiques, doivent être assujettis aux mêmes lois que les combinaisons chimiques en général. Dans chaque combinaison des corps oxydés, combinaison qu'on se figure décomposé dans le travail de la pile électrique, il y a un ou plusieurs oxydes qui se classent autour de pôle négatif, tandis qu'il y en a toujours un (rarement deux) qui est attiré vers le pôle positif; c'est-à-dire avec d'autres mots que dans chaque combinaison d'oxydes (d'une origine fossile ou non) il y a au moins un oxyde qui joue le rôle d'acide, tandis qu'il y en a d'autres qui jouent le rôle de bases; de manière que les minéraux oxydés peuvent être considérés et classés comme des combinaisons salines. J'ai tâché d'indiquer dans ce mémoire les oxydes qui (hors les acides reconnus pour tels) jouent le rôle d'acides et parmi lesquels la silice occupe une place si distinguée. J'ai tâché de faire voir comment il faut juger la composition des minéraux, tant combustibles qu'oxydés, et j'ai cherché à illustrer le

b. Motivations

Berzelius hoped to show that all minerals were inorganic compounds of the same kinds as those encounted in the laboratory, capable of the same kind of quantitative analysis and formulation. Combustible minerals were either radicals or direct compounds of radicals with one another, in the same proportions that they would have assumed in their oxidized combinations. Other minerals were either oxides (or sulfides), or the combinations of oxides (or sulfides). Berzelius's most important innovations were to subsume all minerals composed of oxides (or sulfides) under the rubric of salt, and to bring them into conformity with the models for complex salts developed in the latter parts of the *Versuch.*

The possibility for a mineralogical application of his chemistry had occurred to Berzelius in two connections in the *Versuch:* the potential for a rationalization of mineral systematics in the light of the demonstration of the oxide nature of the earths and the likelihood that the methods for calculation of complex saline composition might be applied to minerals. Berzelius noted the former in the first *Fortsetzung,* and then elaborated it in the 1815 *Mineralogy.* In the first *Fortsetzung,* he argued that some alloys of arsenic were actually compounds in definite proportions of the metallic radicals and that the proportions of the radicals might be determined, in accordance with the principle of the preservation of combining proportions, from those of their oxidized combinations. The metals would have combined with one another in such a proportion that, were they then oxidized, they would have taken up quantities of oxygen dictated by the rule of oxides; that is, either they would take up equal quantities of oxygen or the less negative one (in accordance with Berzelius's original sign convention) would take up an integral multiple of the amount that the other would take up.[2] In this connection, Berzelius remarked in a note that:

> Es ist leicht einzusehen, dass, da die Erdarten als Metalloxyde betrachtet werden müssen, dieses [that is, the statement just summarized here about the proportions of oxides] nach der oben angeführten Regel [of oxides] auch von den Verbindungen der Erdarten und der Metalloxyde in den krystallisirten, d.i. in allen durch chemische Verwandschaft gebildeten Mineralien gelten muss; so dass der Sauerstoff auch in ihnen die

tout par des exemples tirés des ouvrages de nos meilleurs analystes. J'ai ensuite tâché de prouver que dans les combinaisons chimiques fossiles les éléments obéissent aux lois des proportions chimiques, tout aussi bien que dans les expériences de nos laboratoires. Les résultats analytiques n'y sont point toujours concordants, mais cette circomstance tient à des causes dont j'ai fait une courte exposition." Cf. also Berzelius's letter to Berthollet of 7 May 1814 (*Bref,* I,i, pp. 44—48, esp. pp. 46—47), and the opening paragraphs of the 1815 *Mineralogy* (301—302).

For the contention that the division between mineralogy and chemistry was merely pragmatic, see "1814 monograph," 195—196.

[2] Berzelius, *Versuch, die bestimmten und einfachen Verhältnisse aufzufinden, nach welchen die Bestandtheile der unorganischen Natur mit einander verbunden sind* (Leipzig: 1892), pp. 119—120; cited fully above, p. 168, n. 8 and p. 202, n. 6; henceforth cited as *Versuch.*

Proportion [of the radicals] in der Verbindung bestimmt. Alle Analysen der Mineralien müssen also nach dieser Ansicht wiederholt und geprüft werden.[3]

The novelty afforded mineralogy by the demonstration that the earths were metallic oxides was the ability to eliminate the traditional mineralogical class of earthy minerals by classifying the compounds of earths in the same way as those of metals. In the 1815 *Mineralogy,* having concluded his criticism of antecedent mineral systems with a discussion of Haüy's, Berzelius elaborated this conclusion:

Es lässt sich voraussehen, dass wenn Haüy, damals als er sein System aufstellte, Davy's Entdeckung über die Zusammensetzung der Kalien und Erdarten gekannt hatte, auch die Ideen über die Beschaffenheit der Grundmischung erdiger Fossilien, welche die Entdeckungen der neuesten Zeiten veranlassten: so würden wir bereits das vollkommene consequente System erhalten haben, und vielleicht so, wie ich es im Folgenden [that is, in the remainder of the 1815 monograph, where the full classification scheme was set out] vorzuschlagen beabsichtigte. Das neue chemische System kann demnach nicht anders angesehen werden, als eine Anwendung späterer Entdeckungen auf Haüy's System.

Es ist klar, dass der Eintheilungsgrund, welcher im System bei der Mehrheit der Metalle und deren Verbindungen herrscht, auch bei andern Metallen befolgt werden müsse; und dass, so bald eine Klasse von Körpern, die man nicht als aus Metalloxyden bestehend ansah, und die man demnach im System auch nicht als solche anführen konnte, als Metalloxyde befunden werden, die Aufstellungsordnung sich ändern müsse; und dass diese nun nach den nämlichen Gründen angeordnet werden müssen, wie die Körper, von denen man vorher wusste, dass es Metalloxyde waren. Wir finden, dass die meisten Mineralogen einen chemisch richtigen Eintheilungsgrund bei den Metallen, deren Oxyden und übrigen Verbindungen befolgten. Die Chemie entdeckt, dass die Alkalien und Erdarten auch Metalloxyde sind; das giebt sich nun von selbst, dass man sie und ihre Verbindungen, gemäss den Principien für die vorher bekannten Metalle aufführt. Die Salze wurden aufgeführt nach einem chemischen Princip, entweder bei den Säuren oder bei den Basen. Die Chemie entdeckt, dass alle Verbindungen, zwischen oxydirten Körpern nichts anders als salzartige sind (um nicht direct zu sagen Salze), das sie ihre Basen und ihre Säuren haben (oder Körper, so deren Stelle vertretten [*sic*] und dass zu den letztern gehören Kiesellerde, Titanoxyd, Tantaloxyd, Antimonoxyde, und mehr andere. Daraus folgt wieder die Nothwendigkeit, alle Verbindungen zwischen oxydirten Körpern nach denselben Gründen, wie die Salze, zu klassificiren, d.h. entweder alle nach ihrer gemeinsamen Basis, oder nach der gemeinsamen Säure.

Kurz die Veränderung, welche ich vorzuschlagen beabsichtige in den schon zuvor angenommenen chemischen Systemen ist nichts anders als ein Versuch, dieselbe mit den Entdeckungen in Uebereinstimmung zu bringen, womit die Chemie, nach deren Aufstellung, bereichert worden ist; und, ich gestehe es aufrichtig, ich rechne auf vorurtheilsfreier Naturforscher Beifall.[4]

The second inspiration for a mineralogical application emerging in the *Versuch* was the possibility that Berzelius's techniques for analyzing complex salts and rationalizing their composition could be deployed in studying minerals, taken to be complex salts. This notion was adumbrated in the note quoted above and made fully explicit in the concluding section of the *Versuch.* There, Berzelius

[3] *Ibid.*, p. 120n.

[4] "1815 monograph," 361—362.

listed as subject to the law of unitary oxygen not only salts with water of crystallization and double salts, but also,

> Zusammensetzungen aus mehreren Alkalien, Erden und Metalloxyden, wozu die krystallisirten Mineralien gehören. Obgleich wir zu der Anwendung der Regel [of unitary oxygen] auf die krystallisirten Mineralien durch keinen Versuch in diesen Abhandlungen berechtigt sind, lässt sich doch die Richtigkeit derselben nicht bezweifeln; denn warum sollte der Marmor, der Flussspath, der Schwerspath, u.m. nach einer Regel gebildet sein, welche für die andere Mineralien nicht geltend wäre? Dieses lässt sich nicht wohl denken.[5]

The minerals cited in this passage were also chemicals, and, as chemicals, were subject to the laws of Berzelius's chemistry. There was no evident reason to suppose that other minerals were not also chemicals, differing from laboratory chemicals only, perhaps, in their greater complexity. They too ought to have been subject to the law of unitary oxygen. Berzelius was led from considerations of the continuity of mineralogy with chemistry to attempt a subsumption of the former under the latter. From the effort to achieve this was born the *Mineralogy*.

c. Foundations

The constituents of many minerals, however, were known to be very different from the typical acids and bases common in ordinary laboratory experience. If minerals were to be considered chemicals, saline chemicals, their constituents had to be shown to be playing the roles of acids and bases. Berzelius could accomplish this demonstration in terms of his electrochemical generalization of the concepts of acidity and basicity. His chemical studies had revealed that even weakly reactive oxides could play the role of acid or base in saline combinations and that the oxygen content of their constituents entered into calculation in the same way as the oxygen content of the typically acidic or basic oxides. As the passage quoted above (p. 297, accompanied by n. 4) suggests, Berzelius's mineralogical innovation rested on his recognition that weakly reactive substances like the oxides of titanium, tantalum, and silicon (that is, the earth, silica), could play the role of saline constituents.

Though the generalization of the concept of basicity originally had transformed Berzelius's chemistry, the newly recognized acidic components, above all silica, were most significant for mineralogy. The acidic role of silica had been proposed approximately simultaneously by Smithson but had not been recognized previously. In the late eighteenth century, silica had been included among the earths, which had formed one category of salifiable base. If, in the *Traité*, Lavoisier had excluded only silica, of the five earths, from the category of salifiable base, he had done so in recognition not of its acidity, but of its minimal reactivity. That he retained it among the "alkaline earths" in his table of

[5] *Versuch*, p. 208.

elements indicates the persistent force of the standard analogy of silica with the other earths.[6] Citing Smithson's independent proposal in support of his own views, Berzelius argued that a substance previously perceived as largely unreactive but analogous with the alkaline earths played the role of acid in a vast number of minerals.[7]

As the embodiment of the generalization of the ideas of acidity and basicity, the electrochemical theory was the foundation on which this innovation rested:

> Die electrochemische Theorie hat uns gelehrt, dass in jedem zusammengesetzten Körper Bestandtheile von entgegengesetzten electrochemischen Eigenschaften vorhanden sind... Daras folgt, dass in jedem zusammengesetzten Körper ein oder mehrere electropositive mit einem oder mehreren electronegativen Bestandtheilen vorhanden seyn müssen, d.h. im Falle die Verbindung aus Oxyden besteht, dass jedem Stoffe der in einer Verbindung als Base auftritt, ein anderer entsprechen müsse, der dagegen die Rolle einer Säure spielt, selbst wenn letzter in seinem freien Zustande nicht die den starken Säuren eigenen Eigenschaften zeigt, wie den sauren Geschmack und die Reaction gegen Pflanzenfarben. Der Stoff, der in einem Falle electronegative ist gegen einem stärker positiven, d.h., der gegen eine stärkere Basis als Säure reagirt, kann in einem andern electropositiv seyn gegen eine stärker negativen, d.h. ein andermal als Basis gegen eine stärkere Säure sich verhalten. So z.B. vertritt in der Verbindung zweier Säuren die schwächere die Stelle einer Base gegen die stärkere.
>
> Jeder Verbindung aus zwei (oder mehr) Oxyden daher ist von der Natur eines Salzes, d.h. hat ihre Säure und ihre Base, von welcher, wenn die Verbindung der Wirksamkeit der electrischen Säule ausgesetzt wird, erste an dem positiven letzte am negativen Pole sich sammelt. Demnach müssen wir in jedem aus oxydirten Stoffen zusammengesetzten Fossil, es sey von erdiger oder schon voraus erkannter salziger Natur, seine electronegativen und electropositiven Bestandtheile aufsuchen, und wenn diese ihrer Natur und Menge nach bestimmt worden, so sagt uns eine kritische Anwendung der chemischen Theorie, was das untersuchte Mineral ist.[8]

By these techniques, Berzelius went on,

> Wir entdecken sogleich eine starke Klasse von Fossilen, deren den Salzen ähnliche Natur die Chemiker öfters geahnet haben, ohne jedoch dieser Ahnung weitere Ausführung zu

[6] For Smithson's proposal, made in 1811 in connection with an analysis of zeolite, see Partington, *History*, III, p. 710. Two other figures, Winterl and Scheele, also suggested that silica be reckoned among the acids, but their views were not influential. Partington (III, p. 499) notes that Winterl's reputation for fantastic theories probably led contemporaries to ignore whatever reasonable ideas he developed. Scheele expounded his ideas in this connection only in his correspondence, not in his published works; for this, see Olsson, *Kemiens historia*, pp. 326—327. For Lavoisier's exclusion of silica from among the salifiable bases, see his *Traité*, I, Ch. 14; for the place of silica in Lavoisier's table of elements, see the *Traité*, I, p. 192, and cf. Partington, *History*, III, p. 485n.

[7] "1814 monograph," 211. Berzelius noted here that he had suggested the idea of the acidity of silica in the 1811 "Essai sur la nomenclature chimique," *J. de physique*, 73 (1811), 253—286. It is true that in a section devoted to mineralogical nomenclature (270) he had suggested the existence of silicates by including two examples in a list of names of compounds of oxides with one another (*silicias kalicus* and *silicias baryticus*). However, he did not comment explicitly on the acidic properties of silica.

[8] "1814 monograph," 198—199.

geben. Diese Klasse wird durch die Fossile gebildet, in welchen die Kieselerde die Stelle einer Säure vertritt und enthält eine unzählige Mannigfaltigkeit von einfachen, doppelten, drei- und vierfachen Salzen, auf mehrfachen Stufen der Neutralität, oder der Uebersättigung mit der Säure oder Base. Eben so finden wir Klassen von geringerem Umfange, wo Titanoxyd, Tantaloxyd, und andere bisher nicht als Säuren betrachtete Metalloxyde, die Rolle der Säure spielen, so dass das ganze grosse Gebiet der erdartigen Fossile nach denselben Grundsätzen geordnet werden kann, wie die Salze.[9]

These assertions, which read so much like a classic Berzelian profession of faith in dualism, are an extension of the electrochemically based ideas of basicity and acidity, one that permitted the assimilation of mineralogy to chemistry: ontologically, minerals, even earthy ones, were chemicals; in particular, they were salts, because mineral components were the constituents of salts. Accordingly, chemistry specified minerals both qualitatively and quantitatively. Their qualitative specificity rested on the electrochemical characterization of their constituents, which identified them as saline and distinguished them from other ingredients. Their quantitative specificity rested on the conformity of their composition to the stoichiometric laws governing salts.[1]

Given the saline nature of minerals, their composition could be subjected to Berzelius's inorganic stoichiometry. The only additional assumption needed was that the causes operating in the mineral realm were of the same kind as those at work in the laboratory. If the causes were the same, the same visible effects, in particular, submission to identical stoichiometric laws, must and did follow:

In der letzten Hälfte des jüngstverflossenen Jahrzehnds ist die Chemie noch von einer andern Seite zu einer höhern Ausbildung gelangt, durch die Lehre von den chemischen Verhältnissmengen, welche bei der Aufstellung des mineralogischen Systems dieselbe, und ich darf sagen mathematische, Sicherheit gewahren muss, wie sie der Chemie geben wird und zum Theil bereits gegeben hat. Jeder folgerechte Denker konnte sich leicht vorstellen, dass dieselben Gesetze, die durch lange und abgeänderte Erfahrung sich in unsern Laboratorien bewährten, auch in dem grossen Ganzen gelten würden. Es ist überall dieselbe Natur, welche wirket, sind dieselben Gesetze, die diese Wirksamheit beherrschen, es mag entweder der Mensch diese Wirksamheit bisweilen auf ein besonderes Ziel richten, oder sie mag durch den freien Lauf der auf der Erde unendlich sich abändernden Umstände bestimmt werden. Wenn daher die chemische Zerlegung vieler Minerale die Anwendung der chemischen Proportionlehre auf die Mineralogie bis jetzt nicht rechtfertigt, so kann der Grund davon nicht in der Unrichtigkeit dieser Anwendung gesucht werden, sondern in unserem unzureichenden Vermögen, das oft genug, trotz den unsichtigsten Bemühungen, von Schwierigkeiten überwogen wird, die wir bisweilen gewahr werden, nicht selten aber auch übersehen. Indessen giebt es doch eine grosse Menge von Analysen, deren Resultate theils völlig mit der Lehre von den chemischen Verhältnissmengen übereinstimmt, theils sich so sehr annähert, dass die Abweichung mit Recht blos als ein schwer vermeidlicher Beobachtungsfehler angesehen werden kann.[2]

Thus, as its title indicated, the 1814 monograph described the subordination

[9] *Ibid.*, 200—201.
[1] Cf. *ibid.*, 8, 200.
[2] *Ibid.*, 201—202.

of mineralogy to chemistry by the application of the two major theoretical foundations that Berzelius had laid in his inorganic chemistry, the electrochemical theory and the theory of proportions. If the former revealed the ontological identity of mineral components with saline constituents and therefore of minerals with salts, the latter showed minerals to be subject to the same phenomenological restrictions upon the variety and diversity of composition as were chemicals.

One consequence of this point of view was Berzelius's stoichiometric definition of mineral species. Mineral species were individual chemical compounds, as defined by their conformity to definite proportions. All members of a species have "gleiche Zusammensetzung in gleiche Verhältnissmengen," as Berzelius put it in the 1814 *Mineralogy*.[3] In the 1815 monograph, Berzelius offered an even stronger statement of his stoichiometric definition of species, one that symbolized for Wernerians his uncompromising and "unmineralogical" stance:

> Jede besondere Species besteht aus absolut derselben chemischen Verbindung, identisch nach der Qualität und Quantität der Bestandtheile. Die mindeste Particul eines fremden Körpers, oder die vermehrte Anzahl der Particuln irgend eines Bestandtheiles, *welche wesentlich in die chemische Verbindung mit eingehen,* bringt eine neue Species hervor. So is wasserloser Gyps (Anhydrit) eine Species, und wasserhaltiger eine andere.[4]

As might be expected, Berzelius chose the case of alum to illustrate the use of the theory of proportions in establishing the specificity of minerals. In this connection, in the early pages of the 1814 *Mineralogy* Berzelius had

> einige Worte über die Beurtheilung der quantitativen Resultate der Mineral-Analysen und die richtige Zusammenpaarung der gefundenen Bestandtheile zu sagen. Es ist klar, dass, wenn wir bei Vorstellung der Mischung eines Salzes, z.B. des Alauns, nicht weiter gingen, als ihn Kali- und Thonmetall, Schwefel, Wasserstoff und Sauerstoff zusammengesetzt zu betrachten, in wissenschaftlicher Hinsicht wenig gewonnen wäre. Wir kommen der Natur seiner Mischung einen Schritt näher, wenn wir die einzelnen Stoffe zusammenpaaren, und ihn als aus Schwefelsäure, Thonerde, Kali und Wasser zusammengesetzt ansehen. So thaten es die Chemiker auch schon lange und dadurch entstand der Name eines Tripelsalzes, oder eines Salzes aus drei Hauptbestandtheilen. Der nächste Schritt zu einer vollkommeneren Kenntniss war nun, ihn als aus schwefelsauren Kali und schwefelsaurer Thonerde mit Krystallwasser bestehend zu betrachten, wodurch der Name eines Doppelsalzes aufkam. Endlich, dass ich so sagen darf, vollendete die Lehre von den chemischen Verhältnissmengen unsere Vorstellung von der Zusammensetzung dieses

[3] *Ibid.*, 231. Berzelius attributed this species definition to Werner, apparently mindful of the lip service Werner paid to chemistry in the introduction to his 1774 *Aeusserliche Kennzeichen* (for which, see above, p. 108, n. 5 and accompanying text). At the word "Verhältnissmengen," Gehlen signalled this note (231n): "Sollte dies wohl *Werner's* Begriff von Species in der Mineralogie seyn?" Later recognizing that the attribution of a chemical definition to Werner, if correct, was not apt, Berzelius changed the attribution in the French edition (*Nouveau système de minéralogie*, p. 44): "Par minéraux de même espèce, j'entends ici la même chose que Haüy, *la même composition avec la même forme primitive.*"

[4] "1815 monograph," 422; cf. *Nouveau système de minéralogie*, pp. 184—185.

Salzes, in[dem] sie zeigte, dass es aus 1 Partikel schwefelsaurem Kali, 3 Partikeln schwefelsaurer Thonerde und 24 Partikeln Krystallwasser bestehe.[5]

Berzelius went on to affirm that silicates, too, though complex, could be seen as composed quantitatively like salts and named accordingly.

Die Kieselerde, als Säure betrachtet, hat die Eigenschaft, Verbindungen von mehreren ungleichen Sättigungsstufen zu geben. Die gewöhnlichste ist, wo die Kieselerde gleich viel Sauerstoff, wie die Base hat und diese wollen wir in der Folge *Siliciates* nennen. Dann zeigen sich vorzüglich solche, in welchen sie dreimal so viel Sauerstoff hält, und diesen wollen wir den Namen *Trisiliciates* geben. Nicht selten enthält sie in diesen Verbindungen auch zweimal die Sauerstoffmenge der Base, in welchem Fall wir sie *Bisiliciates* heissen. Auch bildet die Kieselerde viele basische Verbindungen von ungleichen Graden, *Subsiliciates*, die, da es immer vortheilhaft seyn wird, die Verschiedenheit in der chemischen Bennenung auszudrucken, durch den Beisatz *bi-*, *tri-*, u.s.w. unterschieden werden können; z.B. *Subsilicias bialuminicus, trialuminicus* etc., je nachdem die Base, (hier die Thonerde), zwei oder dreimal die Sauerstoffmenge der Kieselerde enthält.

Die Kieselerde giebt ferner, gleich andern Säuren, *mehrfache Silicate*, theils mit, theils ohne Krystallwasser. Oft finden wir, dass diejenigen Basen, die gerne mit andern Säuren mehrfache Salze bilden, in den Doppelsilicaten dasselbe Verhältniss der Basen unter sich finden, wie in andern bekannten Doppelsalzen derselben Basen. Würde z.B. in dem gewöhnlichen Feldspath an die Stelle des Siliciums Schwefel gebracht, so wäre die Verbindung wasserfreier Alaun.[6]

Berzelius was aware, however, that some silicates were even more complex and presented chemistry with problems not fully commensurate with those of the laboratory. In this and several other respects, Berzelius gave evidence in the 1814 monograph that he recognized the points at which the chemical subordination of mineralogy would face its sternest tests. Most aware of epistemological problems, he ascribed persistent difficulties to the still imperfect state of chemical analysis and suggested that, with the progress of the latter, the former would be overcome. He also glimpsed, but did not address directly, long-standing ontological problems that his chemistry had not overcome, and, indeed, might not respond to its further progress. Only Hausmann's criticisms induced Berzelius to examine these issues squarely.

d. Obstacles to the consummation of a chemical mineralogy: Berzelius's early perceptions

In the 1814 monograph Berzelius recounted what he considered to have been the principal epistemological obstacles that previously had hindered the attainment of a chemical mineralogy and that he now expected could be overcome by his methods.[7] The import of this discussion was to assimilate the problems confronting analysis of minerals and interpretation of analytical results to the same

[5] "1814 monograph," 209—210.

[6] *Ibid.*, 212.

[7] *Ibid.*, 202—209.

categories of problems confronting quantitative analysis of laboratory chemicals. Though clearly a facet of Berzelius's attempted reduction of mineralogy to inorganic chemistry, this approach tended to ignore the fundamental problems confronting a chemical definition of mineral species.

In this discussion of obstacles to a chemical mineralogy, Berzelius cited in particular three sorts of difficulties. First, the problem of attaining accurate, reproducible results; second, the added difficulties occasioned by the production of impure specimens in natural mineral genesis; third, the issues raised by a particular cause of impurities, a phenomenon that Berzelius called "Zusammenkrystallisirung." The first he considered easiest to overcome—presumably it would be, by the advance of the analytical art.[8] In regard to the second, he divided impure minerals into those that were clearly mixtures and those that were largely pure crystalline specimens.[9] The former he labelled analytically intractable; the latter he deemed accessible to analytical techniques, because they presented difficulties no different in kind from those faced by the chemist in dealing with impure precipitates.[1] Berzelius noted that minerals contained several constituents amounting to one percent or less of their mass; accordingly, he asserted blithely, "Es ist klar, dass diese fremden Bestandtheile abgezogen werden müssen, wenn man das Ergebniss nach der Lehre von den bestimmten Verhältnissmengen beurtheilen will..."[2] A sense for the fundamental ontological problem of discriminating essential from accidental ingredients certainly did not inform these lines. The third obstacle to a chemical mineralogy was the difficulty posed by "Zusammenkrystallisirung," that is, simultaneous crystallization of different compounds into a single crystal, differing in its characters from those of its pure ingredients, but remaining a mixture.[3] To this category he assimilated *inter alia* several famous cases of polymorphism that had

[8] *Ibid.*, 202—203. In part, this claim was directed against the by then notorious variability among the analyses of minerals otherwise determined to belong to the same species. The objective of the statement was to suggest that in mineralogy this variability resulted from the same kinds of problems, though more severe, that had previously encumbered ordinary laboratory analysis. This subject is examined further below.

[9] *Ibid.*, 203—207.

[1] *Ibid.*, 203—204.

[2] *Ibid.*, 206—207. He noted (206) that colors could be considered a sign of impurities. Assuming that mineral genesis proceeded by precipitation from solution, he remarked that "Der Stoff, dessen die Auflösung am meisten enthält und womit sie übersättigt ist, bildet die Krystalle, aber diese [b]ehalten einen Antheil von der Auflösung in ihren Zwischenräumen, werden davon verunreinigt und erhalten dadurch nicht selten eine ihren gar nicht zukommende Farbe. Dieser Umstand verursacht, dass so viele Minerale, die ihren eigentlichen Bestandtheilen nach farblos seyn sollten, roth, grün, gelb, blau u.s.w. gefärbt sind, welche Farben von einer geringen mechanischen Beimischung anderer gefärbter Minerale herrühren, die oft so fein zertheilt sind, dass sie die Durchsichtigkeit des Krystalls nicht merklich beeinträchtigen." Color, then, of principal epistemological importance to the natural-historical mineralogist, had no fundamental significance according to Berzelius. In natural-historical terms, it reflected not the essence of a mineral, its stoichiometry, but the accidents attendant upon its formation.

[3] *Ibid.*, 207—209.

plagued the crystallographers, and he compared them with other similar cases from industry and the chemical laboratory.[4] Just as the man-made cases posed significant but not fundamental problems, Berzelius claimed, their mineralogical counterparts did not reflect on the foundations of his chemical mineralogy. Though the difficulties issuing from the phenomenon of "Zusammenkrystallisierung" were obstacles in the path to a chemical mineralogy, they were not insurmountable.[5]

Supremely confident of the vitality of his chemical achievements and underestimating—because of the analogy he saw between chemicals and minerals—the obstacles in the path to a chemical solution to the species question, Berzelius boldly affirmed that a chemical mineralogy could be made to work. Previously debilitating problems could now be tolerated until they were overcome, as surely they would be. Despite this optimism, an inkling of the complexity of the mineralogical species problem did surface occasionally in Berzelius's prose. He had a glimpse of new difficulties in connection with the most complex compounds: the silicates, aluminates, and fluorosilicates, just those substances that exceeded in complexity anything then found in the laboratory. They represented the point at which Berzelius's initially easy assimilation of minerals to chemicals became difficult, because he had no chemical conceptual techniques for accommodating them that he could import into mineralogy. In the 1814 *Mineralogy*, Berzelius did little to deal with the issues that he raised in connection with these complex substances. After Hausmann's criticisms, he addressed them with new articulations of his chemistry.

The problems Berzelius glimpsed in the original monograph were of three kinds. First, among the most complex minerals, changes in the number of particles of each ingredient produced relatively small changes in the percentage composition of minerals.[6] In this connection, Berzelius produced a chart listing fifteen possible formulae for topazes, taken to be compounds of fluorosilicic acid and alumina, and displaying the percentage composition corresponding to each.[7] The differences among these figures were too small—some of them, Berzelius admitted, smaller than the margin of experimental error then attainable—to permit a decision about the composition of topazes.[8] However, for Berzelius, this was a matter of awaiting improvement in analytical techniques and entailing no issues on the level of fundamentals.[9]

[4] *Ibid.*, 208—209.

[5] This claim was scarcely more than a profession of faith, however, as noted shortly.

[6] *Ibid.*, 47—52. The argument was developed with reference to the topaze genus.

[7] The table reproduced on p. 305 is in *ibid.*, 49. A represented alumina; Fl. fluoric acid; S, silica. Berzelius's use of this mineralogical notation is examined further below (p. 305, n. 1 and p. 308, n. 6).

[8] *Ibid.*, 50.

[9] *Ibid.*, 51. Berzelius indicated that the isolation of "Flussspathsäure" was a prerequisite to the solution of this problem. Like muriatic acid, the acid of fluorspar was then believed to be the oxide of an unknown radical.

Berzelius later pursued the topazes relentlessly, reporting his analytical results in letters

Though he showed himself relatively untroubled about this issue in the 1814 monograph, after Hausmann's critique Berzelius grew more concerned. Even when subject to his quantitative laws, mineral combinations threatened to become too numerous, and, in terms of percentage composition, too similar, to permit discrimination of neighboring species. To Berzelius, the similar composition of proximate species seemed akin to the occurrence of transitions in natural history generally and in organic chemistry in particular. As shown below, Berzelius tried to insure the sharpness of his division between the organic and the inorganic and to retain mineralogy securely on the inorganic side by arguing that apparent transitions occurred only in extreme and therefore exceptional cases. They were still manageable by his chemical techniques, he claimed, and therefore must remain part of inorganic chemistry.

In 1814, however, Berzelius brushed this problem aside. He did likewise with the other problems that he then glimpsed, though they should have seemed to him still less easily dispatched. All of them issued from the magnitude of the number of oxide or sulfide components that formed the most complex minerals. First, in such substances the oxidation stages of the first order components were often uncertain. Berzelius found the oxides of iron particularly troublesome and he devoted separate discussion to them.[1] Generally, for questionable cases

to Gahn during late 1814 and early 1815 (*Bref*, V,ix, pp. 110, 111, 125) and announcing in a letter of 17 April (p. 125) "Topazerna hafva capitulerat." He published his results as "Undersökning af hittills kända fluorsilicater, eller af de till Topazslägtet räknade fossilierna," *Afhandlingar*, 4 (1815), 236—252.

[1] It was this uncertainty that had occasioned the need for the mineralogical symbols. Though in the 1814 monograph (225) Berzelius presented them as a convenience—"leichtere Formel, die blos anzeigen was das Fossil ist"—he affirmed in the 1815 monograph (425—426) that the "chemischen Formuln reichen bis zum Anfang der Radicale von Erdarten. Da fangen die mineralogischen an, aus dem Grunde, weil so grosse Unbestimmtheit, oder beinahe nur Muthmassung herrscht in Hinsicht der Anzahl von Sauerstoffparticeln in den Alkalien und Erdarten; und wenn das, was für eine einzige Erdart angenommen wird, unrichtig ist, so wird die Mehrzahl der Formeln für die Silicate unrichtig. Wir müssen daher hier uns der mineralogischen bedienen, bis unsere Kenntnisse hinreichend sich erweitern, um die chemischen benützen zu können." The mineralogical formulae were constructed from italicized symbols, to distinguish them from the chemical ("1814

For key to symbols, see note 7.

Formel	$AFl^2 + AS^2$	$AFl + AS$	$AFl + 2AS$	$AFl + 3AS$	$AFl + 4AS$
Thonerde	38,62	55,4	53,91	53,07	52,55
Kieselerde	37,68	27,2	35,06	38,08	40,30
Flussspathsäure	23,70	16,4	11,03	8,13	7,15
Formel	$AFl + 5AS$	$AFl + 6AS$	$2AFl + AS$	$A^2Fl + AS$	$A^2Fl + 2AS$
Thonerde	52,21	52,17	47,6	65,36	60,95
Kieselerde	42,44	43,50	23,2	21,26	30,00
Flussspathsäure	5,35	4,33	29,2	13,38	9,07
Formel	$A^2Fl + 3AS$	$A^2Fl + 4AS$	$AFl + AS^2$	$A^2Fl + 2AS^2$	$A^2Fl + 3AS^2$
Thonerde	58,55	57,09	44,56	47,10	55,30
Kieselerde	34,27	31,10	42,40	45,50	40,60
Flussspathsäure	7,18	5,81	13,24	7,40	4,26

he adopted a convention: to calculate the oxidation stages on the assumption that each oxide contained the same number of volumes of oxygen.[2]

Second, Berzelius was attempting to apply the model of alum to all minerals, without, in 1814, perceiving that, because many minerals were more complex than this double salt, the alum model was inadequate. As a double salt, alum had two and only two saline or second order constituents. More complex minerals had more than two—or perhaps had higher order constituents. However, Berzelius had yet to articulate his mature statement of dualism, which called for a dualistic composition for substances even more complex than alum. Lacking this mature expression, which is called here "hierarchical dualism," because it applied even to substances of higher order than alum, Berzelius could not say unequivocally what must be the number and composition of the expected second order constituents of minerals, and therefore he could not provide criteria for partitioning the first order constituents obtained by analyses. Applying the alum model without regard to this question, Berzelius found himself in effect attempting to exhibit the multiple *saline* constituents of complex minerals as he had previously tried to exhibit the multiple *oxide* constituents of complex salts. Just as this latter procedure was inconsistent with the model of double salts and just as it failed to limit the number of first order constituents, so too the multiple saline model failed to delimit the number of second order constituents and was therefore inadequate for mineralogy. Realizing this after Hausmann's criticism, Berzelius produced his classic, mature statement of hierarchical dualism, as shown below. In 1814, however, Berzelius recognized only that his partitionings of first order constituents into multiple saline components were provisional,[3] and he attributed their tentativeness only to the difficulties of analysis, not to the absence of stoichiometric laws.

monograph," 226). Italics are avoided in this study. Instead, mineralogical formulae are explicitly so labelled. The mineralogical symbols ("1814 monograph," 226) are:

S	= silica	K	= potash
A	= alumina	F	= ferric oxide
Z	= zirconia	f	= ferrous oxide
G	= beryllia	Ff	= ferrosoferric oxide
Y	= yttria	Zi	= zinc oxide
M	= magnesia	Mg	= black oxide of manganese
C	= chalk		
St	= strontia	mg	= green oxide of manganese
B	= barytes		
N	= soda	Aq	= water

For Berzelius's discussion of the oxide of iron, see "1814 monograph," 290—296.

[2] *Ibid.*, 221, 18—19.

[3] *Ibid.*, 221, 19. In the former passage, Berzelius was exemplifying the determination of the composition of a complex silicate using the example of byssolith as analyzed by Vauquelin. He remarked (221) that "Nimmt man darin [that is, in each of the metallic-oxide bases] gleichen Sauerstoffgehalt an und theilt man die Kieselerde gleichmässig zwischen allen vier Basen, so besteht das Salz aus 1 Part[ikel] Kalk-Bisilicat, 1 Part. Talk-Bisilicat, 1 Part. Mangan-oxyd-Bisilicat und 2 Part. Eisenoxyd-Silicat." Accordingly, the mineralogical formula for byssolith (227) was $MS^2 + CS^2 + MgS^2 + 2FS$. Though for alum the partitioning of its acid between its two bases was dictated by the model of double salts as compounds of two simple salts, there was no model of more complex compounds that dictated the distribution of the silica among several bases. Berzelius's procedure, then, involved an arbitrary assumption.

Third, lack of additional laws also prevented Berzelius from saying how, or even whether all discernible second order constituents were in fact combined into a mineral as a single compound. After having enumerated the various categories of silicate salts that stoichiometry revealed as so many different degrees of saturation of base with silica (in the passage quoted above, p. 302, accompanied by n. 6), Berzelius suggested the existence of still more complex silicates, those containing several bases that did not correspond to any chemical found in the laboratory and that, if they were single compounds, must be composed of silicates in different degrees of saturation:

Aber die Natur bringt in ihrem Reichtum, durch den schwächsten Verwandtschaftsgrad, eine Menge noch mannigfaltigerer Verbindungen hervor, — und dahin lassen sich ohne den mindesten Widerspruch eine grosse Menge der Silicate bringen, — denen wir nach der Erfahrung in unsern Laboratorien, nur sehr wenige oder gar keine ähnlichen Verbindungen gegenüber zu stellen haben. So, z.B. finden wir Silicate mit drei bis vier, und vielleicht bei erweiterter Erfahrung mit noch mehr, Basen, die alle zusammen zu einer einzigen Verbindung vereinigt sind, und deren reines Krystallgefüge zu erkennen geben scheint [this had been one of Berthollet's criteria of purity], dass man sie als ein chemischen Ganze ansehen müsse; im Fall nicht Umstände in der Folge zeigen, dass sie zu der Klasse von Krystallen gehören, die durch mehrere sich zusammenlegende, aber nicht chemisch verbundene, Stoffe gebildet worden. Dabei findet sich auch oft, dass die verschiedenen mit einander verbundenen Silicate sich nicht auf gleichen Sättigungsstufen befinden, sondern dass eine oder mehrere von den schwachen Basen Subsilicate oder Silicate, eine oder mehrere der stärkern dagegen Bi- oder Trisilicate sind; (ein solches Resultat geben wenigstens für jetzt die Analysen.) Dass in unsern Laboratorien solche Verbindungen nicht erzeugt werden, führt offenbar daher, das wir gewöhnlich unsere Erfolge durch Anwendung solcher Umstände erhalten, deren Wirkung viel zu heftig und gewaltsam ist, als dass sie den Einfluss sehr schwacher Verwandtschaften zuliessen.[4]

Berzelius recognized, however, that these most complex silicates may not have been single compounds of all their supposed second order constituents, but may have amounted instead to cases of "Zusammenkrystallisirung," familiar from his laboratory experience with basic and double salts:

Ich habe mir auch die Frage aufwerfen müssen, ob es wahrscheinlich sey, dass bei einer solchen langsam zusammengewachsenen Verbindung mehrere Silicate derselben Base, aber von ungleichen Sättigungsstufen, sich mit dem Silicat einer andern Base zu einem chemischen Ganzen verbinden können, z.B. 1 Partikel Kalisilicat mit 2 Partikeln Thonsilicat und 2 Partikeln Thonsubsilicat. Theoretische Gründe bestimmen mich indessen, dieses für minder wahrscheinlich zu halten; [besides,] ich habe es bei der Berechnung der Analyse verschiedener krystallisirter Fossile nie gefunden. Wenn daher dieser Fall einträfe, so ist es wahrscheinlich, dass das Fossil als ein erhartetes Gemenge zweier Silicate von ungleicher Sättigungsstufe anzusehen ist, gleich einer eingedickten Auflösung eines neutral und basischen Salzes derselben Base. Von solcher Art ist z.B. die Zusammensetzung des Agalmatholits [a micaceous mineral] nach John's Analyse.[5]

[4] *Ibid.*, 213.

[5] *Ibid.*, 213—214.

Berzelius was clearly troubled by these complex compounds. Suggesting their existence in one breath, he questioned it in the next. His discomfiture evidently derived from his presentiment that his stoichiometric laws did not apply at this level of complexity. Though he thought there was little direct evidence for the existence of such complicated compounds, he was uncertain. The fundamental problem of discriminating complex compounds from mixtures remained. Consequently, in this connection, Berzelius called for the subjection of all minerals to his now reformulated unitary laws: the oxygen of one first order constituent must serve as the unit of the compound; the second order constituent containing the unitary first order constituent must serve as the unitary compound atom for the whole compound. Compounds that did not conform to the second criterion, even if they satisfied the first, were probably mixtures.[6] As shown below, Berzelius offered the mature version of his dualistic doctrine as a way to determine whether extremely complex compounds were compounds or mixtures.

These epistemological problems were compounded by an ontological one, also suggested in the passages just quoted. Berzelius attributed the occurence of extremely complex salts to the operation in nature of circumstances of mineral genesis that did not occur in the laboratory. Coupled with the emergent issues just described, this assertion led Berzelius to see clearly that mineralogy was not confined to the same purview, either in its phenomena or their causes, as the chemistry of the laboratory. In this respect and in regard to the close proximity of neighboring species among complex compounds (as shown by the topazes), minerals came to resemble organic chemicals. This resemblance rendered the boundary between the organic and the inorganic vague, leading Berzelius to articulate more fully his doctrine of their separation. The articulation of the

[6]*Ibid.*, 214, 226—227. These passages, especially the second, show that the two unitary laws had become fused into a single principle, applicable simultaneously to first order compounds and to the multiple saline components of higher order compounds. The fusion was accomplished by the perhaps tacit assumption that the ingredient of which the oxygen was a unit was present as a single particle. This assumption is clear from some of the examples he cited in illustration of the formulae (226—227): "Die mehrfach zusammengesetzten Fossile müssen nach dem obigen als Verbindungen mehrerer einfachen Gemische betrachtet werden; ihre Formeln setzt man demnach aus den Zeichen dieser letzten zusammen. So ist die Formel für den Ichtyophtalm $KS^3 + 5CS^3$. (Der Sauerstoffgehalt des Kali ist hier nämlich als Einheit des Ganzen angenommen und das Kali ist mit einer Menge Kieselerde verbunden, die drei solcher Sauerstoff-Einheiten enthält; und eine Einheit dieses Kali-Trisilica[t]s ist mit fünf Einheiten eben solchen Kalk-Trisilicats verbunden.) Auf solche Weise ist die Formel für den Byssolith $MS^2 + CS^2 + MgS^2 + 2FS$; (also eine Vereinigung von 1 *Partikel* dreier Verbindungen, deren Säure die Sauerstoff-Einheit der Base zweimal enthält, mit zwei *Partikeln* einer vierten, worin auch die Säure nur die Sauerstoff-Einheit der Base enthält.)" (Emphasis added.)

In the other passage noted here (214), which is the continuation of the passage quoted on p. 307, accompanied by n. 5, Berzelius suggested that a complex mineral not conforming to the unitary rules was a mixture: "Dass zwei Silicate auf ungleichen Sättigungsstufen sich nicht ohne Vermittelung des Silicates einer andern Base, welche die *Grundpartikel oder Einheit* der Verbindung ist, vereinigen könne, folgt aus der Lehre von den bestimmten Verhältnissmengen." (Emphasis added.)

dualistic doctrine and the effort to retain mineralogy on the side of the inorganic followed Hausmann's criticisms.

5. Mineralogy and organic chemistry: separating the inorganic and the organic

a. The mineralogical critisicm of chemically defined speciation

Though Berzelius offered his 1814 mineralogy only as a set of illustrated principles, his endeavor to reduce mineralogy to a branch of chemistry seemed more than tentative to mineralogists, who were quick to protest. Though they lodged a variety of complaints against Berzelius's call for the assimilation of mineralogy to chemistry, in this study discussion is confined to the main issue bearing on Berzelius's science: the extent to which stoichiometry alone could attain the specific level. The mineralogists' criticism of stoichiometrically defined speciation led some of them toward a perception of isomorphism, induced Berzelius to formulate the mature version of his dualistic doctrine, and intensified Berzelius's desire to draw a sharp boundary between the inorganic and the organic.

The most prominent of Berzelius's critics was J. F. L. Hausmann (1782—1859), who reviewed the 1814 monograph for the *Göttingische gelehrte Anzeigen.* Hausmann's good reputation and Berzelius's decision to reply to him directly in the 1815 monograph insured that this exchange would become widely known.[7] Less visible than Hausmann's review, but vastly more detailed, was Georg Pusch's lengthy criticism addressed to the audience of specialists.[8] Appearing after the exchange with Hausmann, Pusch's criticism was aimed at the fully articulated mineral system that appeared only in the 1815 monograph. Though his paper exerted no influence on the development of Berzelius's ideas, it is cited here to throw additional light on the debate about chemically defined species. So, too, is J. N. Fuchs's paper of 1815 in which the discovery of isomorphism was strikingly anticipated.[9] Relying on Berzelian precedent, Fuchs introduced a novel method that suggested how, despite the perceptions underlying the Wernerian

[7] For Hausmann, see Gümbel's article in the *Allgemeine deutsche Biographie*, XI (1880), pp. 94—97; and Karl Friedrich Philipp Martius, "[Denkrede auf] Johann Friedrich Ludwig Hausmann," in Martius, *Akademische Denkreden* (Leipzig: 1886), pp. 429—440.

[8] Georg Gottlieb Pusch (1790—1846), "Kritische Betrachtungen über das, auf die electrochemische Theorie und die chemische Proporzionslehre gegründete, Mineral-System des Herrn Professor Berzelius," *Leonhard's Taschenbuch für die gesammte Mineralogie*, 11,ii (1817), 471—522; 12,i (1818), 3—66; henceforth cited as Pusch, "Kritisches."

[9] Johann Nepomuk Fuchs (1774—1856), "Ueber den Gehlenit, ein neues Mineral aus Tirol," *Schweigger's Journal*, 15 (1815), 337—386; henceforth cited as Fuchs, "Gehlenit." For the discovery of isomorphism, see Evan M. Melhado, "Mitscherlich's discovery of isomorphism," *Hist. Stud. Phys. Sci.*, 11,i (1980), 87—123; henceforth cited as Melhado, "Mitscherlich's discovery."

criticism of Berzelius, stoichiometry could gain access to the specific level in mineralogy.

Both Hausmann and Pusch cited the frequent failure of chemically and mineralogically determined species to agree. Working in the immediate aftermath of Berzelius's *Versuch* and mindful of Haüy's insistence upon the unique correlation of composition and form, Hausmann had taken up the correlation question in his own research. Only the year before the appearance of Berzelius's 1814 monograph, he had published a paper addressed to the problem of the dependence of external characters, especially crystal form and crystal structure, on chemical composition.[1] It is scarcely surprising that this problem should have loomed large in his mind while composing his review of Berzelius's mineralogy. As part of his argument that Berzelius's taxonomy was "ganz unmineralogisch," Hausmann asserted of the system that

> das unnatürliche derselben nach einer naturhistorischen Ansicht, leuchtet ganz besonders auch in den Unterscheidungen der Species hervor. Die chemische Ansicht von der Gleich- und Verschiedenartigkeit der Substanzen weicht von der mineralogischen sehr ab; denn wenn man nach jener nur auf die qualitative und quantitative Gleich- oder Verschiedenartigkeit der Bestandtheile siehet, so muss man nach dieser zugleich auch auf die Gleich- oder Verschiedenartigkeit der äusseren Beschaffenheiten Rücksicht nehmen; auf den Einfluss, den die verschiedenen Theile der Mischung auf gewisse constante Beschaffenheiten des Aeussern haben; wodurch man dahin gelangt, in mineralogischer Hinsicht gewisser Bestandtheile für unwesentlich zu halten, die nach einer rein chemischen Ansicht nicht minder wesentlich sind als andere. Der Chemiker wird daher oft gewisse Species trennen, die dem Mineralogen nur als Abänderungen einer und der nähmlichen Substanz erscheinen. Herr Berzelius spricht z.B. zwey in der Mischung etwas verschiedene Varietäten des Eisenkiesels für zwey differente Species an, worin ihm, so wie in vielen ähnlichen Distinctionen, kein wissenschaftlicher Mineralog folgen wird.[2]

According to Hausmann, then, what Berzelius took to be significant differences in composition could have produced insufficient alteration in external characters to justify separation of two substances into two species. External characterization should therefore draw the boundary line between mineralogical species without demanding that it should always coincide with the stoichiometrically defined boundary.[3]

[1] J.F.L. Hausmann, "Specimen de relatione inter corporum naturalium anorganicorum indoles chemicas atque externas," *Commentationes societatis regiae scientiarum gottingensis recentiores*, 2 (1811—1813), (1813), 1—47 (separate pagination). The paper was summarized in *Göttingische gelehrte Anzeigen*, 1813,i (1813), 705—720.

[2] Hausmann, "Review," 2000.

[3] Hausmann's choice of the example of Eisenkiesel was unfortunate, however, because he did not appear to have direct knowledge of one of the two species that Berzelius had distinguished in the 1814 monograph. Ordinary Eisenkiesel, of course, was well known to Hausmann. Though it is now known to be a form of quartz containing a dense mixture of iron oxide, Berzelius concluded from an analysis of Bucholz that it was a distinct compound, which he named "supersilicias ferricus" ("1814 monograph," 34). The other, of which Hausmann seemed to have no direct knowledge, was Hedenbergite, which Berzelius had named for one of his students, A. Ludvig Hedenberg, who had

In part, Pusch shared Hausmann's belief that, however essential certain constituents might have been chemically, they may not have been essential mineralogically, that is, they might not have produced sufficient differences in external characters to justify separation of distinct species. Pusch did recognize the value of the theory of proportions as a control upon analyses,[4] and he acknowledged that the theory permitted distinguishing fine mixtures from pure compounds.[5] However, he denied that every distinct compound formulated in accordance with the theory was *ipso facto* a distinct species. While arguing against Berzelius's contention that mineralogy could be reduced to a branch of chemistry, Pusch remarked:

Wir erlangen freilich die Kenntnis der Verbindungen an sich, ihrer Mischungsweise und chemischen Eigenschaften durch die Chemie, als z.B. die Kenntnis von Schwefelspiesglanz; aber sie kann uns über dessen Krystallvarietäten, die dabei statt findenden mathematischen Verhältnisse, die Mannichfaltigkeit der Textur, Lichtdurchgang, Absonderungsansehen, Vorkommen u. dgl., alles Dinge, welche den Mineralogen so sehr, als die Mischung interessiren, keinen Aufschluss vorschaffen. Sollen wir z.B. zwei Varietäten von Rubellit, welche in nichts als einer kleinen Farbennüanze sich unterscheiden lassen, deswegen wirklich als zwei getrennte Gattungen betrachten, weil der Chemiker ihr ähnliches Mischung-Verhältniss doch so verschieden fand, dass er es durch die zwei Formeln

$$mgS + 2NS + 12AS \text{ und } mgS + NS + 6AS$$

ausdrücken muss? Dagegen möchte sich Vieles einwenden lassen, sobald man die chemische Ansicht nicht mehr vor Augen hat.[6]

analyzed the mineral shortly before his early death; see Hedenberg's "Undersökning af ett manganeshaltigt mineral ifrån Tunaberg," *Afhandlingar,* 2 (1807), 164—170. In the 1814 monograph (218), Berzelius had cited Hedenbergite as an example of a simple silicate, calling it "Eisenoxydul-Trisilicat." Later in the same monograph (34), it appeared with Eisenkiesel as an example of a silicate of iron. A pyroxene, Hedenbergite *does* differ from Eisenkiesel, and, in the 1814 monograph (295—296, n. 16), Berzelius listed its external characters to satisfy Hausmann. For Berzelius's immediate reaction to Hausmann's review and to this point in particular, see his letter to Gahn of 24 October 1814 (*Bref,* V,ix, pp. 104—108, esp. pp. 107—108), and his letter to Hisinger of 11 November 1814 (*Bref,* V,viii, pp. 37—39, esp. pp. 38—39).

Hausmann's example shows that Wernerians considered Eisenkiesel to be a distinct species; (cf. also C.A.S. Hoffmann, *Handbuch der Mineralogie,* 4 vols. (Freyberg: I, 1811; II,a, 1812; II,b—IV,b, fortgesetzt von August Breithaupt, 1815—1818), II,a, pp. 60—64). Haüy considered Eisenkiesel to be an immixed quartz (in his *Tableau comparatif des résultats de la cristallographie et de l'analyse chimique, relativement à la classification des minéraux* [Paris: 1809], p. 25). In the 1815 monograph (429), Berzelius registered a changeover to Haüy's view, listing Eisenkiesel as an impure oxide of silicon.

Though this example was not as well chosen as those adduced by Pusch (discussed below), the use Hoffmann made of it is suggestive of the mineralogist's reaction to Berzelius's innovation. An accomplished student of Werner, Hausmann refused to accede to Berzelius's separation of two silicates of iron into two species merely because of stoichiometric differences.

[4] Pusch, "Kritisches," 477. Hausmann ("Review," p. 1090) also recognized the importance of stoichiometry as a control.

[5] Pusch, "Kritisches," 9.

[6] *Ibid.*, 493. Rubellite is a tourmaline. When discussing more generally Berzelius's classification of tourmalines and garnets, Pusch pursued this argument along slightly different lines. For this, see below, pp. 312—313, nn. 1—6 and accompanying text.

Not merely did chemistry fail to describe the relation between composition and external characters, it failed to recognize that apparently large differences in composition often produced entirely minor differences in external characters. Pusch, therefore, devoted much of his criticism to denying Berzelius's contention, expressed quite forcefully in the 1815 monograph,[7] that each distinct compound was a distinct species. He hoped

> vollkommen darzuthun, dass die vermehrte Anzahl der Partikeln eines Bestandtheils, auch wenn sie wesentlich in die chemische Verbindung mit eingehen, nicht unbedingt eine neue Spezies hervorbringt.[8]

Citing specific examples, Pusch showed himself willing to label as groups of fine mixtures exhibiting transitions in composition the substances containing similar ingredients but having different formulae that Berzelius had distinguished into distinct species.[9]

Pusch took his arguments a step further than Hausmann, however, when examining Berzelius's separation of species with highly variable composition. In particular, he focussed on tourmalines and garnets, substances now known to exhibit isomorphic substitution of ingredients ont a grand scale.[1] The variability in the analyses of these minerals suggested to Pusch, as he asserted in his discussion of garnets, that essential ingredients, in the chemist's sense, perhaps could not be identified. He pursued this notion in regard to tourmalines in terms that apparently brought him to the threshold of isomorphism.[2]

He selected from Berzelius's classification five species of tourmaline:

	in der Familie des Kalks	
den brasilianischen Turmalin	CS + 2FS + 18AS	[Vauquelin]
	in der Familie des Natrons	
den elektrischen Schörl	NS + 9AS	[Klaproth]
hellen violetten Rubellit	mgS + 2NS + 12AS	[Vauquelin]
dunkelvioletten Rubellit	mgS + NS + 6AS	[Vauquelin]
	in der Familie des Kali's	
den Turmaline [gemeine Schörl]	KS + 4FS + 15AS	[Klaproth][3]

[7] See above, p. 301, passage accompanied by n. 4.

[8] Pusch, "Kritisches," 16. The part of the sentence beginning with "dass" was italicized.

[9] E.g., from the lead and copper families (*ibid.*, 16—20).

[1] The modern formula for tourmaline is $WX_3B_3Al_3(AlSi_2O_9)_3(O,OH,F)_4$, where W = Na,Ca and X = Al,Fe(III),Li,Mg. Garnet is a mineral group having the general modern formula $R(II)_3R'(III)_2(SiO_4)_3$, where R = Ca,Mg, Fe(II),Mn, and R' = Al,Fe(III),Ti,Cr. For these formulae, see Dana, *Manual of mineralogy*, 16th ed. (New York: 1952), pp. 374, 384. A slightly different formula for tourmaline is in the *Handbook of chemistry and physics*, ed. Robert C. Weast (53rd ed., Cleveland: 1972), p. B-197: $Na(Mg,Fe,Mn,Li,Al)_3Al_6Si_6O_{18}(BO_3)_3(OH,F)_4$.

[2] For Pusch on garnets, see his "Kritisches," pp. 30—35; on tourmalines, pp. 35—40.

[3] *Ibid.*, 36. The names in brackets are those of the analysts on whose results Berzelius relied to calculate the formulae given; they are drawn from the "1815 monograph," pp. 444—446.

Pusch remarked that the first two items in the list, and the two rubellites, were so similar that there was scarcely any justification in separating them:

> Wie sich brasilianischer Turmalin und elektrischer Schörl äusserlich und physikalisch unterscheiden, wo ihre Grenzen liegen, das kann nur Hr. B. selbst bestimmen, eben so zwischen hellem und dunkeln Rubellit. Der Mährische und Sibirische [not appearing in Berzelius's classification] gehören wahrscheinlich beide hierher und könnten zusammenverbunden allenfalls eine Gattung bilden wegen des Mangangehalts und mancher andern Eigenthümlichkeit. Trennen wird sie aber wegen Farbennuance und einigen vermehrten Partikeln Thonsilikat wohl Niemand.[4]

Of Brazilian tourmaline and common schorl he marked that, should chalk belong exclusively to the former and potash to the latter, they could justly be separated, "gewiss aber nicht weit von einander in ganz verschiedene Familien, denn sie sind sich in allen Kennzeichen und selbst in den Kerngestalten und Krystallformen ganz gleich."[5] In the face of even crystallographic identity, however, Pusch wondered whether these sharply drawn chemical lines were permissible species boundaries. Returning to the first two species in the list, Pusch remarked,

> Wenn Vauquelin bei der Analyse wirklich kein Natron, sondern Kalk statt dessen fand, so haben wir hier in einer philosophisch und naturhistorisch nie zu trennenden Fossiliengattung drei so nahe verwandte positiv elektrische Körper als der Kalk, das Natron und das Kali sind, abwechselnd mit der negativen Kieselerde sich verbinden können, ohne die Natur des Fossils wesentlich zu verändern und dieses Beispiel wird uns zeigen, dass auch der Theil von Hr. B. Gattungsprinzip, welcher sagt: dass die mindeste zu einer chemischen Verbindung hinzutretende Partikel eines fremden Stoffs eine neue Gattung bestimme, sich in der Mineralogie zur Gattirung nicht unbedingt anwenden lässt. Man erinnere hierbei daran, dass die schwefelsaure Thonerde mit Kali sowohl als mit Ammoniak und Natron Alaun bildet, der immer dieselbe Hauptnatur des Alauns behält.[6]

Similar basic ingredients, Pusch noted here, could combine "abwechselnd" with the acidic component and yet produce no significant change in the characters, even in the mathematically exact crystallographic parameters. This apparently close approach to the notion of isomorphic substitution lacked an essential ingredient. Substitution takes place within stoichiometrically defined limits, which Pusch, as a defender of natural history against stoichiometry, was scarcely liable to perceive. Seeing sharp stoichiometric boundaries between species belied by transitions in external characters, Pusch was intent only upon denying the adequacy of stoichiometry as a discriminator of species.[7]

The work of Fuchs stands in striking contrast with that of Pusch. While examining phenomena similar to those explored by Pusch, he carried his reasoning a step further, attaining a clear but undeveloped notion of substitution. Like Pusch, he recognized that characters, including crystallographic ones, might not

[4] *Ibid.*

[5] *Ibid.*, 37.

[6] *Ibid.*, 37—38.

[7] In this connection, see Melhado, "Mitscherlich's discovery."

have undergone significant alteration with certain changes in composition. However, he also realized that the changes in composition were confined within stoichiometric limits. In a paper devoted to describing a new mineral, which he named Gehlenite, Fuchs presented both an external characterization and an analysis.[8] Calculating the oxygen content of the various constituents in accordance with Berzelius's methods, he observed that the oxygen of the chalk present was insufficient to constitute an integral multiple of that containing the least oxygen, in this case, water.[9] However, the deficit could be made up by including in the calculation the oxygen of a small amount of iron oxide found in the analysis. Berzelius would have considered the iron oxide as finely mixed impurity and considered the deficit in the chalk a consequence of faulty analysis. Fuchs accepted neither this conclusion nor Pusch's denial of Berzelius's species definition. Instead, he perceived the possibility of substitution and introduced the term "vicarious constituent" to describe it: his analyses agreed, he remarked,

ziemlich gut mit der Lehre von den bestimmten chemischen Verhältnissmengen überein, von welcher sich, wie der berühmte Berzelius gezeigt, die Mineralogie überhaupt sehr viel zu versprechen hat. — Mit dem Kalk und Eisenoxyd hat es jedoch eine eigene Bewandtniss; weder dieses noch jener passt, für sich genommen, in die Reihe der Verhältnisse, werden aber beide zusammengenommen, so kommt eine Menge von Sauerstoff heraus, welche fast genau das Vierfache von dem Sauerstoff des Wassers, und fast gleich dem Sauerstoff der Thonerde ist.

Dessen ungeachtet halte ich das Eisenoxyd nicht für einen wesentlichen Bestandtheil dieser Gattung, sondern blos für einen *vicariirenden Bestandtheil,* wenn ich mich dieses Ausdruckes bedienen darf, für einen Stellvertreter von fast eben so viel Kalk, welcher bei der Abwesenheit des Eisenoxyds zur Ergänzung noch vorhanden seyn müsste, um mit den übrigen Bestandtheilen in das gehörige Verhältnis zu treten; und ich glaube, dass sich in der Folge Varietäten finden werden, die viel weniger oder gar kein Eisenoxyd, dagegen aber eine grössere Quantität von Kalk enthalten werden. . .

Aus diesem Gesichtspunkte wird man die Resultate mehrerer Analysen von Mineralkörpern betrachten müssen, wenn man sie einerseits mit der chemischen Proportionslehre in Uebereinstimmung bringen, andrerseits verhindern will, dass die Gattungen nicht unnöthiger Weise zu sehr zersplittert werden, was, wenn man immer in kleinen Mischungs-Verschiedenheiten schon einen hinreichenden Grund zur Trennung finden wollte, am Ende so weit gehen würde, dass man bei manchen nicht mehr im Stande wäre, einen bestimmten Gattungscharakter zu fassen. Die schwefelsaure Thonerde liefert mit Ammonium so gut wie mit Kali, oder mit diesen beiden Alkalien zugleich Alaun; wäre es wohl zweckmässig, diese drei verschiedenen Zusammensetzungen, die in ihren physichen Eigenschaften gar nicht von einander abweichen, als drei verschiedene Salzgattungen zu betrachten*)? Das Ammonium kann hier die Stelle des Kali ganz oder zum Theil vertreten, und umgekehrt.[1]

[8]The modern formula of Gehlenite is $(CaO)_3.Al_2O_3.(SiO_2)_2$, with some ferric iron replacing aluminium and some magnesium replacing calcium. See Dana, *System of mineralogy*. (6th ed. New York: 1914).

[9]Fuchs, "Gehlenite," 382.

[1]*Ibid.*, 382—384. The asterisk signalled this note (383n): "Gehlen schrieb mir kurz vor seinem Tode, dass es ihm gelungen sey, auch mit Natrum Alaunkrystalle darzustellen. Dieses erinnert an den Feldspath, welcher Natrum anstatt Kali enthält."

Unlike Pusch, Fuchs was sympathetic with Berzelius's innovations and used his insight to sustain them, even to the advantage of mineral systematics. With the notion of vicarious constituents, he here remarked, excessive splitting of species became unnecessary, and at the same time the chemical definition of species was upheld. Like Pusch, Fuchs concluded with citation of the well-known example of alum, which lay on the boundary line between traditional chemistry and mineralogy. Unlike Pusch, Fuchs showed with this example how the notion of substitution could reduce the number of species, yet keep the mineralogist at the specific level, which Pusch denied was possible.

The promise of this approach, however, would be decisively realized only from the end of the second decade of the century by Mitscherlich. Berzelius himself might have recognized isomorphism, but he appears to have lacked either or both of two prerequisites to seeing it: a knowledge of crystallography, that would permit an observer, like Pusch, Fuchs, and later Mitscherlich, to recognize the persistence of crystallographic characters in defiance of changes in composition; and an ability to shift his focus from the *oxides* of particular basic radicals to the total *oxygen* carried by *all* the basic radicals. Berzelius had trained himself too thoroughly to see each integral dose of oxygen as carried by a *particular* radical in *specific* quantity.[2] His response to the mineralogical critique of the chemical species definition was not the discovery of isomorphism, but a further articulation of his notion of dualism.

b. Hierarchical dualism: further articulation of chemistry to accommodate the complexity of minerals

If, as argued earlier, Berzelius's dualism may be defined as the attribution to compounds of a binary or dualistic composition, so that one of the constituents served as acid and the other as base, up to the 1815 *Mineralogy* he had applied this notion consistently only at the level of simple salts. His model for double salts as compounds of simple salts, though essentially dualistic, coexisted with the multiple oxide model, but did not replace it for lack of laws governing the proportions subsisting between higher order constituents. Berzelius's mature expression of dualism, here termed "hierarchical," explicitly extended dualism to all compounds, no matter how complex, and in particular, to double salts and beyond. According to this view, any compound could be concieved as composed of a pair of electrochemically opposing constituents; these in turn could likewise be considered as composed of two constituents; and so on down to the elements. Each level of composition was to be distinguished by labelling it as belonging to a given "order:" compounds of the elements with one another were first order

[2] As is suggested, for example, by the passages quoted above, p. 284, accompanied by nn. 5 and 6.

compounds, e.g., oxides; those composed of at least one first order compound and a second ingredient of equal or lower order were second order compounds, e.g., salts; those composed of one second order compound and another ingredient of second or lower order were third order compounds, e.g., double salts; double salts with water of crystallization were fourth order compounds. This explicit aspect of hierarchical dualism was expressed in a much quoted *locus classicus* that first appeared in the *Proportions chimiques:*

> Si le point de vue électro-chimique est juste, il s'ensuit que toute combinaison chimique dépend uniquement de deux forces opposées, l'électricité positive et le négative, et qu'ainsi chaque combinaison doit être composée de deux parties constituantes réunies par l'effect de leur réaction électrochimique, attendu qu'il n'existe pas une troisième force. De là découle que tout corps composé, quel que soit d'ailleurs le nombre de ses principes constituants, peut être divisé en deux parties, dont l'une est positivement et l'autre négativement électrique. Ainsi, par example, le sulfate de soude n'est pas composé de soufre, d'oxigène et de sodium, mais d'acide sulfurique et de soude, qui l'un et l'autre, peuvent être encore divisés en deux éléments, l'un positif, l'autre négatif. De même, l'alun ne peut pas être considéré comme directement composé de ses principes simples; mais on doit regarder ce corps comme le produit de la réaction du sulfate d'alumine, élément négatif, sur le sulfate de potasse, élément positif; et c'est ainsi que la manière de voir électrochimique justifie également ce que j'ai déjà exposé des particules composées du premier, du second, du troisième ordre, etc.[3]

Berzelius offered his notion of hierarchical dualism in the 1815 *Mineralogy* as part of a series of "Betrachtungen über die Constitution der Mineralien," apparently motivated by Hausmann's criticisms.[4] Proposed in recognition of the inadequacy of the multiple saline model that Berzelius used in the 1814 monograph, the hierarchical conception of dualism emerged in discussion of three points bearing on the composition of complex compounds and the difficulty of discriminating them from complex mixtures. In this connection, Berzelius introduced the term "constitution" (Sw: *constitution*) to signify the composition of a complex substance portrayed as a higher order compound of preexisting lower order constituents. This conception became a powerful and vexatious motif in nineteenth-century organic chemistry.

The first of the three points was Berzelius's distinction between two kinds of (visually) homogeneous minerals, those that were "zusammenverschmolzen" (Sw: *sammangjutna fossilier*), those containing several compounds that were prevented by the accidents of formation from separating out before the mineral

[3] *Proportions chimiques,* pp. 97—98. The passage continued (pp. 98—99) with these remarks on organic composition: "Ce même point de vue est applicable à la chimie organique, et chaque produit organique peut être considéré comme divisible en oxigène et en un radical composé, bien que la multiplicité des particules simples et la structure plus compliquée qui doit en résulter pour l'atome composé, puissant faire que cette division électrique ne soit possible qu'en idée, parce qu'il est probable que, dans la plupart des cas, le radical électro-positif, hors de l'oxide construit d'une manière déterminée, ne peut pas exister isolément." The organic application of dualism is explored below.

[4] "1815 monograph," 323—336.

hardened, from those that were "ungemengt" (Sw: *oblandade fossilier*), pure compounds. Claiming that most mineralogists had not distinguished the two but referred to them indiscriminately with the word "simple," Berzelius pleaded for acceptance of this innovation intended to improve conceptual precision.[5] He argued that both uncrystallized and crystallized minerals could belong to the category of the "zusammenverschmolzen," the crystallized ones being cases of the "Zusammenkrystallisirung" that he had cited in the original monograph.[6] In the latter, he had merely asserted that the problems posed by "zusammenkrystallisirt" mixtures would be overcome by the advance of the analytical art. In 1815, however, he argued more specifically that the task confronting chemical analysis was to decide whether a given sample was "zusammenverschmolzen" or "ungemengt" by determining whether the (second order) ingredients were present in proportions stoichiometrically permissible for a single compound or could be partitioned in accordance with stoichiometry only among several compounds. In exemplification, he showed how an analysis of an artifically "zusammenverschmolzen" mixture of alum and glauber salt would permit partitioning of the constituent oxides only into the two compounds that were mixed together in the sample.[7] Berzelius thus had equated the problem of discriminating complex mixtures from complex compounds to the problem of partitioning the oxides revealed by analysis either among several compounds or to the lower order ingredients of a higher order compound.

His second point was to suggest that the task of partitioning second order ingredients could be facilitated if there were limits on the number of second order constituents that could enter into a single compound. In particular, he inquired

> 1) Welches ist die grösste Anzahl verschiedener binärer Körper, welche zusammen eine bestimmte chemische Verbindung ausmachen können? und 2) Welches ist das Maximum für die Theilchen eines jeden binären Körpers, welche in eine gewisse Verbindung eingehen können? ist dieses Maximum grösser im Verhältniss als die Anzahl der Bestandtheile zunimmt?[8]

The method he suggested for answering these questions was to impose a further limit on the complex substances under consideration: confining them to samples of a single crystallographic species. Solving the crystallographic problem of discriminating accidental and essential ingredients became a means of limiting the number of stoichiometrically permissible species in mineralogy; for in this manner,

> kann man hoffen, in jedem Fossil von einer complicirten Zusammensetzung durch fortgesetzte Arbeit zu entdecken, welche binäre Bestandtheile desselben die eigentlich Verbindung ausmachen, und welche nur zufällig sind; auch ist unzweifelhaft, dass während

[5] *Ibid.*, 323—324.
[6] *Ibid.*, 325—326.
[7] *Ibid.*, 326—328.
[8] *Ibid.*, 330.

dieser Forschungen die vorgelegten [zwei] Fragen sich durch Vergleichung der gefundene Resultate beantworten.[9]

Though Berzelius now replaced his blithe comment of 1814 that accidental impurities must be subtracted from the analytical results with the concession that variability in mineral analyses "eher zurückschreckt als aufmuntert,"[1] he nevertheless hoped that research of this kind would reveal additional limits on the complexity of mineral species.

As his third point, Berzelius suggested how, given that all the essential second order ingredients of a complex compound were known, to partition them so that the mineral might be considered a higher order compound of lower order constituents. Abandoning the multiple saline model of the 1814 monograph, Berzelius proposed his classic conception of hierarchical dualism in a passage that formed the prototype for the *locus classicus* quoted above:

Denken wir uns ein mannichfaltiger zusammengesetztes Fossil, z.E. Pyrop (worin Kieselerde die Säure ist zu den 4 Basen Kalk, Magnesia, Thonerde und Eisenoxyd) mit der Vorstellung, dass dessen Bestandtheile möglicherweise nur eine einzige ungemischte chemische Verbindung ausmachen. Wir können hier auf dem ersten Anblicke nicht eben so leicht uns vorstellen wie diese 4 Silicate zusammengebunden seyn konnten zu einem gemeinsamen Ganzen, als wir einzusehen glauben, wie Schwefelsäure und Kalkerde im Gipse eine einzige Verbindung ausmachen könne . . .[2]

Wir wissen, dass die Elemente in der *unorganischen Natur* sich zu vereinigen streben in Verhältnisse ihres elektrochemischen Gegensatzes; und daraus folgt, dass, da die Verbindungen eben hierdurch statt haben, nie mehr als 2 Körper sich verbinden können, weil es ein dritte mitwirkende Kraft nicht giebt. Schwefel vereinigt sich mit Sauerstoff; Natrium verbindet sich auch mit Sauerstoff, und jedes von beiden ist nach geschehener Verbindung als *ein einziger* Körper blos zu betrachten, Schwefelsäure oder Natron; diese zwei können sich nun vereinigen, aber nicht als drei, Schwefel, Natrium und Sauerstoff, sondern einzig als zwei, Schwefelsäure und Natron, und nach geschehener Vereinigung, nachdem der elektrische Gegensatz unter ihnen gehoben ist, machen sie wieder blos einen einzigen Körper aus. Dieser vereinigt sich wieder mit Wasser, zu krystallisirtem Glaubersalz, nicht als drei, Schwefelsäure, Natron und Wasser, sondern blos als zwei: schwefelsaures Natron und Wasser[, that is, it is not a compound of multiple oxide components, but a binary combination of a second order compound, sulfate of soda, with a first order one, water of crystallization].

Auf gleiche Weise vereinigt sich schwefelsaures Kali mit schwefelsaurer Thonerde [to form alum], nach des erstern als des elektropositiven Gegensatz zu der letztern als elektronegativer. Diese zwei machen nun einen einzigen Körper aus, welcher wiederum von seiner Seite sich mit Wasser verbinden kann. Um demnach die rein chemische Analyse des krystallisirten Alauns zu machen, wird erfordert, dass man zuerst die zwei nächsten Bestandtheile absondert, aus denen er besteht. Diese sind Wasser und eigentlicher Alaun, dann müssen diese geschieden werden, das erste in seine Elemente, das letztere in seine nächsten Bestandtheile schwefelsaures Kali und schwefelsaure Thonerde, [that is, again, its second order constituents, not its multiple oxide components]; von diesen wiederum

[9] *Ibid.*, 332.
[1] *Ibid.*, 333.
[2] *Ibid.*

jedes in die seinigen, und so fort bis zu den letzten Elementen. Alles was in diesem Fall richtig ist, und bei dem angeführten Beispiel gilt, muss richtig seyn, und gelten bei einem jeden andern Körper, der nach Principien der unorganischen Natur zusammengesetzt ist. [Alum, again, in this revised view, was to be the bridge between chemistry and mineralogy, and, as an object of both sciences, was to assure their commensurability as sciences of the inorganic.]

Kommt uns demnach ein zusammengesetztes Mineral vor, so ist die erste Frage: welches sind seine zwei nächsten Bestandtheile? Diese Frage ist bei einfachern Fossilien leicht beantwortet; aber bei sehr zusammengesetzten dürften wir für jetzt uns einizig von Muthmassungen leiten lassen. Z.B. im Tafelspath = CS^2 sind die nächsten Bestandtheile Kieselerde und Kalkerde. Im Grammatit, dessen Zusammensetzung durch die Formel = $CS^2 + MS^2$ ausgedrückt wird, ist est auch leicht gefunden, dass er besteht aus CS^2 und MS^2, d.i., aus den zwei Verbindungen, deren jede für sich ausmacht Tafelspath und edlen Serpentin [two known lower order compounds].

Nehmen wir nun ein andres mehr zusammengesetztes Fossil vor uns, z.B. Kalkstein und Vesuvian, dessen Zusammensetzung nach Klaproths Analyse durch FS + 4CS + 5AS ausgedrückt wird. Hier verbirgt sich die natürliche Vertheilung mehr, kann aber doch mit Wahrscheinlichkeit gefunden werden. Das Fossil enthält 5AS mit 4CS; demnach können diese zwei nicht die eine Hälfte ausmachen, denn dann müsste deren Anzahl gleich, oder die des einen Antheils ein Vielfaches von dem des andern seyn; wenn wir aber den andern Antheil AS zu FS legen, so erhält man FS + AS als einen Körper, vereinigt mit CS + AS als einem Körper, aber 4 Partic. des letztern verbunden mit einer des erstern. Dieses eine Beispiel dürfte hinreichen mich verständlich zu machen. Es ist klar, dass des Fossils elektrochemische Constitution, wenn ich so diese chemische Entzweiung nennen darf, durch dieselbe Formel dargestellt werden muss, welche die Anzahl der binären Particuln darlegt, welches es enthält; ich sehe es gleichwohl noch für zu frühe an, einen Versuch darüber zu wagen, zumalen die Entwickelung der neu aufgekommenen Probleme sicher mehr als eines Menschen Leben und Arbeit erfordert.[3]

This dualistic conception was not only an aid to rationalizing the composition of complex compounds, but also a stoichiometric constraint upon the proliferation of mineral species, as Berzelius made clear in a paragraph omitted from the preceding quotations. In reference to the difficulty of conceiving how four silicates, unlike the two sulfates in alum, could be united into a single compound, he remarked:

Diess ist wohl wahr, dass nach Daltonschen blos mechanisch atomistischen Gründen, die Sache ganz leicht verstanden würde; denn dort baut man mit einer Anzahl gegebner Atome ein zusammengesetztes Atom, zu gleicher Weise wie man ein Haus baut, ohne auf etwas anders zu sehen, als dass das Mathematische in der Construction richtig ist; aber nirgends leuchtet die Unrichtigkeit solcher halben Ansichten klarer ein, als hier, denn hieraus würde eine unendliche Reihe von Verbindungen hervorgehen in einer beinahe unendlichen Anzahl von Proportionen. Die Erfahrung hingegen bezeugt, dass diess nicht der Fall ist...[4]

The appeal of dualism thus lay in the possibilities it offered for both rationalizing the composition of complex compounds and limiting the number of complex

[3] *Ibid.*, 334—336.

[4] *Ibid.*, 333—334.

species. In the interest of achieving these goals, dualism was a call for the construction of new stoichiometric laws that would govern the proportions subsisting between higher order components and link the composition of adjacent higher orders in the same way as the rule of oxides limited the numbers of salts and linked the composition of salts with that of their oxide constituents. Berzelius never devised such laws, however, and he persisted in relying on the law of unitary oxygen. At least two circumstances were responsible for this. First, shortly after the articulation of hierarchical dualism, Mitscherlich's discovery of isomorphism changed the face of mineralogy: the apparent complexity of many minerals and the resistance of many species to unequivocal chemical characterization could now be seen as the result of isomorphic substitution; strict adherence to classical stoichiometric principles had obscured this effect; minerals henceforth would prove more submissive to chemical techniques. Second, in an age when the distinction between a mineral chemist and an inorganic chemist was becoming sharper, Berzelius fit the latter category more than the former. Having successfully treated virtually all the inorganic chemicals then known and very many minerals, Berzelius had achieved his principal goals. He sensed that to pursue the still greater complexity of many minerals, even if in terms of hierarchical dualism, was to reach beyond not only his own grasp, but that of contemporary chemistry.

It was outside of mineralogy that the dualistic conception had its greatest influence. Berzelius's notion that any compound possessed a "constitution," that it consisted of a hierarchy of pre-existing binary constituents, was an early foundation of nineteenth-century organic chemistry. With the dualistic model in mind, chemists attempted to determine the constitution of organic compounds and express it in a "rational formula"—one that revealed the dualistic constituents of a compound, such as organic radicals and their oxides. However, just as eighteenth-century vegetable analysis was suspect because of the possibly creative role of fire, so in the nineteenth century doubts emerged about whether the supposedly pre-existing dualistic ingredients were not in some fashion the new products of the reactions that revealed them. A unitary view of organic compounds thus came to replace the Berzelian one, but the possibility of establishing rational formulae remained a constant theme of organic chemistry through the establishment of the structural theory.

c. Mineralogy and organic chemistry

The inadequacy of Berzelius's chemistry as it existed in 1814 to reduce mineralogy to a branch of chemistry has already been pointed out. That minerals displayed greater diversity and complexity in patterns of combination than inorganic chemicals had compelled Berzelius's articulation of hierarchical dualism in the 1815 *Mineralogy.* Whether differing forces were responsible for these differences in the phenomena of combination was a question that also demanded an answer from Berzelius. Having argued in 1814 that minerals and chemicals were no different in kind, Berzelius supported this claim by asserting that the forces underlying mineral genesis were the same as those operating in the reactions of the laboratory chemist. However, if the phenomena were indeed different, then so also were their causes. Moreover, if the causes operating in the mineral realm were different from those at the disposal of the chemist, then perhaps mineralogy was not rightly regarded as a branch of chemistry. A closely related question was raised by the attributes that Berzelius assigned to the causes peculiar to the mineral realm and to the character of their perceived effects. As shown below, the ideas Berzelius employed in this connection rendered minerals similar to organic chemicals and marked the point of transition, on the inorganic side, from the inorganic to the organic.

An illustration of these points is provided by the assertions Berzelius made in the passage quoted above (p. 307, accompanied by n. 4). First, Berzelius claimed that the most complex minerals were formed by the operation of the weakest affinities; second, that minerals had been not been synthesized in the laboratory; and third, that the circumstances of laboratory experimentation were insufficiently subtle to achieve such syntheses. These notions, coupled with Berzelius's demonstration, discussed above, in subsection 4.*d.*, that addition of atoms to complex minerals entailed only minor changes in percentage composition, rendered his conception of minerals quite similar to his view of organic compounds: weak affinities, defiance of laboratory synthesis, and transitions were common to both. The similarities presented a potential threat to the distinction between the organic and the inorganic. Now from the mineralogical side as well as from the organic side, the boundaries confining the applicability of his inorganic laws were in danger of becoming vague. To restore their clarity, Berzelius had to assure the full discrimination of mineralogy, as a branch of inorganic chemistry, from organic chemistry. He tried to accomplish this goal in his response to Hausmann.

One issue raised by Hausmann was primarily responsible for evoking Berzelius's defense of the distinction he had drawn between mineralogy as inorganic chemistry and natural history. Reacting to Berzelius's claim that only a chemical mineralogy could be scientific, Hausmann raised the question of the extent to which viable conceptual foundations existed for the natural-historical approach to mineralogy. Hausmann's position, and Pusch's, who approvingly quoted him, was clearly that mineralogy must be natural-historical, that it must differ from chemistry, and that it could remain scientific.[5]

After quoting Hausmann's contentions,[6] Berzelius replied with a discussion of the viability of natural-historical principles of classification in their application to inorganic nature. He pointed out that the objects of natural history were

[5] Hausmann, "Review," 1098—1099.

[6] "1815 monograph," 305—306.

living entities—here again was the primacy of the distinction based upon life functions—and that they were categorized by external characters taken to indicate the essence of a given category.

> In der organischen Naturgeschichte beruht die ganze Classification *einzig und allein* auf äussern Charakteren, und unter diesen beinahe ausschliessend auf *der Form,* ohne die mindeste Berücksichtigung der inneren Zusammensetzung; bei welchen es oft unmöglich wäre die Abänderungen zu erforschen, und aus welchen, für jetzt, nichts geschlossen werden könnte. Die Gegenstände für die organische Naturgeschichte sind künstlich zusammengesetzte Maschinen, beruhend auf einer in ihrem Innern niedergelegten Kraft, welche nur für eine gewisse Zeit wirkt, wobei die Maschine allmälig sich selbst abnützt, und endlich zerstört wird. Die Naturgeschichte sammlet deren äussere Charaktere, stellt die Gegenstände in Klassen zusammen, welche gewisse gemeinsame wesentliche äussere Charaktere haben, wodurch alle zu einer Klasse gehörende eine allgemeine Geschlechtsgleichheit haben, und da man Klassen von ähnlichen wesentlichen Charakteren neben einander stellt, so entsteht der allmälige Uebergang von einer Form in die andere; so dass wie auch immer jedes getrennte Gleid in der Kette dem nächst liegenden gleich seyn mag, sich doch bei weit getrennten Gliedern oft die grössten Ungleichheiten in den Formen finden.[7]

Inquiring first about the applicability of natural-historical methods to organic matter, Berzelius reiterated the sort of argument deployed in his paper on organic analysis to account for the existence of generic categories in organic chemistry and their non-existence in inorganic chemistry. Of organic chemistry he asserted:

> Die organische Natur ist zusammengesetzt aus mehr denn 2 Elementen, gewöhnlich aus 3 oder 4, und bisweilen mehrern, welche beinahe in allen Verhältnissen sich vereinigen zu können scheinen: d.i. so, um gemäss den Ansichten der Corpuscular-Theorie zu sprechen, dass in der Vereinigung zwischen den Particuln oder Atomen der Grundstoffe, kein Atom Einheit zu seyn braucht, sondern dass sie nach allen Verhältnissen, innerhalb gewisser Gränzen eines Maximums und Minimums, sich vereinigen können; z.B. wenn die Elemente sind A, B, C, D so können sie vereinigt seyn in solchen Verhältnissen, dass in folgender Formul $3A + 4B + 6C + 12D$, die Ziffern mit jeder andern beliebigen [with certain exceptions that Berzelius cited in a note, of no import for the example] Ziffer vertauscht werden können; einzig dass sie ein gewisses Maximum nicht übersteigen, das wir indess noch nicht kennen. Durch diesen Umstand kann vermittelst dieser 4 Elemente eine beinahe unberechenbare Anzahl verschiedener Verbindungen hervorgebracht werden, und dadurch dass man ein Atom fortnimmt oder zulegt zu irgend einem der Grundstoffe, entsteht eine neue Verbindung, welche aber nach ihrer Zusammensetzung, wie nach ihren darauf beruhenden Eigenschaften, der vorhergehenden ganz nahe kommt, und mit derselben eine vollkommene Aehnlichkeit hat. So sehen wir z.B. unter den Producten des Gewächsreichs das eine flüchtige Oel sich vom andern unterscheiden; fast alle haben eine gemeinsame Uebereinstimmung in den physischen und chemischen Charakteren; so ist der Zucker von Saccharum offinarum ungleich dem aus Vitis vinifera, und diese beiden ungleich dem Zucker aus Stärke, aber drei sind Zucker. Diese Umstandes wegen liegt das Princip für die systematische Aufstellung der organischen Natur in *dem* für die Zusammensetzung ihrer nähern Bestandtheile.[8]

[7] *Ibid.*, 307.

[8] *Ibid.*, 308—309.

The production of generic categories, resulting from the absence of narrow limitations upon possible patterns of combining proportions, permitted the application of natural-historical method to organic matter. Organic matter, like the ordinary objects of natural history, was to be classified by external characters. Despite his novel concern for the specific level in organic analysis, he granted natural history primacy in this sphere, as the next paragraph indicates:

> Man sieht aus diesen Anschauungen, wie die Chemie indem sie auf der einen Seite das Prinzip für die systematische Aufstellung der organischen Naturhistorie rechtfertigt, auf der andern Seite sich spät in die organische Naturhistorie einmischt, und auch dann blos um aufzuklären, vielleicht[,] aber niemals zu dirigiren; weil in der organischen Natur etwas liegt, welches über die Grenze der Erweiterung gehen dürfte, wohin wir möglicher Weise einmal die Chemie bringen können.[9]

The dividing line between the organic and the inorganic was that between natural history and inorganic chemistry; organic chemistry belonged to the former, mineralogy to the latter; the former was the realm of the naturalist; the latter of the scientist. Berzelius's boundary line therefore lay between natural history and science. The mineralogical branch of traditional natural history had been made to cross that line, but not the others. Moreover, Berzelius did not believe that the organic realm was soon to become scientific, for what was essential to being organic also made the organic inaccessible to science:

> Wenn wir uns die Erdmasse fertig vorstellen mit allen ihren Grundstoffen, aber ohne Bewegung und ohne organische Wesen, so wird eine fremde Einwirking erfordert, um sie zu bewegen und eine, wenn ich so sagen darf, weit unbegreiflichere fremde Einwirkung, um die sich allezeit reproducirende organische Natur hervorzubringen. In dem Umkreis der unorganischen Natur vereinigen sich die Elemente nicht nach dem Princip für Zusammensetzung in der organischen. Gieb dem Chemiker die Grundstoffe, und seine Bemühungen werden sich fruchtlos verzehrten in Versuchen dieselben auf die Weise der organischen Natur zusammen zu binden; die organische Natur scheidet sich demnach von der unorganischen eben darin, dass wenn die Elemente gegeben sind, sie eine primum movens fordert, welches ausser den Elementen liegt, und ohne welches niemals, durch der Elemente ursprüngliche Kräfte, organische Wesen, oder nur mit deren Producten gleichartiger Verbindungen, gebildet werden.[1]

This passage may be compared with the *locus classicus* of hierarchical dualism (quoted above, p. 316, accompanied by n. 3) and its prototype (quoted above, p. 319, accompanied by n. 3), in which the binarism of inorganic chemistry was attributed to the operation in the inorganic realm of two and only two forces, those of the opposing electricities. Organic chemistry, however, was distinguished by the action of a third force, that of the life processes. Because this force was not at the disposal of the scientist, the sphere of the organic was not within the pale of science.

Regarding the application of natural-historical principles to inorganic matter,

[9] *Ibid.*

[1] *Ibid.*, 309—310.

Berzelius restated his leading ideas about inorganic composition and then argued for the conformity of all minerals to the principles of inorganic stoichiometry. He began by citing binary organization and the role of one ingredient as unit:

> Das Princip dagegen für die Zusammensetzung der unorganischen Natur ist gänzlich verschieden. Ihre nähern Bestandtheile sind zusammengesetzt nur aus zwei Elementen; und des Erdglobus unorganische Masse stellt nichts anders dar, als einige wenige elementare Körper ausser dem Verbindungszustand, eine grössere Zahl aus zwei Elementen zusammengesetzter d.h. binärer Körper, und eine ganz grosse Anzahl Verbindungen zwischen binären Körpern mit einander vermischten. In der binären Verbindung muss das eine Element allezeit die Einheit seyn; das will sagen, dass wenn A und B Elemente sind, so können sie sich einzig in solchen Verhältnissen vereinigen, dass das eine A sich verbindet mit 1, 2, 3, 4 . . . B . . . aber keine Verbindungen finden sich zwischen 2A und 3B, oder 4A mit 5B, u.s.w. Daher kommts, dass wenn man ein Atom von B entweder zusetzt oder wegnimmt, so wird die Mischungsveränderung so bedeutend, dass meist alle Aehnlichkeit aufhört, und das A + 3B nicht mehr einige äussere Analogie mit A + 2B hat. Weiter, wenn die binären Körper wechselseitig sich mit einander verbinden, so gilt dasselbe Gesetz, das einer davon allzeit die Einheit seyn muss[, that is, the law of the unitary atom applies to all orders]; und daraus folgt hinweiderum, wegen der grossen Verschiedenheit in den Proportionen der Zusammensetzungen, dass die Verbindung AB + CB meist nicht im geringsten äussere Aehnlichkeit mit AB + 2CB hat.[2]

As a result of the narrow limitations on the possibilities of inorganic combining proportions, variation in external properties was by saltation, not transition:

> Die Mischungsveränderungen in der unorganischen Natur geschehen daher mit so grossem Sprung, dass alles Suchen nach Uebergang dort eine Ungereimtheit ist, und der Grund davon liegt eben in dem Princip für die Zusammensetzung der unorganischen Natur.[3]

The most complex minerals, though exhibiting lesser changes in properties with changes in composition, did not constitute an exception, as Berzelius now stated expressly:

> Doch muss ich erinnern, dass wenn mehrere unterschiedene binäre Körper sich vereinigen, der Sprung minder stark ist im Verhältniss als die Anzahl der verschiedenen binären Körper und mit derselben die Anzahl der zwei oder mehrerer Atome zunimmt; und je näher man dem Maximum dieser Anzahl kommt, desto weniger ungleich werden die verschiedenen Verbindungen, welche um diese Zahl herum möglich sind, so dass man sich überreden könnte etwas zu haben, dass dem Uebergange zwischen den Gliedern in der Kette der organischen Natur ähnlich sey. Aber ungeachtet diese Aehnlichkeit unter den um das Maximum herumliegenden Körpern bisweilen nicht zu verkennen ist, so kann sie gleichwohl nicht angewandt werden als Princip für die Aufstellung des Ganzen, denn die hat bei der grössten Anzahl unorganischer Producte nicht Statt.[4]

Therefore, complex minerals resembled organic compounds in this sense only because they represented the extreme case, just as the vegetable acids represented

[2] *Ibid.*, 310—311.
[3] *Ibid.*, 311.
[4] *Ibid.*

an extreme, similar to inorganic matter, in the organic sphere. The most complex minerals and the simplest vegetable acids—however different in many respects—were the points of contact between the organic and the inorganic.

Berzelius addressed elsewhere the other shared characteristic of organic chemicals and minerals: that neither had been synthesized, for the causes producing them seemed not within the control of the chemist. The point was pursued in a paper of 1818 devoted specifically to this question,[5] in which Berzelius proceeded on two fronts. He distinguished the source of weak affinities in the two spheres, and he provided some examples of mineral syntheses to suggest that, in principle, if not in general fact, mineral synthesis was possible. For organic matter, Berzelius had argued that weak affinities derived both from complexity of composition and, more importantly, from an electrochemical modification imposed on the elements by the life processes. For complex minerals, however, he argued that weak affinities derived from a combination of complexity of composition with the relatively low reactivity of many mineral constituents:

> So long as we confine our attention to those combinations which owe their existence to very strong chemical affinities, we find a striking simplicity and a constant analogy in their composition: but, if we begin to examine those combinations which depend upon weak affinities, we find, that, as we proceed, the modes of combination continue to increase in number and intricacy. Salts formed by the union of strong acids, with salsifiable bases, belong to the former of these classes; whilst a great part of the productions of the mineral kingdom, where bodies slightly electronegative perform the function of acids, belongs to the latter. This class includes also some other very weak combinations; those, namely, which take place among compound molecules of the second or third order, and which form salts of a double, sometimes perhaps of a triple base. It is sufficiently ascertained, that chemical affinity rapidly diminishes as the number of combining atoms augments. Thus, the affinity of compound molecules of the first order is much weaker than that of elementary bodies; and the difference between the affinity exerted by molecules of the second order, and that exerted by molecules of the first, is immensely greater. Now, it is precisely the study of these decreasing affinities which forms the object of mineralogical chemistry, and without which, that science can never arrive at any higher degree of perfection.[6]

Accordingly, the causes underlying the formation of complex minerals were of the same kind as those invoked in laboratory procedure. Mineralogy was still a part of chemistry, for there was no other agency, such as the life force, to move it outside the pale of that science:

> The attempt to prove that the combinations which constitute minerals have been formed under the same laws which regulate the union of elementary substances in our laboratories,

[5] Berzelius, "Examination of some compounds which depend upon very weak affinities," *Edinb. Phil. J.*, 1 (1819), 63—75, 243—253; Swedish, "Undersökning af några föreningar som bero af svagare frändskaper," *Afhandlingar*, 6 (1818), 5—41; henceforth cited from the English text as "Weak affinities."

[6] "Weak affinities," 63—64.

has met with a degree of success, sufficient to remove all doubts concerning the accuracy of its principle; but in order to reduce the results of not a few mineral analyses into conformity with these laws, the supporters of this opinion have been obliged to admit modes of combination, to which the chemistry of our laboratories offers nothing analogous; and it has been found, in general, that compounds in the mineral kingdom consist of a greater number of molecules than we have it in our power to combine by artificial means. The reason of our inability to form such combinations is not, that weak affinities are inert in our experiments; but that they are destroyed by those greater forces of which we are obliged to make use, in order to obtain the combinations in an isolated state. Hitherto few trials have been made to produce artificial compounds analogous to fossils; but I am fully persuaded that such attempts will succeed, to a degree beyond what is hoped for in the present state of science.[7]

In principle, therefore, minerals could be synthesized. Berzelius's paper, an early effort in geochemistry, offered a few exemplary syntheses to illustrate the point. Minerals bordered on the organic in their complex composition, weak affinities, and slow change in properties with composition, but they were not therefore organic. The life functions were absent, and both the causes and the effects were of fundamentally the same nature as those of the laboratory.

d. First order organic compounds: foundations in Berzelius's views for the organic chemistry of the nineteenth century

1) *Berzelius's initial pessimism*

The preceding discussion shows that by 1819 Berzelius's views of first order organic compounds consisted of a cluster of several leading ideas, some of which were precedented in the work of Lavoisier. First order compounds were ternary and higher combinations of their elements, not binary; they could not be synthesized; they could be readily decomposed, but not by procedures subject to fine control; they occurred in generic mixtures of similar species that were difficult to separate from one another; and their properties derived from a peculiar electrochemical modification imposed on their elements by the life processes and could not be correlated with their composition.[8]

Taken together, these ideas amounted to a near renunciation of the possibility of a chemical science of first order organic compounds. If these substances could neither be synthesized nor subtly decomposed, and if the origin of their properties could not be analyzed, there was little to do but discover, analyze, and classify them—that is, treat them as Berzelius believed they should have been treated, in terms of natural history.[9] That he could accept and persist in this

[7] *Ibid.*, 64.

[8] Cf. above, p. 290, n. 1.

[9] Cf. *Lehrbuch,* III,i, p. 147 (fully cited above, p. 287, n. 3). After having listed several well-known examples of the production, from inorganic starting materials, of substances having some properties of organic matter, Berzelius remarked there: "Wenn wir aber auch in Zukunft mehrere solche Producte, aus rein unorganischen Materien und von einer mit den organischen Producten analogen Zusammensetzung, entdecken sollen, so ist doch diese unvollständige Nachahmung immer zu unbedeutend, als dass wir jemals hoffen könnten es zu wagen,

attitude of renunciation is clear from his discussion of organic chemistry in the edition of his textbook after the *Proportions chimiques.* There, Berzelius observed that

> Alles was also unsere Forschungen in diesem geheimnissvollen Theile der Chemie ausrichten können, ist, die chemischen Veränderungen zu beobachten, welche von den lebenden Körpern in den Elementen, worin sie sich befinden, d.h. auf die chemischen Reagentien, von welchen sie umgeben sind, hervorgebracht werden, so wie die Erscheinungen, welche die Lebensprozess begleiten, so weit uns möglich ist, ihn zu verfolgen, dann die organischen Producte von einander zu trennen, ihre Eigenschaften zu studiren und ihre Zusammensetzung zu bestimmen. Aber alles dies ist sehr schwer, die Chemie erhebt sich hier auf einen höheren Werth, wenn ich mich dieses Gleichnisses für mathematische Gegenstände bedienen darf, und auch das schärfste Auge ist beständigen Irrthümern ausgesetzt, während die Entdeckung des Wahren oft eben so sehr vom Zufall als von geistvoller Berechnung abhängt.[1]

One of the principal difficulties confronting even this natural-historical approach to organic matter was precisely that the chemist was still without sure access to the specific level:

> Die erste Schwierigkeit, welche bei Behandlung organischer Materien aufstösst, ist die Unmöglichkeit, zu unterscheiden, ob das, was ich untersuchen will, eine Verbindung, oder sogar nur ein blosses Gemenge aus zwei organischen Stoffen ist, oder ob es wirklich von allen anderen isolirt ist. Es geschieht so oft, dass zwei mit einander verbundene organische Materien sich einander in ihre Verbindungen mit unorganischen Körper [in which form they are submitted to combustion analyses] begleiten, und wenn sie bei keinem von unseren Versuchen getrennt werden, so werden wir dadurch irre geführt und halten das Untersuchte für unvermischt. Auf diese Art sind wir immer sehr unsicher, ob das von uns für einen einzigen Stoff Gehaltene wirklich so ist; und überdem ist es in vielen Fällen absolut unmöglich, einen solchen rein und frei selbst von solchen fremden Materien zu erhalten, deren Gegenwart uns bekannt ist. Ueberall in der organischen Natur sind die Schwierigkeiten zu trennen grösser, und die Wege dafür weniger zahlreich als in der unorganischen, so dass dieser Theile der Chemie als noch gänzlich in seiner Kindheit sich befindend betrachtet werden kann.[2]

organische Stoffe künstlich hervorzubringen, und, wie es in den meisten Fällen in der unorganischen Natur glückt, die Analyse durch die Synthese zu bestätigen." However, second order compounds could be subsumed under chemical science, just like ordinary salts, because they conformed to the rule of oxides. (However, as shown by the passage accompanied by the next note, second order species were as difficult to attain as those of the first order.)

[1] *Ibid.*, p. 147.

[2] *Ibid.*, pp. 147—148; cf. also p. 151, where Berzelius listed this as the first requirement of analysis: "*Muss der zu analysirende Körper von allen anderen brennbaren Körpern getrennt werden.* Dies ist, wie ich schon anführte, sehr schwer; es können dafür keine bestimmten Regeln gegeben werden, und es beruht durchaus auf dem eigenen individuellen Vermögen die Wege aufzufinden, um die Reinheit des Körpers zu prüfen. Viele für eigenthümliche und besondere Stoffe gehaltene Körper möchten wohl Verbindungen aus zweien oder mehreren sein, welche uns der Zufall zu trennen noch nicht gelehrt hat. Diese, die erste Bedingung, wenn eine Analyse ein richtiges und anwendbares Resultat geben soll, ist daher so schwer zu erfüllen, dass man bei der Unsicherheit deshalb gerne eine Reihe von Untersuchungen vermeidet, deren Resultat mit der Zeit unrichtig befunden werden wird." Cf. also p. 174: "... die Hauptsache bei diesen [organischen] Analysen

Berzelius, then, saw organic chemistry as seriously encumbered in two main ways in which inorganic chemistry, including mineralogy, was not. First, the peculiarities of the subject reduced the chemist to a passive observer of vital processes and a taxonomist of organic products. Second, to fulfill the first condition of further progress in organic chemistry—gaining sure access to the level of specific compounds—was an incomparably more difficult task than gaining access to inorganic species. Berzelius's prose did not suggest that no progress on these fronts could be achieved—his was a description of the current state of affairs, not a proscriptive injunction.[3] On the contrary, he explicitly recommended pursuit of research in organic chemistry:

> In der lebenden Natur scheinen die Elemente ganz anderen Gesetzen zu gehorchen, als in der todten, die Producte ihrer gegenseitigen Einwirkungen werden daher ganz anders, als in dem Gebiete der unorganischen Natur. Die Ursache dieser Verschiedenheit zwischen dem Verhalten der Elemente in der todten Natur und den lebenden Körpern zu entdecken, würde der Schlüssel zur Theorie der organischen Chemie sein. Sie ist indessen auf eine solche Weise verborgen, dass wir, wenigstens gegenwärtig, ohne alle Hoffnung sind, sie auszumitteln. Dessen ungeachtet müssen wir streben, dieser Kenntniss näher zu kommen; denn einmal wird uns glücken, entweder sie ganz zu erreichen, oder bis an einer bestimmten Grenze stehen zu bleiben, über welche das menschliche Forschungsvermögen nicht weiter hinaus erstreckt werden kann.[4]

By pressing on in organic chemistry, the burdens would be lifted, or at least, the extent to which they must be accepted would become known. However, Berzelius evidently believed that there was to be no early significant progress.

2) *Foundations for organic chemistry in the early nineteenth century*

Though Berzelius's pessimism of the 1820's was justified, the obstacles to progress in organic chemistry began to be cleared in his own lifetime. During the two decades following the *Proportions chimiques,* practitioners mindful of Berzelius's concern for the specific level devoted most of their energies to building a growing store of specific compounds, analyzing them quantitatively, and characterizing them qualitatively. Advances on this front were recorded by Thomas Thomson in his 1838 *Chemistry of organic bodies: vegetables.* The object of the tome, he remarked in its preface,

ist, dass man den Stoff, welchen man untersucht, genau von allen anderen organischen Stoffen trenne, und dass man hernach seine Sättigungscapacität bestimme. Ein jedes aufgeführte analytische Resultat, das nicht von genauen Untersuchungen über die Reinheit des Stoffes und über Bestimmung seines Sättigungscapacität begleitet ist, verdient kein Vertrauen, und hat es dessen ungeachtet der Untersuchende versucht, nach seinem Resultate die relative Menge der Atome zu berechnen, so beweist dies, dass er nicht eingesehen hat, was zur Gewinnung eines zuverlässigen Resultates erfordert wird."

[3] As several scholars have suggested: see Jørgensen, "Lebenskraft;" John H. Brooks, "Wöhler's urea, and its vital force?—a verdict from the chemists," *Ambix,* 15 (1968), 84—114; and Benton, "Vitalism." Benton's paper is particularly enlightening.

[4] *Lehrbuch,* III,i, p. 135.

is to lay before the British chemical public, a pretty full view of the present state of the chemistry of Vegetable Bodies. This branch of the science has made so much progress of late years, that a very wide and inviting field has been laid open. Several hundred new substances have either been discovered, or their characters have been determined with such precision, and their composition investigated with such accuracy, as to give a pretty accurate idea of their constitution, and of their connexion with each other.[5]

This and similar works recorded the construction of one of the two principal experimental foundations of the older radical theory, a body of exact, quantitative analyses of specific compounds and the translation of analytical results into formulae.

A second experimental pillar of the older radical theory, one that Berzelius did not foresee, was the recognition of a new class of organic reactions. Appearing in the early 1830's, the newly recognized reactions were neither decompositions nor syntheses, but "metamorphoses," as they came to be called by contemporaries. Metamorphoses were reactions providing genetic links between compounds of comparable complexity. The justly celebrated study of benzaldehyde and its reactions (1832) by Liebig and Wöhler is perhaps the most prominent early example of the novel reaction pattern of which more and more examples rapidly accumulated.[6] Entailing the perception of stable groups of atoms that passed intact from one compound to another in a fashion comparable to inorganic displacements, studies of metamorphoses gave the concept of a compound organic radical a practical foundation that it previously lacked. Berzelius himself remarked in his textbook that the idea of organic radicals was still, in the mid-1820's, entirely speculative.[7] The perception of groups of atoms moving intact

[5]Thomas Thomson, *Chemistry of organic bodies: vegetables* (London: 1838), p. v. The passage here quoted continued with this remark: "These ultimate analyses, with a very few exceptions, have been all made upon the continent, and chiefly in Germany and France. British chemists have scarcely entered upon the investigations." Another comparable source that recorded progress in discovering, analyzing, and formulating specific organic compounds is Franz Döbereiner, *Supplement zu J.W. Döbereiner's Grundriss der Chemie: Tabellarische Darstellung der organischen Stoffe in alphabetischer Ordnung für Aertzte, Chemiker und Pharmaceuten* (Stuttgart: 1837). For this reference, the author is indebted to Dr. Frank H. Stodola, formerly of the United States Department of Agriculture.

[6]Justus Liebig and Friedrich Wöhler, "Untersuchungen über das Radikal der Benzoesäure," *Ann. Chem.*, 3 (1832), 249—282. Another pioneering study in the history of the older radical theory was Robert Wilhelm Bunsen's study of cacodyl compounds (1837—1843), for which see Partington, *History*, IV, pp. 284—286.

[7]*Lehrbuch*, III,i, pp. 139—140: "Da in der organischen Natur der Sauerstoff ein wesentlichen Bestandtheile [in ternary and higher compounds] ist, so können auch die organischen Producte als Oxyde von zusammengesetzten Radikalen betrachtet werden. Diese Radikale existiren nicht ausser Vereinigung mit Sauerstoff, wenigstens kennen wir kein einziges derselben, und sind folglich ganz hypothetisch, denn sonst wären, was wir nun zusammengesetzte Atome der ersten Ordnung nennen, eigentlich Atome der zweiten Ordnung, und ihre hypothetischen Radikale würden die wirklichen Atome der ersten Ordnung ausmachen. Auch kennen wir von diesen Radikalen nicht mehrere Oxydationsstufen, sondern es wird die Zusammensetzung des Radikales immer durch Zukommen von Sauerstoff verändert." Cf. the passage from the *Proportions chimiques* quoted above, p. 316, in n. 3.

through a succession of metamorphoses transformed the idea of radicals from a speculation into a plausible interpretation of experimental observations.

That interpretation, the older radical theory, completed the organic deployment of Berzelius's theory of salts. It was constructed entirely in the spirit of Berzelius's inorganic chemistry. The difference between organic chemistry and inorganic chemistry rested wholly in the differences in their radicals. Inorganic radicals were elementary; organic radicals were compound. In both cases, specificity resided in the radicals, and higher order compounds were saline in nature, containing one substance serving as base and another serving as acid. Though radicals apparently could not exist independently of their compounds, their presence and their composition could be discerned by reflection upon the course of metamorphoses. The older vocabulary that described organic compounds as ternary (and higher combinations) of the elements gave way to a vocabulary of radicals and their compounds. Accordingly, the older radical theory was an advance over Berzelius's original pessimism of 1819, because it permitted a description of the constitution of organic compounds.

That advance was limited, however. The theory did not permit a description of the constitution of the radicals themselves. Still perceived as the peculiar product of the life functions, the radicals remained as mysterious as ever. For a while, this was no disadvantage. Chemists could table discussion of fundamental questions about organic matter while they exploited the older radical theory to organize and interpret a growing body of experimental material. Only when experiment revealed a chemistry of the radicals themselves, a chemistry manifest in the laboratory rather than just in the organism, did the older radical theory prove unsatisfactory. Phenomena such as isomerism and substitution forced chemists to address the fundamentals that the older radical theory ignored.

Though these events led to a rapid demise for the older radical theory, its legacy was never outlived. Later chemists soon found that, though radicals were insufficient to serve as the foundation of organic chemistry, they were also indispensable. Organic chemistry could neither rest content with radicals nor do without them. What role was to be attributed to the radicals and what to their constituents and how those roles were played were the chief questions debated by organic chemists during the three decades from 1840 to 1870. As the chemistry of groups of atoms, Berzelian science was an important resource for ideas about these issues. Despite the fall of the formal theory, the Berzelian legacy remained influential in the organic chemistry of subsequent decades.

Abbreviations

1. *Books*

Some books are cited in the notes by abbreviation only. The abbreviations and the full citations they represent are as follows:

Bref	Berzelius, Jöns Jacob, *Jac. Berzelius Bref, utgifna af Kungl. Svenska Vetenskapsakademien genom H. G. Söderbaum,* 6 vols., suppl. (Stockholm and Uppsala: 1912—1935).
Davy, *Works*	Davy, Humphrey, *The collected works of Sir Humphrey Davy, Bart. Ll.D. F.R.S. edited by his brother, John Davy, M.D. F.R.S.* 9 vols. (1839—1840; rpt., New York: 1972).
DSB	*Dictionary of Scientific Biography,* ed. Charles C. Gillispie, 16 vols. (New York: 1970—1980).
Lavoisier *et al.*, *Méthode*	Lavoisier, Antoine Laurent, Louis Bernard Guyton de Morveau, Claude Louis Berthollet, and Antoine François de Fourcroy, *Méthode de nomenclature chimique, proposé par MM. de Morveau, Lavoisier, Bertholet [sic] et de Fourcroy. On y a joint un nouveau système de caractères chimiques, adaptées à cette nomenclature, par MM. Hassenfratz et Adet* (Paris: 1787).
Lavoisier, *Traité*	Lavoisier, Antoine Laurent, *Traité élémentaire de chimie, présenté dans un ordre nouveau et d'après les découvertes modernes; avec figures,* 2 vols. (1789; rpt., Brussels: 1965).
Lavoisier, *Oeuvres*	Lavoisier, Antoine Laurent, *Oeuvres de Lavoisier, publiées par les soins de son excellence le ministre de l'instruction publique et des cultes,* 6 vols. (1862—1893; rpt., New York: 1965).
Olsson, *Kemiens historia*	Olsson, Hugo, *Kemiens historia i Sverige intill år 1800,* Lychnos-Bibliotek: studier och källskrifter utgivna av Lärdomshistoriska samfundet, Vol. XVII (Uppsala: 1971).
Partington, *History*	Partington, J.R., *A history of chemistry,* 4 vols. (I,i, II—IV, London: 1961—1970; I,ii, not published).

2. *Journals*

Most journals are cited in abbreviated form. The abbreviations and the fuller citations they represent are as follows:

Abhl. Akad. Berlin	*Abhandlungen.* (Akademie der Wissenschaften, Berlin).
Afhandlingar	*Afhandlingar i fysik, kemi, och mineralogie.* Stockholm.

Ann. chim.	*Annales de chimie et de physique.* Paris.
Ann. Phil.	*Annals of philosophy, or magazine of chemistry, mineralogy, mechanics, natural history, agriculture, and the arts.* London.
Ann. Sci.	*Annals of science.*
Arch. int. d'hist. sci.	*Archives internationales d'histoire des sciences.*
Chemisch Weekblad	*Chemisch Weekblad.* (Nederlandsche Chemische Vereeniging, Amsterdam).
Chymia	*Chymia: annual studies in the history of chemistry.*
Edinb. Phil. J.	*Edinburgh philosophical journal.*
Hist. Stud. Phys. Sci.	*Historical studies in the physical sciences.*
Janus	*Janus: revue internationale de l'histoire des sciences, de la médicine, de la pharmacie, et de la technique.*
J. de physique	*Journal de physique, de chimie, d'histoire naturelle, et des arts.*
Gehlen's Journal	*Neues allgemeines Journal der Chemie,* ed. A.F. Gehlen. 8 vols. (Leipzig: 1803—1806); *Journal für die Chemie*[, *Physik* from vol. 4] *und Mineralogie,* 9 vols. (Berlin: 1806—1810).
Gilbert's Annalen	*Annalen der Physik,* ed. L.W. Gilbert, vols. 21—60 (Halle: 1809—1818).
K.S.V.A. Economiska Annaler	*Economiska annaler.* (Kungliga Svenska Vetenskapsakademien, Stockholm).
K.S.V.A. Handlingar	*Handlingar.* (Kungliga Svenska Vetenskapakademien, Stockholm).
Med. Hist.	*Medical History.*
Mém. de l'Acad.	*Histoire de l'Académie royale des sciences avec les mémoires de mathématique et physique.* (Académie des sciences, Paris).
Nicholson's Journal	*Journal of natural philosophy, chemistry, and the arts,* ed. W. Nicholson.
Mém. soc. Arcueil	*Mémoires de physique et de chimie.* (Société d'Arcueil).
Obs. sur la physique	*Observations sur la physique, sur l'histoire naturelle, et sur les arts,* vols. 1—43 (Paris: 1773—1793).
Phil. Trans.	*Philosophical transactions.* (Royal Society of London).
Rev. d'hist. sci.	*Revue d'histoire des sciences et de leurs applications.*
Samml. chem. u. chem.-tech. Vorträge	*Sammlung chemischer und chemisch-technischer Vorträge* (Stuttgart: 1896—1935).
Schweigger's Journal	*Journal für Chemie und Physik,* ed. J.S.C. Schweigger, 69 vols. (Nürnberg: 1811—1833).
Stud. HPS	*Studies in history and philosophy of science.*
Trans. Am. Phil. Soc.	*Transactions.* (American Philosophical Society, Philadelphia).

Bibliography

The bibliography lists all sources, primary and secondary, cited in the notes (except for a handful of items having minor significance). It also includes a few sources that, although not cited in the notes, illuminate topics explored in the text. Items in the bibliography are arranged alphabetically by author. Items by the same author are listed in chronological order by publication date of the text cited, except for 1) reprints, which are cited by date of original publication; 2) articles in journals of certain learned societes, which are cited by the years of the volumes (which often precede the years of publication); and 3) items cited by date supplied (as in the case of Bergman's dissertions). Works by a single author appearing in the same year are listed in strict alphabetical order by title. The alphabetization invokes all words in each title, including articles, particles, and prepositions, and it treats the letters *å, ä, ö,* and *ü,* as if they were *a, a, o,* and *u,* respectively. Works written by an author in collaboration are listed after works by the author alone (if any are cited). Where several collaborative works of an author are listed, they are arranged in alphabetical order by name of first collaborator. Several works by the same author and collaborator are arranged by the same principles governing works by single authors.

Baumé, Antoine, *Manuel de chymie ou exposé des opérations de la chymie et de leurs produits* (2nd ed.; Paris: 1766).

Beck, Richard, *Abraham Gottlob Werner: eine kritische Würdigung des Begründers der modernen Geologie; zu seinem hundertjährigen Todestage* (Berlin: 1918).

Beekman, E. H. M., *Geschiedenis der systematische mineralogie* ([the Hague: 1906]).

Benton, E., "Vitalism in nineteenth-century scientific thought: a typology and reassessment," *Stud. HPS,* 5 (1974), 17—48.

Bergman, Torbern Olof, "Vorbericht zur Abhandlung über Luft und Feuer" (1777), in Scheele, *Werke,* I, pp. 3—27.

— *Opuscula physica et chymica,* 6 vols. (I—III, 1779—1783; rpt., Leipzig: 1786—1792; IV—VI, Leipzig: 1779—1790).

— *De acidis metallicis* (1781), in Bergman, *Opuscula,* III, pp. 124—131.

— *De analysi ferri* (1781), in Bergman, *Opuscula,* III, pp. 1—108.

— *De causa fragilitatis ferri frigidi* (1781), in Bergman, *Opuscula,* III, pp. 109—123.

— *De diversa phlogisti quantitate in metallis* (1782), in Bergman, *Opuscula,* III, pp. 132—156.

— *Meditationes de systemate fossilium naturale* (1784), in Bergman, *Opuscula,* IV, pp. 180—278.

— *Torbern Bergman's foreign correspondence,* eds. Göte Carlid and Johan Nordström, Lychnos-Bibliotek: studier och källskrifter utgivna av Lärdomshistoriska samfundet, Vol. XXIII,i (Stockholm: 1965).

Berthollet, A. B., "Mémoire sur la combination du soufre avec l'oxygène et l'acide muriatique," *Mém. soc. Arcueil,* 1 (1807), 161—179.

Berthollet, Claude Louis, *Essai de statique chimique,* 2 vols. (Paris: 1803).

Berzelius, Jöns Jacob, *Afhandling om Galvanismen* (Stockholm: 1802).

— *Föreläsningar i djurkemien,* 2 vols. (Stockholm: 1806—1808).

— "Alkaliernas decomposition," *K.S.V.A. Economiska Annaler,* 6 (April, 1808), 114—121. (Published anonymously.)

— "Undersökning af islands-mossans sammansättning och dess användande såsom näringsmedel," *K.S.V.A. Economiska Annaler,* 8 (November, 1808), 53—94.

— "Einige prüfende Versuche über die Theorie der Volta'schen Säule, und Nachricht seinen [Berzelius's] Untersuchungen über die Metallisirung der Alkalien, und über die Zahlenverhältnisse, wonach Sauerstoff und Schwefel sich mit den Metallen verbinden . . . In einem Briefe an den Prof. Gilbert. Stockholm, den 20 Apr. 1810," *Gilbert's Annalen,* 35 (1810), 269—277.

— "Försök till kiseljordens analys," *Afhandlingar,* 3 (1810), 117—128.

— "Försök rörande de bestämda proportioner, hvari den oorganiska naturens beståndsdelar finnas förenade," *Afhandlingar,* 3 (1810), 162—276; 5 (1818), 94—128.

— "Ueber die Streitigkeiten zwischen Davy und den HH. Gay-Lussac und Thenard, die Metalle aus den Alkalien, das Ammoniak und den Stickstoff betreffend (in einem Schreiben an den Prof. Gilbert) Stockholm, den 1 Aug. 1810," *Gilbert's Annalen,* 36 (1810), 198—203.

— "Vermischte Notizen (aus den Briefen an den Herausgeber [Gehlen] vom 19ten April 1808, 2ten September und 8ten October 1809)," *Gehlen's Journal,* 9 (1810), 585—589.

— "Zerlegung der Kieselerde durch gewöhnliche chemische Mittel," *Gilbert's Annalen,* 36 (1810), 89—102.

— "Auszug aus einem Briefe des Hrn. Prof. Berzelius an den Prof. Gilbert. Stockholm, d. 20 Mai 1811," *Gilbert's Annalen,* 38 (1811), 227—232.

— "Ein spätere Zusatz aus der Handschrift des Verfassers. Neue Untersuchungen über das Schwefel, die Bleioxyd und die Schwefelsäure," in *Versuch,* pp. 47—53, and *Gilbert's Annalen,* 37 (1811), 324—334.

— "Essai sur la nomenclature chimique," *J. de physique,* 73 (1811), 253—286.

— "Om de bestämda proportioner hvari den oorganiska naturens beståndsdelar finnas förenade; summariskt resultat af de försök som häröfver blifvit anställde," *K.S.V.A. Handlingar,* [2] 32 (1811), 169—197.

— "Schreiben des Herrn. Prof. Berzelius, Mitgl. der schwed. Akad. d. Wiss. an den Prof. Gilbert, über einige Gegenstände, welche zwischen Davy und den HH. Gay-Lussac und Thénard [*sic*] streitig sind, und über ein zweites neues Gesetz, welches er im Verfolge seiner Untersuchungen aufgefunden hat. Stockholm, d. 1. Oct. 1810," *Gilbert's Annalen,* 37 (1811), 208—220.

— "Spätere Zugabe, die organischen Körper betreffend. Aus der Handschrift des Verfassers," in *Versuch,* pp. 82—87 and *Gilbert's Annalen,* 37 (1811), 465—472.

— "Ueber das Verhältniss der Sauerstoffmengen etc. zu einander, welche die Körper auf verschiedenen Oxydationsstufen aufnahmen, etc. (Aus einem Schreiben an den Akademiker Gehlen vom 14. Aug. 1810)," *Schweigger's Journal,* 1 (1811), 257—262.

— "Ueber die bestimmten und einfachen Verhältnisse nach welchen die Bestandtheile der unorganischen Natur verbunden sind. Summarischer Bericht über diesen Gegenstand an den königl. Acad. in Stockholm erstattet," *Schweigger's Journal,* 2 (1811), 297—326.

— "Versuch, die bestimmten und einfachen Verhältnisse aufzufinden, nach welchen die Bestandtheile der unorganischen Natur mit einander verbunden sind," *Gilbert's Annalen,* 37 (1811), 249—334, 415—472; 38 (1811), 161—226; 40 (1812), 162—208, 235—320.

— "Försök till en chemisk nomenclatur," *K.S.V.A. Handlingar.* [2] 33 (1812), 28—74.

— "Försök till ett rättfärdigande af de theoretiskt chemiska åsigter, på hvilka den systematiska uppställningen i mitt försök till en förbättring af den chemiska nomenclaturen grundar sig," *K.S.V.A. Handlingar.* [2] 33 (1812), 166—198, 223—244; 34 (1813),

39—62, 175—215.
— *Öfversigt af djur-kemiens framsteg och närvarande tillstånd; tal, hållet för Kungl. Vetenskaps-Akademien vid Praesidii nedläggande den 13 Augusti 1810* (Stockholm: 1812).
— "Versuch, die chemischen Ansichten, welche die systematische Aufstellung der Körper, in meinem Versuch einer Verbesserung der chemischen Nomenclatur begründen, zu rechtfertigen," *Schweigger's Journal,* 6 (1812), 119—144, 144—176, 284—322; 7 (1813), 43—78.
— "Zwei Schreiben des Hrn. Prof. Berzelius, Mitgl. d. schw. Ak. d. W., an den Professor Gilbert (Wahre Natur der Eisenoxyde und der salpetrigen Säure; Entdeckungen neuer Metall-Oxyde und Schwefel-Metalle; Kritik von Davy's Lehre von der Chlorine; eine neue und merkwürdige Erscheinung von Feuer; Druckfehler.)," *Gilbert's Annalen,* 42 (1812), 276—298.
— *A view of the progress and present state of animal chemistry,* tr. Gustavus Brunmark (London: 1813; reissued: 1818).
— "Essay on the cause of chemical proportions, and on some circumstances relating to them; together with a short and easy method of expressing them," *Ann. Phil.,* 2 (1813), 443—454; 3 (1814), 51—62, 93—106, 244—257, 353—364.
— "On the composition of animal fluids," *Ann. Phil.,* 2 (1813), 19—26, 195—208, 377—387, 415—425.
— "Versuche über die Mischung des isländischen Mooses und seine Anwendung als Nährungsmittel," *Schweigger's Journal,* 7 (1813), 317—352.
— "Experiments to determine the definite proportions in which the elements of organic nature are combined," *Ann. Phil.,* 4 (1814), 323—331, 401—409; 5 (1815), 93—101, 174—184, 260—275.
— *Försök att genom användandet af den electrokemiska theorien och de kemiska proportionerna grundlägga ett rent vettenskapligt system för mineralogien* (Stockholm: 1814).
— "Neue Untersuchungen über die Natur des Stickstoffs, des Wasserstoffs, und des Ammoniaks; und einige Bemerkungen über die wahre Zusammensetzung der salpetrigen Säure und der Salpetersäure: Erster Nachtrag zu seinem Versuche, die bestimmten und einfachen Verhältnisse aufzufinden, nach welchen die Bestandtheile der unorganischen Natur mit einander verbunden sind," *Gilbert's Annalen,* 46 (1814), 131—175.
— "Versuch, durch die Anwendung der electrisch-chemischen Theorie und der chemischen Proportion-Lehre ein rein wissenschaftliches System der Mineralogie zu begründen," *Schweigger's Journal,* 11 (1814), 193—233; 12 (1814), 17—62; 15 (1815), 277—300.
— "An address to those chemists who wish to examine the laws of chemical proportions, and the theory of chemistry in general," *Ann. Phil.,* 5 (1814), 122—131.
— "Försök till ett rent kemiskt mineralsystems," *Afhandlingar,* 4 (1815), 1—148.
— "Versuch eines rein chemischen Mineralsystems," *Schweigger's Journal,* 15 (1815), 301—363, 419—451; 22 (1818), 274—302.
— "Untersuchungen über die Zusammensetzung der Phosphorsäure, der phosphorigen Säure und ihrer Salze: Zweiter Nachtrag zu seinem Versuche, die bestimmten und einfachen Verhältnisse aufzufinden, nach welchen die Bestandtheile der unorganischen Natur mit einander verbunden sind," *Gilbert's Annalen,* 53 (1816), 393—446; 54 (1816), 31—55.
— "Nouvelles recherches sur les proportions chimiques," *Ann. chim.,* 5 (1817), 174—181.

— "Allmän öfversigt af hittils vunna resultat af undersökningarna om bestämda proportioner," *Afhandlingar,* 5 (1818), 190—197.

— "Försök att närmare bestämma åtskilliga oorganiska kroppars sammansättning, till vinnande af en närmare utveckling af läran om de kemiska proportionerna," *Afhandlingar,* 5 (1818), 379—520.

— "Försök öfver de bestämda förhållanden, hvari elementen för den organiska naturen äro förenade," *Afhandlingar,* 5 (1818), 520—646.

— "Försök öfver vattnets förhållande till syror, saltbaser, och salter; öfver basiska och dubbla salters sammansättning, samt öfver salpetersyrliga salter," *Afhandlingar,* 5 (1818), 132—189.

— "Undersökning af några föreningar som bero af svagare frändskaper," *Afhandlingar,* 6 (1818), 5—41.

— "Undersökning af quäfvets, vätets, och ammoniakens natur; jemte några anmärkningar om salpetersyrans och salpetersyrlighetens rätta sammansättning," *Afhandlingar,* 5 (1818), 198—238.

— "Versuche die Zusammensetzung verschiedener unorganischer Körper näher zu bestimmen,zur genaueren Entwickelung der Lehre von den chemischen Proportionen," *Schweigger's Journal,* 23 (1818), 98—122, 129—202, 279—299.

— *Essai sur la théorie des proportions chimiques et de l'influence chimique de l'électricité* (Paris: 1819).

— "Examination of some compounds which depend upon very weak affinities," *Edinb. Phil. J.,* 1 (1819), 63—75, 243—253.

— *Nouveau système de minéralogie* (Paris: 1819).

— *Om blåsrörets användande i kemien och mineralogien* (Stockholm: 1820).

— *Lehrbuch der Chemie nach des Verfassers schwedischer Bearbeitung der Blöde-Palmstedt'schen Auflage übersetzt von F. Wöhler,* 4 vols. in 8 (Dresden: 1825—1831).

— "Ueber die Veränderungen in dem chemischen Mineral-Systeme, eine nothwendige Folge der Eigenschaft isomorpher Körper, sich einander, in unbestimmten Verhältnissen, zu ersezzen," *Leonhard's Taschenbuch für die gesammte Mineralogie,* 1825,ii (1825), 169—182, 193—218.

— "Proportions, determinate," in the *Edinburgh encyclopedia, conducted by David Brewster,* XVII (1830), pp. 175—190.

— *Versuch, die bestimmten und einfachen Verhältnisse aufzufinden, nach welchen die Bestandtheile der unorganischen Natur mit einander verbunden sind,* ed. Wilhelm Ostwald, Ostwald's Klassiker der exakten Wissenschaften, No. 35 (Leipzig: 1892).

— *Jac. Berzelius: självbiografiska anteckningar, utgivna af Kongl. Svenska Vetenskapsakademien genom H. G. Söderbaum* (Stockholm: 1901).

— *Jac. Berzelius Bref, utgifna af Kungl. Svenska Vetenskapsakademien genom H. G. Söderbaum,* 6 vols., suppl. (Stockholm and Uppsala: 1912—1935).

— *Autobiographical notes,* ed. H. G. Söderbaum, tr. Olof Larsell (Baltimore: 1934).

Berzelius, Jöns Jacob, and Wilhelm Hisinger, "Versuch, betreffend die Wirkung der elektrischen Säule auf Salze und auf einige von ihren Basen," *Gehlen's Journal,* 1 (1803), 115—149.

— "Försök med elektriska stapelns verkan på salter och på några af deras baser," *Afhandlingar,* 1 (1806), 1—38.

Berzelius, Jöns Jacob, and Magnus Martin Pontin, "Försök med alkaliernas och jordarternas sönderdelning," *K.S.V.A. Economiska Annaler,* 6 (May, 1808), 110—130; (June, 1808), 113—118.

— "Electrisch-chemische Versuche mit der Zerlegung der Alkalien und Erden," *Gilbert's Annalen,* 36 (1810), 247—280.

Blanck, Anton, "Berzelius som medicine studerande," *Lychnos,* 1948—1949 (1950), 168—205.
Blomstrand, C. W., *Die Chemie der Jetztzeit vom Standpunkte der electrochemischen Auffassung aus Berzelius Lehre entwickelt* (Heidelberg: 1869).
Brooks, John H., "Wöhler's urea, and its vital force?—a verdict from the chemists," *Ambix,* 15 (1968), 84—114.
Bucholz, Christian Friedrich, "Versuche über Thomson's schwefelhaltige Salzsäure (Verbindung von Schwefel, Sauerstoff, und Salzsäure); zur nähern Kenntniss dieser merkwürdigen Verbindung," *Gehlen's Journal,* 9 (1810), 172—188.
Buffon, George Louis Leclerc, comte de, "Seconde vue de la nature" (1765), in Buffon, *Oeuvres philosophiques,* ed. Jean Pivetau, pp. 35—41.
— *Oeuvres philosophiques de Buffon,* ed. Jean Pivetau, Corpus général des philosophes français, Vol. XLI,i, ed. Raymond Bayer (Paris: 1954).
Cardwell, D. S. L., ed., *John Dalton and the progress of science: papers presented to a conference of historians of science held in Manchester, September 19—24, 1966, to mark the bicentenary of Dalton's birth* (Manchester: 1968).
Chevreul, Michel Eugène, "Faits et observations pour servir à l'histoire des combinaisons de l'oxide de plomb jaune avec les acides nitrique et nitreux," *Ann. chim.,* 83 (1812), 67—105.
Coleby, L. J. M., *The chemical studies of P. J. Macquer* (London: 1938).
Crosland, Maurice P., *Historical studies in the language of chemistry* (London: 1962).
— "Lavoisier's theory of acidity," *Isis,* 64 (1973), 306—317.
Cuvier, Georges, "Eloge historique d'Abraham Gottlob Werner," in Cuvier, *Recueil des éloges historiques,* II, pp. 305—336.
— *Recueil des éloges historiques lus dans les séances publiques de l'Institut de France,* 2 vols. (Paris: 1819).
— "Eloge historique de René Just Haüy," in Cuvier, *Eloges historiques,* pp. 247—283.
— *Eloges historiques, précédés par l'éloge de l'auteur par M. Flourens* (Paris: n.d.).
Dalton, John, "Remarks on the essay of Dr. Berzelius on the cause of chemical proportions," *Ann. Phil.,* 3 (1814), 174—180.
Daubenton, Louis Jean Marie, "Histoire naturelle," *Encyclopédie,* VIII (1765), pp. 225—230.
— "Introduction à l'histoire naturelle," in his *Encyclopédie méthodique, histoire naturelle des animaux,* I (Paris: 1782), pp. i—x.
Daudin, Henri, *De Linné à Jussieu: méthodes de classification et idée de série en botanique et en zoologie (1740—1790)* (Paris: [1926]).
Davy, Humphrey, "On some chemical agencies of electricity" (read 20 Nov. 1806), in Davy, *Works,* V, pp. 1—56.
— "On some new phenomena of chemical changes produced by electricity, particularly the decomposition of the fixed alkalies, and the exhibition of the new substances which constitute their bases; and on the general nature of alkaline bodies" (read, 19 Nov. 1807), in Davy, *Works,* V, pp. 57—101.
— "Electrochemical researches on the decomposition of the earths, with observations on the metals obtained from the alkaline earths, and on the amalgam procured from ammonia" (read, 30 June 1808), in Davy, *Works,* V, pp. 102—139.
— *Elements of chemical philosophy* (London: 1812), republished as Davy, *Works,* IV.
— *The collected works of Sir Humphrey Davy, Bart, Ll.D. F.R.S., edited by his brother, John Davy, M.D. F.R.S.,* 9 vols. (1839—1840; rpt., New York: 1972).
Debus, Allen G., "Solution analysis prior to Boyle," *Chymia,* 8 (1962), 41—61.
— "Fire analysis and the elements in the sixteenth and seventeenth centuries," *Ann.*

Sci., 23 (1967), 127—147.
Dictionary of scientific biography, ed. Charles C. Gillispie, 16 vols. (New York: 1970—1980).
Duncan, A. M., "Some theoretical aspects of eighteenth-century tables of affinity," *Ann. Sci.*, 18 (1962), 177—194, 217—232, plates VI—VII, VIII—IX.
Duveen, Denis I., and Herbert S. Klickstein, "A letter from Guyton de Morveau to Macquart relating to Lavoisier's attack against the *phlogiston* theory (1778); with an account of de Morveau's conversion to Lavoisier's doctrines in 1787," *Osiris*, 12 (1956), 342—367.
Encyclopédie méthodique, chymie, pharmacie et métallurgie, 6 vols., atlas (Paris: 1786—1815).
Eriksson, Gunnar, "Berzelius och atomteorin: den idéhistoriska bakgrunden," *Lychnos*, 1965—1966 (1967), 1—37.
Fichman, Martin, "French Stahlism and chemical studies of air, 1750—1770," *Ambix*, 18 (1971), 94—122.
Florkin, Marcel, *A history of biochemistry*, Comprehensive biochemistry, Vol. XXX (Amsterdam: 1972).
Fourcroy, Antoine François de, *Systême des connaissances chimiques, et de leurs applications aux phénomènes de la nature et de l'art*, 11 vols. (Paris: An 9—10 [1801—1802]).
Fourcroy, Antoine François de, François Chaussier, and Jean Pierre François Guillot Duhamel, *Encyclopédie méthodique, chimie, pharmacie et métallurgie* (III, La chimie, par le citoyen Fourcroy. La pharmacie, par le citoyen Chaussier. La métallurgie, par le citoyen Duhamel. Paris: An 4 [1796]).
Freund, Ida, *The study of chemical composition: an account of its method and historical development* (1904; rpt., New York: 1968).
Fuchs, Johann Nepomuk, "Ueber den Gehlenit, ein neues Mineral aus Tirol," *Schweigger's Journal*, 15 (1815), 377—386.
Gay-Lussac, Joseph Louis, "Mémoire sur la combinaison des substances gazeuses les unes avec les autres," *Mém. soc. Arcueil*, 2 (1809), 207—234.
— "Sur la vapeur nitreuse, et sur le gaz nitreux comme moyen eudiométrique," *Mém. soc. Arcueil*, 2 (1809), 235—251.
Gay-Lussac, Joseph Louis, and Louis Jacques Thenard, "De la nature et des propriétés de l'acide muriatique et de l'acide muriatique oxygéné. (lu [à l'Institut] le 9 janvier 1809)," in Gay-Lussac and Thenard, "Mémoires lus," 339—342.
— "Extrait des mémoires lus à l'Institut national, depuis le 7 mars 1808 jusqu'au 27 février 1809," *Mém. soc. Arcueil*, 2 (1809), 295—358.
— "Propriétés du métal de la potasse," in Gay-Lussac and Thenard, "Mémoires lus," 302—311.
— "Sur la production d'un amalgame par l'ammoniaque et les sels ammoniacaux, au moyen de la pile électrique," *Ann. chim.*, 73 (1810), 197—214.
Geike, Archibald, *The founders of geology* (2nd. ed., 1905; rpt., New York: 1962).
Gillispie, Charles C., *Genesis and geology: a study in the relations of scientific thought, natural theology, and social opinion in Great Britain, 1790—1850* (1951; rpt., New York: 1959).
Goodman, D. C., "The application of chemical criteria to biological classification in the eighteenth century," *Med. Hist.*, 15 (1971), 23—44.
Grimaux, Edouard, and Charles Gerhardt [fils], *Charles Gerhardt [père], sa vie, son oeuvre, sa correspondance, 1816—1856: document d'histoire de la chimie* (Paris: 1900).

Groth, Paul, *Entwicklungsgeschichte der mineralogischen Wissenschaften* (1926; rpt., Wiesbaden: 1970).
Guerlac, Henry, "The continental reputation of Stephen Hales," *Arch. int. d'hist. sci.*, 4 (1951), 393—404.
— *Lavoisier—the crucial year: the background and origin of his first experiments on combustion in 1772* (Ithaca, N.Y.: 1961).
Gümbel, [Wilhelm von,] "Hausmann: Joh. Friedr. Ludwig," in *Allgemeine deutsche Biographie*, XI (1880), pp. 94—97.
Guyton de Morveau, Louis Bernard, "Conciliation des principes de Sthaal [*sic*] avec les expériences modernes sur l'air fixe," *Obs. sur la physique*, 7 (1776), 380—395.
Guyton de Morveau, Louis Bernard, Hughes Maret, and Jean Pierre François Guillot Duhamel, *Encyclopédie méthodique, chymie, pharmacie et métallurgie* (I, La chymie, par M. de Morveau. La pharmacie, par M. Maret. La métallurgie, par M. Duhamel; Paris and Liège: 1786[—Dijon: 1789]).
Guyton de Morveau, Louis Bernard, Hughes Maret, and J. F. Durande, *Elémens de chymie, théorique et pratique, rédigés dans un nouvel ordre, d'après les découvertes modernes, pour servir aux cours publics de l'académie de Dijon*, 3 vols. (Dijon: 1777—1778).
[Hall], Marie Boas, "The establishment of the mechanical philosophy," *Osiris*, 10 (1952), 413—541.
— *Robert Boyle and seventeenth-century chemistry* (Cambridge: 1958).
Hasse, T. L., ed., *Denkschrift zur Erinnerung an die Verdienste des in Dresden am 30. Juni 1817 verstorbenen K. S. Bergrath's Werner und an die Fortschritte bei der Bergakademie zu Freiberg* (Dresden and Leipzig: 1848).
Hausmann, J.F.L., "[Review of] *Försök att genom användandet af den electrokemiska theorien och de kemiska proportionerna grundlägga ett rent vetenskapligt system för mineralogien*, af. J. Jacob Berzelius, M.D. Chemiae Prof. etc. 1814. 103 S. in Octav," *Göttingische gelehrte Anzeigen*, 1814,ii (1814), 1089—2001 (1100 and 1101 being mispaginated as 2000 and 2001).
Haüy, René Just, *Traité de minéralogie*, 4 vols., atlas (1801; rpt., Brussels: 1968).
— *Tableau comparatif des résultats de la cristallographie et de l'analyse chimique, relativement à la classification des minéraux* (Paris: 1809).
Hjelt, E., "Berzelius—Liebig—Dumas. Ihre Stellung zur Radikaltheorie 1832—1840," *Samml. chem. u. chem.-techn. Vorträge*, 12 (1908), 447—482.
— *Geschichte der organischen Chemie von ältester Zeit bis zur Gegenwart* (Brunswick: 1916).
Hoffmann, Carl August Siegfried, *Handbuch der Mineralogie*, 4 vols. (Freyberg: I, 1811; II,a, 1812; II,b—IV,b, fortgesetzt von August Breithaupt, 1815—1818).
Holbach, Paul Thiry, baron d', "Minéral règne," *Encyclopédie*, X (1765), p. 534.
Holmberg, Arne, *Bibliografi över J. J. Berzelius* (I, *Tryckta arbeten av och om Berzelius*, Stockholm and Uppsala: 1933; I,suppl. i, Stockholm and Uppsala: 1936; I,suppl. ii, Stockholm: 1953; I,suppl. iii, Stockholm: 1967; II, *Manuskript*, Stockholm: 1936; II,suppl., Stockholm: 1953).
Holmes, Frederick L., "From elective affinities to chemical equilibria: Berthollet's law of mass action," *Chymia*, 8 (1962), 105—145.
— "Analysis by fire and solvent extractions: the metamorphosis of a tradition," *Isis*, 62 (1971), 129—148.
Hooykaas, R., "De invloed der biologie op de anorganische natuurwetenschappen," *Geloof en wetenschap (Christlijke vereeniging van natuur- en geneeskundigen. orgaan)*, 101 (1936), 101—114.

— "De natuurlijke klassificatie der chemische substanties," *Chemisch Weekblad*, 33 (1936), 599—610.
— "Die Elementenlehre der Iatrochemiker," *Janus*, 41 (1937), 1—28.
— "The discrimination between 'natural' and 'artificial' substances and the development of the corpuscular theory," *Arch. int. d'hist. sci.*, 1 (1947—1948), 640—651.
— "De kristallografie van J. B. de Romé de l'Isle (1783)," *Chemisch Weekblad*, 47 (1951), 848—855.
— "The species concept in eighteenth-century mineralogy," *Arch. int. d'hist. sci.*, 5 (1952), 45—55.
— "Torbern Bergman's crystal theory," *Lychnos*, 1952 (1952), 21—54.
— *La naissance de la cristallographie en France au XVIIIe siècle* (Paris: 1953).
— "Les débuts de la théorie cristallographique de R. J. Haüy, d'après les documents originaux," *Rev. d'hist. sci.*, 8 (1955), 319—337.
— "Haüy, René Just," *DSB*, VI, pp. 178—183.
Jørgensen, Bent Søren, "Berzelius und die Lebenskraft," *Centaurus*, 10 (1964), 258—281.
Jorpes, J. Erik, *Jac. Berzelius: his life and work* (Berkeley: 1970).
Kapoor, S. C., "Berthollet, Proust, and proportions," *Chymia*, 10 (1965), 53—110.
Kerker, Milton, "Herman Boerhaave and the development of pneumatic chemistry," *Isis*, 46 (1955), 36—49.
Kobell, Franz von, *Geschichte der Mineralogie von 1650—1860*, Geschichte der Wissenschaften in Deutschland. Neure Zeit. Vol. II (Munich: 1864).
Kopp, Hermann, *Die Entwicklung der Chemie in der neueren Zeit* (Munich: 1873).
Kuhn, Thomas S., "Robert Boyle and structural chemistry in the seventeenth century," *Isis*, 43 (1952), 12—36.
Ladenburg, Albert, *Vorträge über die Entwicklungsgeschichte der Chemie in den letzten Hundert Jahren* (2nd ed., Brunswick: 1887).
Larson, James L., *Reason and experience: the representation of the natural order in the work of Carl von Linné* (Berkeley: 1971).
Lavoisier, Antoine Laurent, "Mémoire sur la nature du principe qui se combine avec les métaux pendant leur calcination et qui en augment le poids" (1775), in Lavoisier, *Oeuvres*, II, pp. 122—128.
— "Mémoire sur l'existence de l'air dans l'acide nitreux et sur les moyens de décomposer et de recomposer cet acide" (1776; rev., 1779), in Lavoisier, *Oeuvres*, II, pp. 129—138.
— "Expériences sur la combustion de l'alun avec les matières charbonneuses et sur les altérations qui arrivent à l'air dans lequel on fait brûler du pyrophore" (1777), in Lavoisier, *Oeuvres*, II, pp. 199—208.
— "Mémoire sur la combustion des chandelles dans l'air atmosphérique et dans l'air éminemment respirable" (1777), in Lavosier, *Oeuvres*, II, pp. 184—193.
— "Mémoire sur la combustion en général" (1777), in Lavoisier, *Oeuvres*, II, pp. 225—233.
— "Mémoire sur la dissolution du mercure dans l'acide vitriolique et sur la résolution de cet acide en acide sulfureux aériforme et en air éminemment respirable" (1777), in Lavoisier, *Oeuvres*, II, pp. 194—198.
— "Considérations générales sur la nature des acides, & sur les principes dont ils sont composés" (1778), in Lavoisier, *Oeuvres*, II, pp. 248—260.
— "Mémoire dans lequel on a pour objet de prouver que l'eau n'est point une substance simple, un élément proprement dit, mais qu'elle est susceptible de décomposition et de recomposition" (1781), in Lavoisier, *Oeuvres*, II, pp. 334—359.
— "Mémoires sur la formation de l'acide nommé aire fixe ou acide crayeux, et que je

désignerai désormais sous le nom d'*acide du charbon*" (1781), in Lavoisier, *Oeuvres,* II, pp. 403—422.
— "Mémoire sur l'union du principe oxygine avec le fer" (1782), in Lavoisier, *Oeuvres,* II, pp. 557—574.
— "Mémoire sur la combinaison du principe oxygine avec l'esprit-de-vin, l'huile et différents corps combustibles" (1784), in Lavoisier, *Oeuvres,* II, pp. 586—600.
— "Réflexions sur la décomposition de l'eau par les substances végétales et animales" (1786), in Lavoisier, *Oeuvres,* II, pp. 656—670.
— *Traité élémentaire de chimie, présenté dans un ordre nouveau et d'après les découvertes modernes; avec figures,* 2 vols. (1789; rpt., Brussels: 1965).
— *Oeuvres de Lavoisier, publiées par les soins de son excellence le ministre de l'instruction publique et des cultes,* 6 vols. (1862—1893; rpt., New York: 1965).
Lavoisier, Antoine Laurent, Louis Bernard Guyton de Morveau, Claude Louis Berthollet, and Antoine François de Fourcroy, *Méthode de nomenclature chimique, proposée par MM. de Morveau, Lavoisier, Bertholet [sic], et de Fourcroy. On y a joint un nouveau système de caractères chimiques, adaptés à cette nomenclature, par MM. Hassenfratz et Adet* (Paris: 1787).
Lavoisier, Antoine Laurent, and Jean Baptiste Marie Charles Meusnier de la Place, "Mémoire où on prouve, par la décomposition de l'eau, que ce fluide n'est point une substance simple, et qu'il y a plusieurs moyens d'obtenir en grand l'air inflammable qui y entre comme principe constituant" (1781), in Lavoisier, *Oeuvres,* II, pp. 360—373.
Le Grand, E. H., "Lavoisier's oxygen theory of acidity," *Ann. Sci.,* 29 (1972), 1—18.
— "A note on fixed air: the universal acid," *Ambix,* 20 (1973), 88—94.
Lindroth, Sten, "Berzelius och hans tid," *Vetenskaps-Societeten i Lund. Årsbok,* 1964 (1964), 15—40.
— *Kungl. Svenska Vetenskapsakademiens historia 1739—1818, utgiven av Kungl. Vetenskapsakademien,* 3 vols. in 2 (Stockholm: 1967).
Lough, John, *Contributors to the* Encyclopédie (London: 1973).
Love, Rosaleen, "Some sources of Herman Boerhaave's concept of fire," *Ambix,* 19 (1972), 157—174.
— "Herman Boerhaave and the element-instrument concept of fire," *Ann. Sci.,* 31 (1974), 547—559.
Lovejoy, Arthur O., *The great chain of being: a study in the history of an idea* (1936; rpt., New York: 1960).
Lundgren, Anders, *Berzelius och den kemiska atomteorin,* Acta Universitatis Upsaliensis. Studier i idé- och lärdomshistoria, Vol. II, utgivna av institutionen för idé- och lärdomshistoria vid Uppsala Universitet (Uppsala: 1979).
Macquer, Pierre Joseph, *Elémens de chymie théorique* (Paris: 1749).
— *A dictionary of chemistry,* [tr. James Keir,] 2 vols. (London: 1771).
Martius, Karl Friedrich Philipp, *Akademische Denkreden* (Leipzig: 1866).
— "[Denkrede auf] Johann Friedrich Ludwig Hausmann," in Martius, *Akademische Denkreden,* pp. 429—440.
Mauskopf, Seymour H., "Crystals and compounds: molecular structure and composition in nineteenth-century French science," *Trans. Am. Phil. Soc.,* [NS], 66,iii (1976).
Mélanges Alexandre Koyré, 2 vols. (Paris: 1964).
Melhado, Evan M., "Jac. Berzelius: foundations and development of his chemistry" (Doctoral dissertation, Princeton University: 1977).
— "Mitscherlich's discovery of isomorphism," *Hist. Stud. Phys. Sci.,* 11,i (1980), 87—123.

— "Phlogiston, oxygen, caloric: an interpretation," forthcoming.
Metzger, Hélène, *La genèse de la science des cristaux* (1918; rpt., Paris: 1969).
— *Les doctrines chimiques en France du début du XVIIe à la fin du XVIIIe siècle* (1923; rpt., Paris: 1969).
— *Les concepts scientifiques* (Paris: 1926).
— *Newton, Stahl, Boerhaave et la doctrine chimique* (1930; rpt., Paris: 1974).
— "La philosophie de la matière chez Lavoisier," *Actualités scientifiques et industrielles,* no. 218 (1935), 1—48 (separate pagination).
Meyer, Ernst von, *Geschichte der Chemie von den ältesten Zeit bis zur Gegenwart* (Leipzig: 1889).
Mitscherlich, Eilhard, "Ueber die Krystallization der Salze, in denen das Metall der Basis mit zwei Proportionen Sauerstoff verbunden ist," *Abhl. Akad. Berlin,* 1818—1819 (1819), 427—437.
Moström, Birgitta, *Torbern Bergman: a bibliography of his works* (Stockholm: 1957).
Multhauf, Robert P., *The origins of chemistry* (New York: 1967).
Oldroyd, D. R., "A note on the status of A. F. Cronstedt's simple earths and his analytical methods," *Isis,* 65 (1974), 506—512.
— "Some phlogistic mineralogical schemes, illustrative of the evolution of the concept of 'earth' in the seventeenth and eighteenth centuries," *Ann. Sci.,* 31 (1974), 269—305.
Olsson, Hugo, *Kemiens historia i Sverige intill år 1800,* Lychnos-Bibliotek: studier och källskrifter utgivna av Lärdomshistoriska samfundet, Vol. XVII (Uppsala: 1971).
Partington, J. R., "Jeremias Benjamin Richter and the law of equivalent proportions," *Ann. Sci.,* 7 (1951), 173—198; 9 (1953), 289—314.
— *A history of chemistry,* 4 vols. (I,i, II—IV, London: 1961—1970; I,ii not published).
Prandtl, Wilhelm, *Humbrey Davy, Jöns Jacob Berzelius: zwei führende Chemiker aus der ersten Hälfte des 19. Jahrhunderts,* Grosse Naturforscher, H. W. Frickhinger, ed., Vol. III (Stuttgart: 1948).
Proust, Louis Joseph, "Remarques sur le Système des connoissances chimiques de Fourcroy," *J. de physique,* 56 (1802), 200—220.
Pusch, Georg Gottlieb, "Kritische Betrachtungen über das, auf die electrochemische Theorie und die chemische Proporzionslehre gegründete, Mineral-System des Herrn Professor Berzelius," *Leonhard's Taschenbuch für die gesammte Mineralogie,* 11,ii (1817), 417—522; 12,i (1818), 3—66.
Rappaport, Rhoda, "G.-F. Rouelle: an eighteenth-century chemist and teacher," *Chymia,* 6 (1960), 68—101.
— "Rouelle and Stahl—the phlogistic revolution in France," *Chymia,* 7 (1961), 73—102.
Rau, Albrecht, *Die Theorien der modernen Chemie,* 3 pts. (Brunswick: 1877—1884).
Richter, Jeremias Benjamin, *Ueber die neueren Gegenstände der Chymie,* 11 pts. (Breslau and Hirschberg, and, from pt. iv, Lissa: 1791—1802).
— *Anfangsgründe der Stöchyometrie oder Messkunst chymischer Elemente,* 3 vols. in 4 (Breslau and Hirschberg: 1792—1794).
Rouelle, Guillaume François, "Mémoire sur les sels neutres, dans lequel on propose une division méthodique de ces sels, qui facilite les moyens pour parvenir à la théorie de leur cristallisation," *Mém. de l'Acad.,* 1744 (rpt., Amsterdam: 1750), 480—495.
Russell, Colin A., "The electrochemical theory of Berzelius," *Ann. Sci.,* 19 (1965), 117—145.
— "Berzelius and the development of the atomic theory," in D. S. L. Cardwell, ed., *John Dalton and the progress of science,* pp. 259—273.
— "Introduction to the reprint edition [of Berzelius's *Proportions chimiques* (1819)]," in Berzelius, *Proportions chimiques* (1819; rpt., New York: Johnson Reprint Corporation,

1972), pp. [v]—[xlix].
Sachs, Julius von, *History of botany (1530—1860),* rev. Isaac Bayley Balfour, tr. Henry E. F. Garnsey (Oxford: 1890).
St. Clair, Charles S., "The classification of minerals: some representative mineral systems from Agricola to Werner" (Doctoral dissertation, University of Oklahoma: 1966).
Scheele, Carl Wilhelm, "On manganese or magnesia, and its properties" (1774), in Scheele, *Papers,* pp. 17—49.
— "On arsenic and its acid" (1775), in Scheele, *Papers,* pp. 53—77.
— *Chemische Abhandlung von der Luft und dem Feuer* (Uppsala: 1777), republished as Scheele, *Werke,* I.
— "Experiments with lead ore: molybdaena" (1778), in Scheele, *Papers,* pp. 186—194.
— "The constituents of tungsten" (1781), in Scheele, *Papers,* pp. 225—229.
— *Sämmtliche physische und chemische Werke, nach dem Tode des Verfassers gesammlet, und in deutscher Sprache herausgegeben von Sigismund Friedrich Hermbstädt,* 2 vols. (Berlin: 1793).
— *The collected papers of Carl Wilhelm Scheele; translated from the Swedish and German originals by Leonard Dobbin* (London: 1931).
Schwab, Richard N., Walter E. Rex, and John Lough, *Inventory of Diderot's* Encyclopédie, 6 vols., *Studies on Voltaire and the eighteenth century,* ed. Theodore Besterman, nos. 80, 83, 85, 91—93 (I, no. 80, Geneva: 1971; II—IV, nos. 83, 85, 91—93, Banbury, England: 1971—1972).
Siegfried, R., and B. J. Dobbs, "Composition, a neglected aspect of the chemical revolution," *Ann. Sci.,* 24 (1968), 275—293.
Smeaton, W. A., "L. B. Guyton de Morveau (1737—1816): a bibliographical study," *Ambix,* 6 (1957—1958), 18—34.
— "Guyton de Morveau's course of chemistry in the Dijon Academy," *Ambix,* 9 (1961), 53—69.
— *Fourcroy: chemist and revolutionary, 1775—1809* (Cambridge: 1962).
— "Guyton de Morveau and chemical affinity," *Ambix,* 11 (1963), 55—64.
— "Guyton de Morveau and the phlogiston theory" (1964), in *Mélanges Alexandre Koyré,* I, pp. 522—540.
— "Macquer, Pierre Joseph," *DSB,* VIII, pp. 618—624.
Söderbaum, H. G., *Berzelius' Werden und Wachsen, 1779—1821,* Kahlbaum's Monographien aus der Geschichte der Chemie, Vol. III (Leipzig: 1899).
— *Jac. Berzelius: Levnadsteckning,* 3 vols. (Uppsala: 1929—1931).
Svenson, H. K., "On the descriptive method of Linnaeus," *Rhodora,* 47 (1945), 273—302, 363—388.
Szabadváry, Ferenc, *History of analytical chemistry,* tr. Gyula Shvela, International series of monographs in analytical chemistry, Vol. XXVII (New York: 1966).
Thackray, Arnold, *Atoms and powers: and essay on Newtonian matter-theory and the development of chemistry* (Cambridge, Mass.: 1970).
Thomson, Thomas, "On the compounds of sulphur and oxygen," *Nicholson's Journal,* 6 (1803), 92—109.
— "On the Daltonian theory of definite proportions in chemical combination," *Ann. Phil.,* 2 (1813), 32—52, 109—115, 167—171, 293—301; 3 (1814), 134—140, 375—378; 4 (1814), 11—18, 83—89.
— "Some observations in answer to Mr. Chevenix's attack upon Werner's mineralogical method," *Ann. Phil.,* 1 (1813), 241—258.
— *An attempt to establish the first principles of chemistry by experiment,* 2 vols. (London: 1825).

— *Chemistry of organic bodies: vegetables* (London: 1838).
Theorell, Hugo, "Berzelius och livskraften," *Nordisk Medicin,* 60 (1958), 1537—1540.
Walden, P., "Die Lösungstheorien in ihrer geschichtlichen Aufeinanderfolge," *Samml. chem. u. chem.-tech. Vorträge,* 15 (1910), 277—454.
Weeks, Mary Elvira, and Henry M. Leicester, *Discovery of the elements* (7th ed., Easton, Pa.: 1968).
Werner, Abraham Gottlob, *Von den äusserlichen Kennzeichen der Fossilien* (1774; rpt., Amsterdam: 1965).
Wertime, T. A., "The discovery of the element carbon," *Osiris,* 11 (1954), 211—220.
Wollaston, William Hyde, "On superacid and subacid salts," *Phil. Trans.,* 98 (1808), 96—102.
Wurtz, Charles Adolph, *Histoire des doctrines chimiques depuis Lavoisier jusqu'à nos jours,* Dictionnaire de chimie pure et appliquée, Discours préliminaire (Paris: 1868).

Index